Selected Titles in This Series

198 **V. A. Smirnov,** Simplicial and operad methods in algebraic topology, 2001

197 **Kenji Ueno,** Algebraic geometry 2: Sheaves and cohomology, 2001

196 **Yu. N. Lin′kov,** Asymptotic statistical methods for stochastic processes, 2001

195 **Minoru Wakimoto,** Infinite-dimensional Lie algebras, 2001

194 **Valery B. Nevzorov,** Records: Mathematical theory, 2001

193 **Toshio Nishino,** Function theory in several complex variables, 2001

192 **Yu. P. Solovyov and E. V. Troitsky,** C^*-algebras and elliptic operators in differential topology, 2001

191 **Shun-ichi Amari and Hiroshi Nagaoka,** Methods of information geometry, 2000

190 **Alexander N. Starkov,** Dynamical systems on homogeneous spaces, 2000

189 **Mitsuru Ikawa,** Hyperbolic partial differential equations and wave phenomena, 2000

188 **V. V. Buldygin and Yu. V. Kozachenko,** Metric characterization of random variables and random processes, 2000

187 **A. V. Fursikov,** Optimal control of distributed systems. Theory and applications, 2000

186 **Kazuya Kato, Nobushige Kurokawa, and Takeshi Saito,** Number theory 1: Fermat's dream, 2000

185 **Kenji Ueno,** Algebraic Geometry 1: From algebraic varieties to schemes, 1999

184 **A. V. Mel′nikov,** Financial markets, 1999

183 **Hajime Sato,** Algebraic topology: an intuitive approach, 1999

182 **I. S. Krasil′shchik and A. M. Vinogradov, Editors,** Symmetries and conservation laws for differential equations of mathematical physics, 1999

181 **Ya. G. Berkovich and E. M. Zhmud′,** Characters of finite groups. Part 2, 1999

180 **A. A. Milyutin and N. P. Osmolovskii,** Calculus of variations and optimal control, 1998

179 **V. E. Voskresenskiĭ,** Algebraic groups and their birational invariants, 1998

178 **Mitsuo Morimoto,** Analytic functionals on the sphere, 1998

177 **Satoru Igari,** Real analysis—with an introduction to wavelet theory, 1998

176 **L. M. Lerman and Ya. L. Umanskiy,** Four-dimensional integrable Hamiltonian systems with simple singular points (topological aspects), 1998

175 **S. K. Godunov,** Modern aspects of linear algebra, 1998

174 **Ya-Zhe Chen and Lan-Cheng Wu,** Second order elliptic equations and elliptic systems, 1998

173 **Yu. A. Davydov, M. A. Lifshits, and N. V. Smorodina,** Local properties of distributions of stochastic functionals, 1998

172 **Ya. G. Berkovich and E. M. Zhmud′,** Characters of finite groups. Part 1, 1998

171 **E. M. Landis,** Second order equations of elliptic and parabolic type, 1998

170 **Viktor Prasolov and Yuri Solovyev,** Elliptic functions and elliptic integrals, 1997

169 **S. K. Godunov,** Ordinary differential equations with constant coefficient, 1997

168 **Junjiro Noguchi,** Introduction to complex analysis, 1998

167 **Masaya Yamaguti, Masayoshi Hata, and Jun Kigami,** Mathematics of fractals, 1997

166 **Kenji Ueno,** An introduction to algebraic geometry, 1997

165 **V. V. Ishkhanov, B. B. Lur′e, and D. K. Faddeev,** The embedding problem in Galois theory, 1997

164 **E. I. Gordon,** Nonstandard methods in commutative harmonic analysis, 1997

163 **A. Ya. Dorogovtsev, D. S. Silvestrov, A. V. Skorokhod, and M. I. Yadrenko,** Probability theory: Collection of problems, 1997

162 **M. V. Boldin, G. I. Simonova, and Yu. N. Tyurin,** Sign-based methods in linear statistical models, 1997

(*Continued in the back of this publication*)

Simplicial and Operad Methods in Algebraic Topology

Translations of
MATHEMATICAL
MONOGRAPHS

Volume 198

Simplicial and Operad Methods in Algebraic Topology

V. A. Smirnov

American Mathematical Society
Providence, Rhode Island

В. А. Смирнов

СИМПЛИЦИАЛЬНЫЕ И ОПЕРАДНЫЕ МЕТОДЫ В АЛГЕБРАИЧЕСКОЙ ТОПОЛОГИИ

Translated from the Russian manuscript by G. L. Rybnikov

2000 *Mathematics Subject Classification.* Primary 55P15, 55S20.

Abstract. In recent years, for solving problems of algebraic topology and, in particular, difficult problems of homotopy theory, algebraic structures more complicated than just a topological monoid, an algebra, etc. have been repeatedly used. A convenient language for describing various structures arising naturally on topological spaces and on their cohomology and homotopy is the language of operads and algebras over an operad; this language was proposed by May to describe the structures on n-fold and infinite loop spaces. This language is developed in the book.

The book is useful for research mathematicians and graduate students working in algebraic topology.

Library of Congress Cataloging-in-Publication Data

Smirnov, V. A. (Vladimir Alekseevich), 1949–

Simplicial and operad methods in algebraic topology / V. A. Smirnov.

p. cm. — (Translations of mathematical monographs, ISSN 0065-9282 ; v. 198)

Includes bibliographical references and index.

ISBN 0-8218-2170-9 (acid-free paper)

1. Algebraic topology. 2. Ordered algebraic structures. I. Title. II. Series.

QA612.S58 2001

514′.2—dc21 00-068957

∞ The paper used in this book is acid-free and falls within the guidelines
established to ensure permanence and durability.
Visit the AMS home page at URL: http://www.ams.org/

10 9 8 7 6 5 4 3 2 1 06 05 04 03 02 01

Contents

Preface

In recent years, in solving various problems of algebraic topology and, in particular, difficult problems of homology theory and homotopy theory, the usage of algebraic structures more complicated than just a topological monoid, an algebra, a coalgebra, etc. has been required more and more often.

Among the first examples of these structures were the Massey products introduced by W. S. Massey in 1954, which are partial, multivalued, and many-placed operations. The Massey products naturally arise on homology of differential algebras; they are used for computing differentials of various spectral sequences.

Another example of many-placed operations is provided by the A_∞-structure, which was introduced by J. D. Stasheff in 1963 with the goal of describing loop spaces. He proved that a connected CW-complex possesses an A_∞-structure if and only if it is homotopy equivalent to a loop space.

A general method for describing many-placed operations on topological spaces was proposed by J. P. May in 1972. He introduced the concept of an operad $\mathcal{E}$ as a family of topological spaces $\mathcal{E}(j)$, $j \geq 0$, whose points should be thought of as abstract j-ary operations. An action of an operad $\mathcal{E}$ on a topological space X is a family of operations $\mathcal{E}(j) \times X^{\times j} \to X$. By restricting these operations to points of the spaces $\mathcal{E}(j)$, we obtain many-placed operations $X^{\times j} \to X$ on X.

May showed that on any n-fold loop space there is an action of the little n-cubes operad E_n, and each arcwise connected E_n-space has the weak homotopy type of an n-fold loop space.

The concept of an operad turned out to be useful not only in describing structures on topological spaces, but also in investigating various algebraic structures. In 1981 the author generalized the notion of operad to the case of the category of chain complexes. It was shown that the singular chain complex $C_*(X)$ of a topological space X admits an action of the operad E_∞, and, in the case of a simply connected space, this action determines the weak homotopy type of this space. Thus many problems of homotopy theory can be reduced to the study of the E_∞-structure on the chain complex.

In this book we present a detailed study of the concept of an operad in the categories of topological spaces and of chain complexes. We introduce the notions of an algebra and a coalgebra over an operad and investigate their properties. We elucidate the algebraic structure of the singular chain complex of a topological space and show how the problem of homotopy classification of topological spaces can be solved using this structure. For algebras and coalgebras over operads, we define standard constructions, in particular, the bar and cobar constructions. We apply operad methods to computing the homology of iterated loop spaces, investigating the algebraic structure of generalized cohomology theories, describing cohomology of groups and algebras, computing differentials in the Adams spectral sequence for the homotopy groups of the spheres, and some other problems.

Vladimir Smirnov
Moscow, 2000

CHAPTER 1

Operads in the Category of Topological Spaces

1.1. A_∞-structure on a loop space

We shall start with the study of a structure on a loop space which served as an impetus for creating the concepts of an operad and an algebra over an operad.

Let X be a topological space with base point $* \in X$. The *loop space* ΩX is the space of all continuous maps $f\colon [0,1] \to X$ from $[0,1]$ to X such that the endpoints of the segment are mapped onto the base point, $f(0) = f(1) = *$.

Sometimes it is convenient to define the loop space in a different way, i.e., as the space $\Omega_l X$ consisting of all continuous maps $f\colon [0,T] \to X$ from $[0,T]$ to X, $0 \le T$, $f(0) = f(T) = *$. Intuitively, T can be regarded as the time in which the point traverses the loop. In the first case (for the space ΩX), this time is the same for all loops and is equal to 1. In the second case (for the space $\Omega_l X$), this time is varying.

Let us define the multiplication $\cdot\colon \Omega_l X \times \Omega_l X \to \Omega_l X$ in the space $\Omega_l X$ by assigning to each pair (f_1, f_2) of loops $f_1\colon [0,T_1] \to X$ and $f_2\colon [0,T_2] \to X$ the loop $f_1 \cdot f_2\colon [0,T] \to X$, where $T = T_1 + T_2$ and

$$f_1 \cdot f_2(t) = \begin{cases} f_1(t), & 0 \le t \le T_1, \\ f_2(t - T_1), & T_1 \le t \le T. \end{cases}$$

Thus the loop $f_1 \cdot f_2$ consists of the loops f_1 and f_2. Moving along this loop, the point first traverses f_1 and then f_2. Besides, the time in which the point traverses the entire loop is equal to the sum of the times in which it covers f_1 and f_2.

Clearly, the multiplication so defined is associative, i.e., for any loops f_1, f_2, and f_3, the following relation holds: $(f_1 \cdot f_2) \cdot f_3 = f_1 \cdot (f_2 \cdot f_3)$. Besides, there is a neutral element $e\colon [0,0] \to X$ mapping the point 0 to the base point $* \in X$. It is clear that for any loop f, we have $e \cdot f = f \cdot e$. Thus the space $\Omega_l X$ is a monoid.

Now let us define the multiplication of loops in ΩX. An attempt to do it in the same way as for the case of the space $\Omega_l X$ fails. Indeed, the product of two loops traversed in time 1 will be covered in time $T = 2$, while we require that this time be equal to 1. Hence the multiplication in ΩX must be defined differently. Namely, let $f_1, f_2\colon [0,1] \to X$. We define $f_1 \cdot f_2\colon [0,1] \to X$ by setting

$$f_1 \cdot f_2(t) = \begin{cases} f_1(2t), & 0 \le t \le \frac{1}{2}, \\ f_2(2t - 1), & \frac{1}{2} \le t \le 1. \end{cases}$$

Thus the loop $f_1 \cdot f_2$ consists of the loops f_1 and f_2. The point first traverses the loop f_1 and then the loop f_2, moving with a velocity twice as great as the initial one.

Note that the multiplication $\cdot\colon \Omega X \times \Omega X \to \Omega X$ defined above is not associative. Indeed, although both loops $(f_1 \cdot f_2) \cdot f_3$ and $f_1 \cdot (f_2 \cdot f_3)$ consist of f_1, f_2, and f_3, in the first case the point traverses the loops f_1 and f_2 with 4-fold velocity and the loop f_3 with doubled velocity,

$$((f_1 \cdot f_2) \cdot f_3)(t) = \begin{cases} f_1(4t), & 0 \le t \le \frac{1}{4}, \\ f_2(4t-1), & \frac{1}{4} \le t \le \frac{1}{2}, \\ f_3(2t-1), & \frac{1}{2} \le t \le 1, \end{cases}$$

and in the second case the point traverses the loop f_1 with doubled velocity and the loops f_2 and f_3 with 4-fold velocity,

$$(f_1 \cdot (f_2 \cdot f_3))(t) = \begin{cases} f_1(2t), & 0 \le t \le \frac{1}{2}, \\ f_2(4t-2), & \frac{1}{2} \le t \le \frac{3}{4}, \\ f_3(4t-3), & \frac{3}{4} \le t \le 1. \end{cases}$$

Let us show that the multiplication in ΩX is *homotopy associative*, i.e., there is a homotopy $h: I \times \Omega X \times \Omega X \times \Omega X \to \Omega X$, $I = [0,1]$, such that

$$h(0, f_1, f_2, f_3) = (f_1 \cdot f_2) \cdot f_3, \quad h(1, f_1, f_2, f_3) = f_1 \cdot (f_2 \cdot f_3).$$

We define this homotopy h by setting

$$h(u, f_1, f_2, f_3)(t) = \begin{cases} f_1(\frac{4t}{u+1}), & 0 \le t \le \frac{u+1}{4}, \\ f_2(4t-u-1), & \frac{u+1}{4} \le t \le \frac{u+2}{4}, \\ f_3(\frac{4t-u-2}{2-u}), & \frac{u+2}{4} \le t \le 1. \end{cases}$$

It can be thought of as a path in ΩX connecting the loops $(f_1 \cdot f_2) \cdot f_3$ and $f_1 \cdot (f_2 \cdot f_3)$. The required relations are easily verified.

Now let us consider four loops $f_1, f_2, f_3, f_4 \in \Omega X$, and let us arrange parentheses in their product (reducing it to pairwise products) in all possible ways. We obtain five vertices

$$\begin{gathered} ((f_1 \cdot f_2) \cdot f_3) \cdot f_4, \quad (f_1 \cdot (f_2 \cdot f_3)) \cdot f_4, \\ (f_1 \cdot f_2) \cdot (f_3 \cdot f_4), \qquad f_1 \cdot ((f_2 \cdot f_3) \cdot f_4), \\ f_1 \cdot (f_2 \cdot (f_3 \cdot f_4)). \end{gathered}$$

By using the homotopy associativity of the multiplication of loops, adjacent vertices can be joined by the paths

$$\begin{gathered} h(f_1, f_2, f_3) \cdot f_4, \\ h(f_1 \cdot f_2, f_3, f_4), \qquad h(f_1, f_2 \cdot f_3, f_4), \\ h(f_1, f_2, f_3 \cdot f_4), \qquad f_1 \cdot h(f_2, f_3, f_4), \end{gathered}$$

which, taken together, constitute the boundary of a pentagon.

Denote the pentagon by K and its boundary by $\operatorname{Fr} K$. Then we have the map $\operatorname{Fr} K \times \Omega X^{\times 4} \to \Omega X$. Stasheff proved **[45,46]** that this map can be extended to the entire pentagon: $K \times \Omega X^{\times 4} \to \Omega X$. Moreover, continuing the process, he constructed CW-complexes K_i, $i \geq 2$, of dimension $i-2$ and the maps $K_i \times \Omega X^{\times i} \to \Omega X$ which define higher homotopies corresponding to various arrangements of parentheses in the product of i factors.

The cells of codimension $k \geq 0$ in the CW-complex K_i are in one-to-one correspondence with all arrangements of k pairs of parentheses in the product of i

factors. A cell of codimension $k+1$ is a part of the boundary of a cell of codimension k if and only if the latter can be obtained from the former by deleting a pair of parentheses. For example, K_2 is a point, K_3 is a segment, K_4 is a pentagon, etc.

Stasheff called the structure on ΩX consisting of the maps $K_i \times \Omega X^{\times i} \to \Omega X$, as described above, the *A_∞-structure.* He showed that this structure completely characterizes loop spaces in the sense that a connected space possesses an A_∞-structure if and only if it is homotopy equivalent to a loop space.

The construction of the A_∞-structure on a loop space can also be approached in another way. Namely, note that the space ΩX is homotopy equivalent to the monoid $\Omega_l X$. We have the inclusion map $\xi\colon \Omega X \to \Omega_l X$ and the projection $\eta\colon \Omega_l X \to \Omega X$ mapping each loop $f\colon [0,T] \to X$ to the loop $\eta(f)\colon [0,1] \to X$ defined by the formula $\eta(f)(t) = f(T\cdot t)$.

It is clear that $\eta \circ \xi = \mathrm{Id}$. Let us show that $\xi \circ \eta \simeq \mathrm{Id}$. To this end, we define the homotopy $h\colon \Omega_l X \to \Omega_l X$ by setting

$$h(u,f)(t) = f(((T-1)u+1)t).$$

Thus this is a special case of the following general situation.

Let M be a monoid; by $\pi\colon M \times M \to M$ we denote the multiplication in M, which is assumed to be associative. The product of elements x_1, x_2 in M will also be denoted by $x_1 \cdot x_2$. Further, suppose that the space Y is homotopy equivalent to M; let $\xi\colon Y \to M$, $\eta\colon M \to Y$, $h\colon Y \to Y$ be the corresponding maps and the homotopy satisfying the relations $\eta \circ \xi = \mathrm{Id}$, $\xi \circ \eta \simeq \mathrm{Id}$.

It turns out that in all such situations the space Y possesses the structure of an A_∞-space. It can be constructed as follows.

We define a multiplication $\widetilde{\pi}\colon Y \times Y \to Y$ by setting $\widetilde{\pi} = \eta\pi(\xi \times \xi)$. This multiplication will also be denoted by a dot. Then, for $y_1, y_2 \in Y$, we have $y_1 \cdot y_2 = \eta(\xi(y_1)\cdot \xi(y_2))$.

Similarly, for any elements $y_1, \dots, y_n \in Y$, we define their product $y_1 \cdot \dots \cdot y_n$ by setting

$$y_1 \cdot \dots \cdot y_n = \eta(\xi(y_1) \cdot \dots \cdot \xi(y_n)).$$

The elements $(y_1 \cdot y_2)\cdot y_3$ and $y_1 \cdot y_2 \cdot y_3$ are connected by the homotopy

$$\eta\pi(h\pi \times 1)(\xi(y_1) \times \xi(y_2) \times \xi(y_3)).$$

The elements $y_1 \cdot (y_2 \cdot y_3)$ and $y_1 \cdot y_2 \cdot y_3$ are connected by the homotopy

$$\eta\pi(1 \times h\pi)(\xi(y_1) \times \xi(y_2) \times \xi(y_3)).$$

Combining these homotopies, we obtain a homotopy between the elements

$$(y_1 \cdot y_2)\cdot y_3, \quad y_1 \cdot (y_2 \cdot y_3).$$

Continuing this process, we can construct CW-complexes $\widetilde{K}_i$ consisting of $(i-2)$-dimensional cubes and maps $\widetilde{K}_i \times Y^{\times i} \to Y$ that form the required structure of an A_∞-space on Y.

1.2. Operads and spaces over operads

In the previous section we endowed a loop space with a structure which is not just a binary operation but a whole family of n-ary operations, $n \geq 2$, representing the higher homotopies for the associativity of the multiplication in the loop space. A space possessing such a structure was called an A_∞-space. As was already noted, the A_∞-structure provides higher homotopies for the associativity of multiplication, but it says nothing about the commutativity relation. Thus, in the double loop space $\Omega^2 X = \Omega\Omega X$ of the space X, the multiplication of loops is not only homotopy associative but also homotopy commutative. Hence in the description of the structure of this space, not only the higher homotopies for the associativity, but also the higher homotopies for the commutativity and the relations between them must be included.

An appropriate language for describing the structures of this kind is the language of operads and spaces over operads, which was introduced by May [**29**].

An *operad* $\mathcal{E}$ is a family of spaces $\mathcal{E}(j)$, $j \geq 1$, (whose points should be thought of as j-ary operations) such that the following conditions hold:

1. The space $\mathcal{E}(0)$ contains exactly one point $*$.
2. Continuous operations

$$\gamma\colon \mathcal{E}(k) \times \mathcal{E}(j_1) \times \cdots \times \mathcal{E}(j_k) \to \mathcal{E}(j), \quad j = \sum_{i=1}^{k} j_i,$$

are given that satisfy the associativity relation

$$\gamma(\gamma(u \times v_1 \times \cdots \times v_k) \times w_1 \times \cdots \times w_j) = \gamma(u \times u_1 \times \cdots \times u_k),$$

where $u \in \mathcal{E}(k)$, $v_i \in \mathcal{E}(j_i)$, $w_s \in \mathcal{E}(i_s)$, and

$$u_i = \begin{cases} \gamma(v_i \times w_{j_1+\cdots+j_{i-1}+1} \times \cdots \times w_{j_1+\cdots+j_i}), & j_i \neq 0, \\ *, & j_i = 0. \end{cases}$$

3. There is a distinguished element $1 \in \mathcal{E}(1)$ (the *identity element*) such that $\gamma(1 \times v) = v$ for any $v \in \mathcal{E}(j)$ and $\gamma(u \times 1 \times \cdots \times 1) = u$ for any $u \in \mathcal{E}(k)$.

4. Right actions of the symmetric groups Σ_j on the spaces $\mathcal{E}(j)$ are given, and the following equivariance relations hold:

$$\gamma(u\sigma \times v_1 \times \cdots \times v_k) = \gamma(u \times v_{\sigma^{-1}(1)} \times \cdots \times v_{\sigma^{-1}(k)})\sigma(j_1, \ldots, j_k),$$
$$\gamma(u \times v_1\tau_1 \times \cdots \times v_k\tau_k) = \gamma(u \times v_1 \times \cdots \times v_k)(\tau_1 \times \cdots \times \tau_k),$$

where $\sigma(j_1, \ldots, j_k)$ is the permutation of j elements which is defined by partitioning the set of these elements into k blocks of $j_1, \ldots, j_k$ elements and acting on the blocks by the permutation σ; $\tau_1 \times \cdots \times \tau_k$ is the image of $(\tau_1, \ldots, \tau_k)$ under the natural embedding of the direct product $\Sigma_{j_1} \times \cdots \times \Sigma_{j_k}$ into Σ_j.

A *map of operads* $f\colon \mathcal{E} \to \mathcal{F}$ is a family $f = \{f(j)\}$ of Σ_j-equivariant continuous maps $f(j)\colon \mathcal{E}(j) \to \mathcal{F}(j)$ such that the diagrams

$$\begin{array}{ccc} \mathcal{E}(k) \times \mathcal{E}(j_1) \times \cdots \times \mathcal{E}(j_k) & \xrightarrow{\gamma} & \mathcal{E}(j) \\ {\scriptstyle f(k)\times f(j_1)\times\cdots\times f(j_k)}\big\downarrow & & \big\downarrow{\scriptstyle f(j)} \\ \mathcal{F}(k) \times \mathcal{F}(j_1) \times \cdots \times \mathcal{F}(j_k) & \xrightarrow{\gamma} & \mathcal{F}(j) \end{array}$$

are commutative and $f(1)(1) = 1$.

Suppose that X is an arbitrary pointed space. We shall define an operad, which will be denoted by $\mathcal{E}_X$, by letting $\mathcal{E}_X(j)$ be the space of all pointed maps $X^{\times j} \to X$ ($X^{\times 0} = *$).

The operation $\gamma\colon \mathcal{E}_X(k) \times \mathcal{E}_X(j_1) \times \cdots \times \mathcal{E}_X(j_k) \to \mathcal{E}_X(j)$ is given by

$$\gamma(g \times g_1 \times \cdots \times g_k) = g(g_1 \times \cdots \times g_k),$$

where $g \in \mathcal{E}_X(k)$ and $g_i \in \mathcal{E}_X(j_i)$.

The identity element $1 \in \mathcal{E}_X(1)$ is the identity map of the space X.

The action of the symmetric groups is defined by

$$(g\sigma)(x) = g(\sigma x), \quad g \in \mathcal{E}_X(j), \quad \sigma \in \Sigma_j, \quad x \in X^{\times j},$$

and Σ_j acts on $X^{\times j}$ by permuting the components, i.e.,

$$\sigma(x_1 \times \cdots \times x_j) = (x_{\sigma^{-1}(1)} \times \cdots \times x_{\sigma^{-1}(j)}).$$

Clearly, thus we obtain an operad. It is called the *endomorphism operad of the space* X.

A space X is called a *space over an operad* $\mathcal{E}$ or just an $\mathcal{E}$*-space* if an operad map $\mu: \mathcal{E} \to \mathcal{E}_X$ is given.

Note that giving such an operad map is equivalent to giving a family of maps

$$\mu(j): \mathcal{E}(j) \times X^{\times j} \to X.$$

Each point $u \in \mathcal{E}(2)$ defines a multiplication on X, i.e., $\pi = \mu(2)(u): X \times X \to X$. If the space $\mathcal{E}(3)$ is arcwise connected, then $\gamma(u \times u \times 1)$ and $\gamma(u \times 1 \times u)$ can be connected by a path in it, which means that the multiplication π is homotopy associative. If the space $\mathcal{E}(2)$ is arcwise connected, then u and uT, where T is the generator of Σ_2, can be connected by a path, which means that the multiplication π is homotopy commutative. The connectivity of the spaces $\mathcal{E}(j)$ in higher dimensions leads to higher conditions of homotopy commutativity and associativity and to homotopy coordination between them.

The simplest examples of operads are provided by the following operads E_0 and A.

The operad E_0 consists of singletons $E_0(j)$. Denote the unique point of $E_0(j)$ by $e(j)$. The operad structure γ is defined by

$$\gamma(e(k) \times e(j_1) \times \cdots \times e(j_k)) = e(j_1 + \cdots + j_k).$$

The action of the symmetric group Σ_j on the space $E_0(j)$ is trivial.

The operad A consists of the spaces $A(j)$, where $A(j)$ is the Σ_j-free space generated by $E_0(j)$. The operad structure on A is induced by the operad structure on E_0.

It is clear that giving an A-space structure on a space X means giving a monoid structure on it, and giving an E_0-space structure is equivalent to giving a commutative monoid structure.

Let us reformulate the definition of an operad by using a specific product of families of topological spaces.

Suppose that $\mathcal{E}' = \{\mathcal{E}'(j)\}$ and $\mathcal{E}'' = \{\mathcal{E}''(j)\}$ are families of topological spaces on which the symmetric groups Σ_j act. Such families are said to be *symmetric*. For the symmetric families $\mathcal{E}'$ and $\mathcal{E}''$, let us define the symmetric family $\mathcal{E}' \times \mathcal{E}''$ by setting $(\mathcal{E}' \times \mathcal{E}'')(j)$ to be the quotient space of the Σ_j-free space generated by

$$\bigcup_{j_1+\cdots+j_k=j} \mathcal{E}'(k) \times \mathcal{E}''(j_1) \times \cdots \times \mathcal{E}''(j_k)$$

modulo the equivalence relation generated by the following relations:

$$\begin{aligned} u'\sigma \times v_1'' \times \cdots \times v_k'' &\sim u' \times v_{\sigma^{-1}(1)}'' \times \cdots \times v_{\sigma^{-1}(k)}'' \times \sigma(j_1, \ldots, j_k), \\ u' \times v_1''\tau_1 \times \cdots \times v_k''\tau_k &\sim u' \times v_1'' \times \cdots \times v_k'' \times (\tau_1 \times \cdots \times \tau_k), \end{aligned}$$

where $u' \in \mathcal{E}'(k)$, $v_i'' \in \mathcal{E}''(j_i)$, $1 \le i \le k$.

Thus we have

$$(\mathcal{E}' \times \mathcal{E}'')(j) = \bigcup_{j_1+\cdots+j_k=j} \mathcal{E}'(k) \times \mathcal{E}''(j_1) \times \cdots \times \mathcal{E}''(j_k)/\sim.$$

It is readily seen that the multiplication of symmetric families defined above is associative in the sense that for symmetric families $\mathcal{E}$, $\mathcal{E}'$, and $\mathcal{E}''$, there is a natural isomorphism

$$(\mathcal{E} \times \mathcal{E}') \times \mathcal{E}'' \cong \mathcal{E} \times (\mathcal{E}' \times \mathcal{E}'').$$

Note that even the ordinary direct product of the topological spaces X, X', and X'' is not associative in the sense of the equality

$$(X \times X') \times X'' = X \times (X' \times X''),$$

since the elements of the space on the left-hand side are the collections $((x, x'), x'')$, while on the right-hand side the elements are $(x, (x', x''))$. Associativity of the direct product is understood in the sense of a natural isomorphism

$$(X \times X') \times X'' \cong X \times (X' \times X'').$$

For any space X with an associative multiplication $\pi\colon X \times X \to X$, commutativity of the diagram

$$\begin{array}{ccc} X \times X \times X & \xrightarrow{\pi\times 1} & X \times X \\ {\scriptstyle 1\times\pi}\downarrow & & \downarrow{\scriptstyle \pi} \\ X \times X & \xrightarrow{\pi} & X \end{array}$$

is understood as commutativity of the diagram

$$\begin{array}{ccc} (X \times X) \times X & \xrightarrow{\pi\times 1} & X \times X \\ {\scriptstyle \cong}\downarrow & & \downarrow{\scriptstyle =} \\ X \times (X \times X) & & X \times X \\ {\scriptstyle 1\times\pi}\downarrow & & \downarrow{\scriptstyle \pi} \\ X \times X & \xrightarrow{\pi} & X \end{array}$$

Associativity for the product of symmetric families will be understood in a similar way.

It is readily seen that a symmetric family $\mathcal{E}$ is an operad if there is given a map $\gamma: \mathcal{E} \times \mathcal{E} \to \mathcal{E}$ satisfying the following conditions:

– associativity, i.e., $\gamma(\gamma \times 1) = \gamma(1 \times \gamma)$, which can be expressed as the commutative diagram

$$\begin{array}{ccc} \mathcal{E} \times \mathcal{E} \times \mathcal{E} & \xrightarrow{\gamma \times 1} & \mathcal{E} \times \mathcal{E} \\ {\scriptstyle 1\times\gamma}\downarrow & & \downarrow{\scriptstyle \gamma} \\ \mathcal{E} \times \mathcal{E} & \xrightarrow{\gamma} & \mathcal{E} \end{array}$$

– existence of an identity $1 \in \mathcal{E}(1)$ satisfying the relations $\gamma(u \times 1) = \gamma(1 \times u) = u$, which can be expressed as the commutative diagram

$$\begin{array}{ccccc} \mathcal{E}(.2) & \xrightarrow{i'} & \mathcal{E} \times \mathcal{E} & \xleftarrow{i''} & \mathcal{E} \\ & \searrow^{=} & \downarrow{\scriptstyle \gamma} & \swarrow^{=} & \\ & & \mathcal{E} & & \end{array}$$

where i' and i'' are embeddings, $i'(u) = u \times 1 \times \cdots \times 1$, $i''(u) = 1 \times u$.

Thus an operad is nothing else but a monoid in the category of symmetric families with the multiplication defined above.

1.3. Little cubes operads and loop spaces

The main examples of operads are the little n-cubes operads E_n introduced by Boardman and Vogt [**7**] and studied by May [**29**]. Let us recall their definition.

Let J^n be the open unit n-dimensional cube. An (*open*) *little n-cube* is an arbitrary linear embedding u of the cube J^n into itself parallel to the axes. Such an embedding has the form $u = u_1 \times \cdots \times u_n$, where $u_i: J \to J$ are linear functions of the form

$$u_i(t) = (y_i - x_i)t + x_i, \quad 0 \le x_i < y_i \le 1.$$

Denote by $E_n(j)$ the set of j-tuples $(u^1, \ldots, u^j)$ of little n-cubes whose images are pairwise disjoint. These tuples $(u^1, \ldots, u^j)$ can be regarded as maps ${}^jJ^n \to J^n$ of the disjoint union of j copies of the unit cube into the unit cube.

By definition, the set $E_n(0)$ consists of the unique embedding of the empty set into the cube J^n.

Thus the set $E_n(j)$ is a subset of the space of all continuous maps ${}^jJ^n \to J^n$. Let us endow $E_n(j)$ with the topology of this space.

The operad structure on the family $E_n = \{E_n(j)\}$ is defined as follows:

1. If $u \in E_n(k)$, $u: {}^kJ^n \to J^n$, and $v_i \in E_n(j_i)$, $v_i: {}^{j_i}J^n \to J^n$, $1 \le i \le k$, then the element $\gamma(u \times v_1 \times \cdots \times v_k) \in E_n(j_1 + \cdots + j_k)$ is defined as the composition

$$u \circ (v_1 \cup \cdots \cup v_k): {}^{j_1}J^n \cup \cdots \cup {}^{j_k}J^n \to J^n.$$

2. The element $1 \in E_n(1)$ is the identity map $J^n \to J^n$.
3. The action of the symmetric groups is defined by

$$(u^1, \ldots, u^j)\sigma = (u^{\sigma(1)}, \ldots, u^{\sigma(j)}), \quad \sigma \in \Sigma_j.$$

Note that since the little cubes constituting a point of the space $E_n(j)$ have disjoint images, the symmetric group Σ_j acts freely on $E_n(j)$.

Let us show that any n-fold loop space $\Omega^n X$ is an E_n-space. We shall interpret the n-fold loop space $\Omega^n X$ as the space of all continuous maps of the unit cube I^n into X that take the boundary of this cube to the base point $* \in X$. Let us define maps

$$\mu_n(j)\colon E_n(j) \times (\Omega^n X)^{\times j} \to \Omega^n X$$

by setting

$$\mu_n(j)(u; y_1, \dots, y_j)(t) = \begin{cases} y_i((u^i)^{-1}(t)), & t \in \operatorname{Im} u^i, \\ * & \text{otherwise}, \end{cases}$$

for $u = (u^1, \dots, u^j) \in E_n(j)$, $y_i \in \Omega^n X$, $1 \le i \le j$, and $t \in I^n$. Direct verification shows that the required relations hold.

The E_n-space structure on an n-fold loop space is natural in the sense that if $f\colon X' \to X''$ is a continuous map of topological spaces, then the following diagram is commutative:

$$\begin{array}{ccc} E_n(j) \times (\Omega^n X')^{\times j} & \xrightarrow{\mu'_n(j)} & \Omega^n X' \\ \Big\downarrow{\scriptstyle \mathrm{Id} \times (\Omega^n f)^{\times j}} & & \Big\downarrow{\scriptstyle \Omega^n f} \\ E_n(j) \times (\Omega^n X'')^{\times j} & \xrightarrow{\mu''_n(j)} & \Omega^n X'' \end{array}$$

The embeddings $i_n(j)\colon E_n(j) \to E_{n+1}(j)$ given by

$$i_n(j)(u^1, \dots, u^j) = (u^1 \times 1, \dots, u^j \times 1),$$

where $1\colon J \to J$, define the operad map $i_n\colon E_n \to E_{n+1}$. Thus the following operad is defined:

$$E_\infty = \{E_\infty(j)\}, \quad E_\infty(j) = \varinjlim E_n(j).$$

It is clear that since the operads E_n have free actions of the symmetric groups, the operad E_∞ is also an operad with a free action of the symmetric groups.

Let us show that an infinite loop space is an E_∞-space. Indeed, if the space X is a loop space, i.e., $X = \Omega X'$, then, as can be readily seen, we have the commutative diagrams

$$\begin{array}{ccc} E_n(j) \times (\Omega^n X)^{\times j} & \xrightarrow{\mu_n(j)} & \Omega^n X \\ \Big\downarrow{\scriptstyle i_n(j) \times \mathrm{Id}^{\times j}} & & \Big\downarrow{\scriptstyle \mathrm{Id}} \\ E_{n+1}(j) \times (\Omega^{n+1} X')^{\times j} & \xrightarrow{\mu_{n+1}(j)} & \Omega^{n+1} X' \end{array}$$

Hence if the space X is an infinite loop space, i.e., there is a sequence of spaces X_n, $n \ge 0$, such that $X_0 = X$ and $X_n = \Omega X^{n+1}$, then X is an E_∞-space.

Now let us return to the little n-cubes operads E_n acting on the n-fold loop spaces $\Omega^n X$ by the maps

$$\mu(j)\colon E_n(j) \times (\Omega^n X)^{\times j} \to \Omega^n X.$$

Note that the existence of the higher relations of homotopy associativity and homotopy commutativity for the multiplication in n-fold and infinite loop spaces is determined by the connectivity of the spaces $E_n(j)$. The greater the order of connectivity is, the more of these relations hold.

Now let us elucidate the question of connectivity of the spaces $E_n(j)$. To do this, consider the configuration spaces $F(\mathbb{R}^n; j)$.

Recall that if M is a manifold of dimension n, then its *jth configuration space* is the space

$$F(M;j) = \{(x_1, \ldots, x_j) : \ x_i \in M, \ x_i \neq x_k \text{ if } i \neq k\} \subset M^j.$$

It is a jn-dimensional manifold, $F(M;1) = M$. The group Σ_j acts freely on $F(M;j)$ by permuting the components

$$(x_1, \ldots, x_j)\sigma = (x_{\sigma(1)}, \ldots, x_{\sigma(j)}).$$

Let us show that for any $n \geq 1$ and $j \geq 1$, the space $E_n(j)$ is Σ_j-equivariantly homotopy equivalent to $F(\mathbb{R}^n; j)$ or, what is the same, to $F(J^n; j)$.

We define the map $g: E_n(j) \to F(J^n; j)$ by setting

$$g(u) = \left(u^1(c), \ldots, u^j(c)\right),$$

where

$$u = \left(u^1, \ldots, u^j\right) \in E_n(j), \quad c = \left(\frac{1}{2}, \ldots, \frac{1}{2}\right) \in J^n.$$

Let us define the map $f: F(J^n; j) \to E_n(j)$. Let $(x_1, \ldots, x_j) \in F(J^n; j)$. Consider disjoint little n-cubes $u^1, \ldots, u^j \in J^n$ centered at the points $x_1, \ldots, x_j$, respectively, and having sides of the same length d. Clearly, for a sufficiently small d, such little cubes exist. Choose $(u^1, \ldots, u^j)$ so that d is as large as possible. Let $f(x_1, \ldots, x_j) = (u^1, \ldots, u^j)$.

It is clear that the maps f and g are continuous, Σ_j-equivariant, and satisfy the relation $g \circ f = \mathrm{Id}$. In [**29**] the homotopy $h: E_n(j) \times I \to E_n(j)$ between $f \circ g$ and Id is defined as follows.

Let $u = (u^1, \ldots, u^j) \in E_n(j)$, $u^k = u^k_1 \times \cdots \times u^k_n$, where the linear functions $u^k_i: J \to J$ are of the form $u^k_i(t) = (y^k_i - x^k_i)t + x^k_i$. We set

$$h(u, s) = \left(\prod_{i=1}^{n} h^1_i(s), \ldots, \prod_{i=1}^{n} h^j_i(s)\right), \quad 0 \leq s \leq 1,$$

where

$$h^k_i(s)(t) = [(1-s)(y^k_i - x^k_i) + sd]t + \frac{1}{2}[sy^k_i + (2-s)x^k_i - sd]$$

and d is the side of the equilateral element $fg(u)$.

The homotopy h stretches or contracts each coordinate interval (x^k_i, y^k_i) linearly into an interval of length d, leaving the middle point of the interval fixed. It is Σ_j-equivariant and continuous. However, in the process of this homotopy the little cubes may intersect. To avoid this, we must improve the homotopy, expressing it as a composition of two homotopies. First, the initial little cubes are contracted to equilateral little cubes with side d' equal to the minimum of the sides of the initial little cubes, and then the resulting little cubes are stretched to obtain little cubes with side d. Clearly, in this process the little cubes cannot intersect.

Thus the space $F(J^n, j)$ is a Σ_j-equivariant deformation retract of $E_n(j)$.

Consider the space

$$F_0 = \{(x_1, \ldots, x_j) : \ x_1 < \cdots < x_j\} \subset F(\mathbb{R}^1; j).$$

It is homeomorphic to the interior of the n-dimensional simplex and hence is contractible. On the other hand, it is an arcwise connected component of the space

$F(\mathbb{R}^1; j)$, and the action of Σ_j on $F(\mathbb{R}^1; j)$ defines a homeomorphism

$$F_0 \times \Sigma_j \cong F(\mathbb{R}^1; j).$$

Thus the arcwise connected components of the space $F(\mathbb{R}^1; j)$ are in one-to-one correspondence with the elements of the symmetric group Σ_j, and each component is contractible. The projection $F(\mathbb{R}^1; j) \to \Sigma_j$ defines an operad map $E_1 \to A$, which is a homotopy equivalence.

For $n \geq 2$, consider the projection $p: F(\mathbb{R}^n; j) \to \mathbb{R}^n$ onto the first coordinate, so that the fiber over a point y_1 is the space $F(\mathbb{R}^n \setminus \{y_1\}; j-1)$.

We define the map

$$f: \mathbb{R}^n \times F(\mathbb{R}^n \setminus \{y_1\}; j-1) \to F(\mathbb{R}^n; j)$$

by setting

$$f(x_1, (x_2, \dots, x_j)) = (x_1, x_2 + x_1 - y_1, \dots, x_j + x_1 - y_1).$$

Its inverse is the map

$$g: F(\mathbb{R}^n; j) \to \mathbb{R}^n \times F(\mathbb{R}^n \setminus \{y_1\}; j-1)$$

defined by

$$g(x_1, \dots, x_j) = (x_1, (x_2 + y_1 - x_1, \dots, x_j + y_1 - x_1)).$$

Thus we obtain a homeomorphism $F(\mathbb{R}^n; j) \cong F(\mathbb{R}^n \setminus \{y_1\}; j-1)$. In particular, we have $F(\mathbb{R}^n; 2) \cong \mathbb{R}^n \setminus \{y_1\} \cong S^{n-1}$.

1.4. Operads and monads

Let $\mathcal{T}$ be an arbitrary category. A *monad* (C, γ, i) *on the category* $\mathcal{T}$ is a (covariant) functor $C: \mathcal{T} \to \mathcal{T}$ together with natural transformations of functors $\gamma: C \circ C \to C$ and $i: \mathrm{Id} \to C$ such that the following diagrams are commutative:

$$\begin{array}{ccc} C^3 & \xrightarrow{\gamma\circ 1} & C^2 \\ {\scriptstyle 1\circ\gamma}\downarrow & & \downarrow{\scriptstyle \gamma} \\ C^2 & \xrightarrow{\gamma} & C \end{array} \qquad \begin{array}{ccccc} C & \xrightarrow{C\circ i} & C^2 & \xleftarrow{i\circ C} & C \\ & \searrow^{=} & \downarrow{\scriptstyle \gamma} & {}^{=}\swarrow & \\ & & C & & \end{array}$$

A *morphism of monads* $f: (C', \gamma', i') \to (C'', \gamma'', i'')$ is a natural transformation $f: C' \to C''$ such that the following diagrams are commutative:

$$\begin{array}{ccc} C'^2 & \xrightarrow{f^2} & C''^2 \\ {\scriptstyle \gamma'}\downarrow & & \downarrow{\scriptstyle \gamma''} \\ C' & \xrightarrow{f} & C'' \end{array} \qquad \begin{array}{ccc} & \mathrm{Id} & \\ {}^{i'}\swarrow & & \searrow^{i''} \\ C' & \xrightarrow{f} & C'' \end{array}$$

An *algebra* (X, μ) *over a monad* (C, γ, i) or just a C-*algebra* is an object $X \in \mathcal{T}$ together with a morphism $\mu: CX \to X$ such that the following diagrams are commutative:

$$\begin{array}{ccc} C^2X & \xrightarrow{\gamma X} & CX \\ {\scriptstyle C\mu}\downarrow & & \downarrow{\scriptstyle \mu} \\ CX & \xrightarrow{\mu} & X \end{array} \qquad \begin{array}{ccc} X & \xrightarrow{i} & CX \\ & \searrow{\scriptstyle =} & \downarrow{\scriptstyle \mu} \\ & & X \end{array}$$

A *morphism of C-algebras* $f{:}\,(X',\mu') \to (X'',\mu'')$ is the morphism $f{:}\,X' \to X''$ in $\mathcal{T}$ such that the following diagram is commutative:

$$\begin{array}{ccc} CX' & \xrightarrow{Cf} & CX'' \\ {\scriptstyle \mu'}\downarrow & & \downarrow{\scriptstyle \mu''} \\ X' & \xrightarrow{f} & X'' \end{array}$$

Let us discuss some examples of monads and algebras over monads.

1. Let G be a topological monoid. The functor on the category of topological spaces that assigns the space $G \times X$ to a space X will also be denoted by G. The multiplication in G determines a natural transformation $\pi{:}\,G \circ G \to G$, and the identity element in G determines natural maps $i(X){:}\,X \to G(X)$.

It is clear that (G,π,i) is a monad on the category of topological spaces. Algebras over this monad are topological spaces acted on by the monoid G.

2. Let X be a topological space with base point $*$, and denote by TX the free monoid generated by X,

$$TX = \bigcup_n X^{\times n}/\sim,$$

where the equivalence relation $\sim$ is generated by the relations

$$x_1 \times \cdots \times x_{i-1} \times * \times x_{i+1} \times \cdots \times x_n \sim x_1 \times \cdots \times x_{i-1} \times x_{i+1} \times \cdots \times x_n.$$

The map $\gamma X{:}\,T^2X \to TX$ is defined by removing the parentheses, i.e.,

$$\gamma X((x_1 \times \cdots \times x_n) \times \cdots \times (y_1 \times \cdots \times y_m)) = x_1 \times \cdots \times x_n \times \cdots \times y_1 \times \cdots \times y_m.$$

It is readily seen that the functor T is a monad on the category of pointed topological spaces, and algebras over this monad are topological monoids.

3. In the category of pointed topological spaces, consider a correspondence assigning to a space X the n-fold loop space $\Omega^n S^n X$ of the n-fold suspension of X, $X \longmapsto \Omega^n S^n X$. There is a natural transformation $\eta{:}\,S^n\Omega^n X \to X$ defined by

$$\eta(t,u) = u(t), \quad u \in \Omega^n X.$$

Let us define a natural transformation $\gamma{:}\,\Omega^n S^n \Omega^n S^n X \to \Omega^n S^n X$ by setting $\gamma = \Omega^n \eta S^n X$. The maps $i{:}\,X \to \Omega^n S^n X$ are induced by the identity maps $S^n X \to S^n X$.

It is easy to check that $(\Omega^n S^n, \gamma, i)$ is a monad.

4. Suppose that $F{:}\,\mathcal{T}' \to \mathcal{T}''$ and $G{:}\,\mathcal{T}'' \to \mathcal{T}'$ are adjoint functors, i.e., there are natural transformations $\xi{:}\,\mathrm{Id} \to G \circ F$ and $\eta{:}\,F \circ G \to \mathrm{Id}$ such that the composite maps

$$F \xrightarrow{F\xi} F \circ G \circ F \xrightarrow{\eta F} F, \quad G \xrightarrow{\xi G} G \circ F \circ G \xrightarrow{G\eta} G$$

are identity transformations.

Denote the composition $G \circ F$ by C. Then the natural transformation η induces a transformation $\gamma = G\eta F{:}\,C \circ C \to C$. It is readily seen that (C,γ,ξ) is a monad.

A special case of this monad is the monad $\Omega^n S^n$, since the functors of n-fold suspension S^n and n-fold loop space Ω^n are adjoint.

Let us show that any operad defines a monad on the category of pointed topological spaces, and a space is a space over the operad if and only if it is an algebra over the corresponding monad.

Let $\mathcal{E} = \{\mathcal{E}(j)\}$ be an operad; consider the maps

$$\sigma^i\colon \mathcal{E}(j) \to \mathcal{E}(j-1), \quad 0 \le i < j,$$

defined by

$$\sigma^i(u) = \gamma(u \times 1^{\times i} \times * \times 1^{\times j-i-1}), \qquad 1 \in \mathcal{E}(1), \quad * \in \mathcal{E}(0).$$

For any space X, consider the maps $s_i\colon X^{\times j-1} \to X^{\times j}$ given by

$$s_i(x_1 \times \cdots \times x_{j-1}) = x_1 \times \cdots \times x_i \times * \times x_{i+1} \times \cdots \times x_{j-1}.$$

We now define a space $\underline{\mathcal{E}}X$ by setting

$$\underline{\mathcal{E}}X = \bigcup_{j \ge 0} \mathcal{E}(j) \times X^{\times j} / \sim,$$

where the equivalence relation $\sim$ is generated by the relations

$$\sigma^i(u) \times x \sim u \times s_i(x), \quad u \in \mathcal{E}(j), \quad x \in X^{\times j-1};$$
$$u\sigma \times x \sim u \times \sigma x, \quad u \in \mathcal{E}(j), \quad \sigma \in \Sigma_j, \quad x \in X^{\times j}.$$

The correspondence $X \longmapsto \underline{\mathcal{E}}X$ defines a functor $\underline{\mathcal{E}}$ on the category of pointed topological spaces. The operad structure on $\mathcal{E}$ defines the monad structure on $\underline{\mathcal{E}}$. A space X is a $\mathcal{E}$-space if and only if it is an algebra over the monad $\underline{\mathcal{E}}$.

The space $\underline{\mathcal{E}}X$ can be regarded as a free $\mathcal{E}$-space generated by X in the sense that if Y is a $\mathcal{E}$-space, then for any continuous map $f\colon X \to Y$ there is a unique map of $\mathcal{E}$-spaces $\tilde{f}\colon \underline{\mathcal{E}}X \to Y$ such that the following diagram is commutative:

$$\begin{array}{ccc} X & \xrightarrow{i} & \underline{\mathcal{E}}X \\ & \searrow^{f} & \downarrow{\tilde f} \\ & & Y \end{array}$$

Consider the space $\Omega^n S^n X$. The little n-cubes operad E_n acts on this space and, therefore, we have a map

$$\mu_n\colon \underline{E}_n(\Omega^n S^n X) \to \Omega^n S^n X.$$

Besides, the inclusion $i\colon X \to \Omega^n S^n X$ induces a map

$$\underline{E}_n(i)\colon \underline{E}_n(X) \to \underline{E}_n(\Omega^n S^n X).$$

The composition of the maps μ_n and $\underline{E}_n(i)$ yields a map $\underline{E}_n(X) \to \Omega^n S^n X$. Direct verification shows that this map is a morphism of monads, and the following diagram is commutative:

$$\begin{array}{ccc} \underline{E}_n & \longrightarrow & \Omega^n S^n \\ \downarrow & & \downarrow \\ \underline{E}_{n+1} & \longrightarrow & \Omega^{n+1} S^{n+1} \end{array}$$

Hence we can define the limit morphism of monads, i.e., $\underline{E}_\infty \to \Omega^\infty S^\infty$.

May showed (see [**29**]) that for any arcwise connected space X and for $1 \leq n \leq \infty$, the map $\underline{E}_n(X) \to \Omega^n S^n X$ is a weak homotopy equivalence. Besides, each arcwise connected E_n-space is weakly homotopy equivalent to the n-fold loop space $\Omega^n Y$ of a certain space Y. These assertions are proved by induction over n.

For $n = 1$, the space ΩSX is weakly homotopy equivalent to the free monoid TX generated by X. The map $\xi\colon TX \to \Omega SX$ defining this equivalence is given by the following formulas: $\xi(x) = i(x)$ for $x \in X$, and $\xi(x_1 \times \cdots \times x_n) = \xi(x_1) \cdot \ldots \cdot \xi(x_n)$.

To prove that this map is a homotopy equivalence, consider the cone CX of the space X and the space defined in the same way as TX with one copy of X replaced by CX, i.e.,

$$T(X, CX) = \bigcup_n X^{n-1} \times CX/\sim$$

This space is contractible; besides, we have the short exact sequence

$$0 \to TX \xrightarrow{i} T(X, CX) \xrightarrow{p} SX \to 0$$

and the commutative diagram

$$\begin{array}{ccccccccc}
0 & \longrightarrow & TX & \xrightarrow{i} & T(X,CX) & \xrightarrow{p} & SX & \longrightarrow & 0 \\
 & & \xi\downarrow & & \downarrow & & =\downarrow & & \\
0 & \longrightarrow & \Omega SX & \xrightarrow{i} & PSX & \xrightarrow{p} & SX & \longrightarrow & 0
\end{array}$$

where PSX is the path space of SX.

In this diagram, the top row is a cofibration, and the bottom row is a fibration. The middle terms of the rows are contractible. If the top row were a fibration, the required assertion would follow by comparing the long exact sequences of homotopy groups. Fortunately, it turns out that in spite of the fact that the upper row is not a fibration, it is a quasifibration, which makes it possible to compare the long exact sequences of homotopy groups and to obtain the required result.

1.5. Bioperads in the category of topological spaces

It turns out that the concept of an operad does not cover all natural structures arising on families of topological spaces. We shall introduce the concept of a bioperad and consider some situations in which this structure appears.

Suppose that $\mathcal{E} = \{\mathcal{E}(n)\}_{n\geq 1}$ and $\mathcal{F} = \{\mathcal{F}(n)\}_{n\geq 1}$ are families of topological spaces. Let

$$\mathcal{E}(k) * \mathcal{F}(n) = \mathcal{E}(k) \times \mathcal{F}(n)^{\times k}.$$

It is readily seen that the following permutation maps are defined:

$$T\colon (\mathcal{E}'(k) \times \mathcal{E}''(k)) * (\mathcal{F}'(n) \times \mathcal{F}''(n)) \to (\mathcal{E}'(k) * \mathcal{F}'(n)) \times (\mathcal{E}''(k) * \mathcal{F}''(n)).$$

Let us consider the family $\Sigma = \{\Sigma_n\}$ of symmetric groups Σ_n with discrete topology, and let us find out the structure existing on this family.

There are interior multiplications

$$\cdot\colon \Sigma_n \times \Sigma_n \to \Sigma_n,$$

induced by the group structure. Besides, there are exterior multiplications

$$\pi\colon \Sigma_n \times \Sigma_m \to \Sigma_{n+m}, \quad \gamma\colon \Sigma_k * \Sigma_n \to \Sigma_{kn}.$$

The multiplication π assigns to permutations $\sigma \in \Sigma_n$, $\tau \in \Sigma_m$ the permutation $\sigma \times \tau$ in Σ_{n+m} which acts on the first n elements as σ and on the last m elements as τ.

The multiplication γ assigns to permutations $\sigma \in \Sigma_k$ and $\tau_i \in \Sigma_n$, $1 \leq i \leq k$, the permutation $\sigma(\tau_1 \times \cdots \times \tau_k) \in \Sigma_{k \cdot n}$, acting on $(1, \ldots, k \cdot n)$ as follows. First, the set $(1, \ldots, k \cdot n)$ is partitioned into k blocks of n elements each; then, the elements of each block are permuted by permutations $\tau_1, \ldots, \tau_k$, respectively; finally, the blocks themselves are permuted by the permutation σ.

Direct verification shows that the multiplications π and γ defined above are associative and distributive, which means that the following diagrams are commutative:

$$\begin{array}{ccc}
(\Sigma_{k'} \times \Sigma_{k''}) * \Sigma_n & \xrightarrow{\pi * 1} & \Sigma_{k'+k''} * \Sigma_n \\
\downarrow & & \downarrow \gamma \\
(\Sigma_{k'} * \Sigma_n) \times (\Sigma_{k''} * \Sigma_n) & & \Sigma_{(k'+k'')n} \\
\gamma \times \gamma \downarrow & & \downarrow = \\
\Sigma_{k'n} \times \Sigma_{k''n} & \xrightarrow{\pi} & \Sigma_{k'n+k''n}
\end{array}$$

Consider the diagrams

$$\begin{array}{ccc}
\Sigma_k * (\Sigma_n \times \Sigma_m) & \xrightarrow{1 * \pi} & \Sigma_k * \Sigma_{n+m} \\
\nabla * 1 \downarrow & & \downarrow \gamma \\
(\Sigma_k \times \Sigma_k) * (\Sigma_n \times \Sigma_m) & & \Sigma_{k(n+m)} \\
T \downarrow & & \uparrow \pi \\
(\Sigma_k * \Sigma_n) \times (\Sigma_k * \Sigma_m) & \xrightarrow{\gamma \times \gamma} & \Sigma_{kn} \times \Sigma_{km}
\end{array}$$

where $\nabla \colon \Sigma_k \to \Sigma_k \times \Sigma_k$ is the diagonal map, $\nabla(\sigma) = \sigma \times \sigma$.

These diagrams are not commutative. However, they are commutative up to certain isomorphisms $\xi \colon \Sigma_{kn+km} \to \Sigma_{kn+km}$. To define these isomorphisms, we fix a partition of $(1, \ldots, k(n+m))$ into alternating blocks of n and m elements each and consider the permutation $\xi \in \Sigma_{k(n+m)}$ which places first the blocks of n elements and then the blocks of m elements. We define an isomorphism ξ by setting $\xi(x) = \xi^{-1} x \xi$. Straightforward computations show that if this isomorphism is put in the diagram after the map π, then the resulting diagram is commutative.

We shall now give the main definition.

DEFINITION. A symmetric family $\mathcal{E}$ is called a *bioperad* if maps

$$\pi \colon \mathcal{E}(n) \times \mathcal{E}(m) \to \mathcal{E}(n+m), \quad \gamma \colon \mathcal{E}(k) * \mathcal{E}(n) \to \mathcal{E}(kn)$$

are given that satisfy the following conditions:

1. The maps π and γ are compatible with the actions of the symmetric groups, that is,

$$\pi(x\sigma \times y\tau) = \pi(x \times y)\pi(\sigma \times \tau),$$

where $x \in \mathcal{E}(n)$, $y \in \mathcal{E}(m)$, $\sigma \in \Sigma_n$, $\tau \in \Sigma_m$;

$$\gamma(x\sigma \times y_1\tau_1 \times \cdots \times y_k\tau_k) = \gamma(x \times y_{\sigma^{-1}(1)} \times \cdots \times y_{\sigma^{-1}(k)})\gamma(\sigma \times \tau_1 \cdots \times \tau_k),$$

where $x \in \mathcal{E}(k)$, $y_i \in \mathcal{E}(n)$, $\sigma \in \Sigma_k$, $\tau_i \in \Sigma_n$.

2. The maps π and γ are associative, i.e., the following diagrams are commutative:

$$\begin{array}{ccc}
\mathcal{E}(k) \times \mathcal{E}(n) \times \mathcal{E}(m) & \xrightarrow{\pi \times 1} & \mathcal{E}(k+n) \times \mathcal{E}(m) \\
{\scriptstyle 1\times\pi}\downarrow & & \downarrow{\scriptstyle \pi} \\
\mathcal{E}(k) \times \mathcal{E}(n+m) & \xrightarrow{\pi} & \mathcal{E}(k+n+m)
\end{array}$$

$$\begin{array}{ccc}
\mathcal{E}(k) * \mathcal{E}(n) * \mathcal{E}(m) & \xrightarrow{\gamma * 1} & \mathcal{E}(kn) * \mathcal{E}(m) \\
{\scriptstyle 1*\gamma}\downarrow & & \downarrow{\scriptstyle \gamma} \\
\mathcal{E}(k) * \mathcal{E}(nm) & \xrightarrow{\gamma} & \mathcal{E}(knm)
\end{array}$$

3. There is given an identity element $1 \in \mathcal{E}(1)$ such that

$$\gamma(1 * x) = \gamma(x * 1) = x, \quad x \in \mathcal{E}(n).$$

4. The distributivity relations hold, which means that the following diagrams are commutative:

$$\begin{array}{ccc}
(\mathcal{E}(k') \times \mathcal{E}(k'')) * \mathcal{E}(n) & \xrightarrow{\pi * 1} & \mathcal{E}(k'+k'') * \mathcal{E}(n) \\
\downarrow & & \downarrow{\scriptstyle \gamma} \\
(\mathcal{E}(k') * \mathcal{E}(n)) \times (\mathcal{E}(k'') * \mathcal{E}(n)) & & \mathcal{E}((k'+k'')n) \\
{\scriptstyle \gamma\times\gamma}\downarrow & & \downarrow{\scriptstyle =} \\
\mathcal{E}(k'n) \times \mathcal{E}(k''n) & \xrightarrow{\pi} & \mathcal{E}(k'n+k''n)
\end{array}$$

The diagrams

$$\begin{array}{ccc}
\mathcal{E}(k) * (\mathcal{E}(n) \times \mathcal{E}(m)) & \xrightarrow{1 * \pi} & \mathcal{E}(k) * \mathcal{E}(n+m) \\
{\scriptstyle \nabla * 1}\downarrow & & \downarrow{\scriptstyle \gamma} \\
(\mathcal{E}(k) \times \mathcal{E}(k)) * (\mathcal{E}(n) \times \mathcal{E}(m)) & & \mathcal{E}(k(n+m)) \\
\downarrow & & \uparrow{\scriptstyle \xi\circ\pi} \\
(\mathcal{E}(k) * \mathcal{E}(n)) \times (\mathcal{E}(k) * \mathcal{E}(m)) & \xrightarrow{\gamma\times\gamma} & \mathcal{E}(kn) \times \mathcal{E}(km)
\end{array}$$

are commutative up to certain homeomorphisms

$$\xi \colon \mathcal{E}(k(n+m)) \to \mathcal{E}(k(n+m))$$

such that $\xi(x\sigma) = \xi(x)\xi(\sigma)$, $x \in \mathcal{E}(k(n+m))$, $\sigma \in \Sigma_{k(n+m)}$.

A bioperad E is called an E_∞-*bioperad* if the spaces $E(n)$ are contractible and the symmetric groups act freely on them.

Note that the concept of a bioperad differs from that of an operad. In the case of an operad $\mathcal{E}$, it is required that the operations

$$\gamma \colon \mathcal{E}(k) \times \mathcal{E}(n_1) \times \cdots \times \mathcal{E}(n_k) \to \mathcal{E}(n_1 + \cdots + n_k)$$

be defined for arbitrary positive integers $n_1, \dots, n_k$, and not just for equal ones as in the case of bioperads. On the other hand, for an operad it is not required in general that the operations π be given nor that the distributivity relations hold.

DEFINITION. Let $\mathcal{F} = \{\mathcal{F}(n)\}$ be a family of topological spaces, and let $\mathcal{E}$ be a bioperad. We say that the family $\mathcal{F}$ is *$\mathcal{E}$-multiplicative* if there are given maps

$$\pi: \mathcal{F}(n) \times \mathcal{F}(m) \to \mathcal{F}(n+m), \quad \mu: \mathcal{E}(k) * \mathcal{F}(n) \to \mathcal{F}(kn)$$

compatible with the actions of the symmetric groups and satisfying the associativity and distributivity relations.

A topological space X is said to be *$\mathcal{E}$-multiplicative* if the corresponding family $\{X(n)\}$, $X(n) = X$, is $\mathcal{E}$-multiplicative.

If E is an E_∞-bioperad, then an E-multiplicative family F is called an *E_∞-multiplicative family*, and an E-multiplicative space X is called an *E_∞-multiplicative space.*

In 1966 Milgram [**33**] introduced polyhedra $C(k)$ and used them to construct an approximation of the iterated loop spaces $\Omega^n S^n X$ of a connected CW-complex X.

Later, approximations of iterated loop spaces of arcwise connected topological spaces were constructed by May with the help of the little n-cubes operads E_n. However, the operads E_n are not CW-complexes.

Thus, on the one hand, we have the Milgram polyhedra $C(k)$, on which there is no natural operad structure. On the other hand, we have the little n-cubes operads E_n, but they are not CW-complexes. We shall show that there is a bioperad structure on the Milgram polyhedra.

Let us recall the definition of the Milgram polyhedra $C(k)$. We denote the set of real numbers by $\mathbb{R}$. The symmetric groups Σ_n act on $\mathbb{R}^n$ by permuting the components,

$$\sigma(x_1, \dots, x_n) = (x_{\sigma^{-1}(1)}, \dots, x_{\sigma^{-1}(n)}), \quad \sigma \in \Sigma_n.$$

Then $C(n-1)$ is the convex hull of the set of points $\sigma(1, 2, \dots, n) \in \mathbb{R}^n$, $\sigma \in \Sigma_n$. It is more convenient to denote $C(n-1)$ by $E^1(n)$. The symmetric group Σ_n acts on $E^1(n)$ as a group of homeomorphisms $E^1(n) \to E^1(n)$.

The space $E^1(n)$ can also be defined as the set of all points $(x_1, \dots, x_n) \in \mathbb{R}^n$ satisfying the relations

$$\begin{gathered} x_1 + \dots + x_n = \frac{n(n+1)}{2}; \\ x_i \geq 1, \quad 1 \leq i \leq n; \\ x_{i_1} + x_{i_2} \geq 3, \quad 1 \leq i_1 < i_2 \leq n; \\ \dots \\ x_{i_1} + \dots + x_{i_{n-1}} \geq \frac{(n-1)n}{2}, \quad 1 \leq i_1 < \dots < i_{n-1} \leq n. \end{gathered}$$

Unfortunately, the family $E^1 = \{E^1(n)\}$ has no natural operad structure. However, the following assertion holds.

THEOREM. *The family $E^1 = \{E^1(n)\}$ of the Milgram polyhedra is a bioperad.*

PROOF. Let us define operations $\pi\colon E^1(n)\times E^1(m)\to E^1(n+m)$ by setting

$$\pi((x_1,\dots,x_n)\times(y_1,\dots,y_m))=(x_1,\dots,x_n,y_1+n,\dots,y_m+n).$$

It is readily seen that these operations are associative and compatible with the actions of the symmetric groups.

We define operations $\gamma\colon E^1(k)*E^1(n)\to E^1(kn)$ by setting

$$\begin{aligned}\gamma(x\times y^1\times\cdots\times y^k)&=(z_1,\dots,z_{kn})\\&=\big((x_1-1)n+y^1_1,\dots,(x_1-1)n+y^1_n,\\&\qquad(x_2-1)n+y^2_1,\dots,(x_2-1)n+y^2_n,\\&\qquad\dots\\&\qquad(x_k-1)n+y^k_1,\dots,(x_k-1)n+y^k_n\big),\end{aligned}$$

where $x=(x_1,\dots,x_k)\in E(k)$, $y^i=(y^i_1,\dots,y^i_n)\in E(n)$, $1\le i\le k$.

It is clear that the operations γ are compatible with the actions of the symmetric groups and are linear maps which assign vertices of the polyhedron $E(kn)$ to the vertices of the polyhedron $E(k)*E(n)$. Direct verification shows that the associativity and distributivity relations hold. □

Note that if $x=(1,2,\dots,k)\in E(k)$, then $\gamma(x\times y^1\times\cdots\times y^k)$ is equal to the iterated product of $y^1,\dots,y^k$. In particular, for $x=(1,2)\in E(2)$, we have $\gamma((1,2)\times y^1\times y^2)=\pi(y^1\times y^2)$.

Let us consider another example of a bioperad, also from Milgram's paper [**33**].

First, we introduce the following notation: $I(n)=[0,n]$, $I=I(1)=[0,1]$, I^n is the n-dimensional cube, and $P(n)$ is the space of the paths $f\colon I(n)\to I^n$ of length n that begin at the origin $(0,\dots,0)$ and end at the final point $(1,\dots,1)$.

THEOREM. *The family $P=\{P(n)\}$ is a bioperad.*

PROOF. Let us define operations $\pi\colon P(n)\times P(m)\to P(n+m)$ by setting

$$\pi(f\times g)(\tau)=\begin{cases}(y_1,\dots,y_n,0,\dots,0), & 0\le\tau\le n,\\(1,\dots,1,z_1,\dots,z_n), & n\le\tau\le n+m,\end{cases}$$

where $f\in P(n)$, $g\in P(m)$, $(y_1,\dots,y_n)=f(\tau)$, $(z_1,\dots,z_m)=g(\tau-n)$.

It is clear that the operations π satisfy the associativity relation and are compatible with the action of the symmetric groups.

Next we define the operations $\gamma\colon P(k)*P(n)\to P(kn)$. Let $f\colon I(k)\to I^k$ be a map. It induces a map $nf\colon I(kn)\to I^{kn}$ defined by the formula $(nf)(\tau)=n\cdot f(\tau/n)$.

Further, for any maps $g_1,\dots,g_k\colon I(n)\to I^n$, we set

$$\gamma(f\times g_1\times\cdots\times g_k)=(g_1\times\cdots\times g_k)\circ(nf).$$

The correspondence $f\longmapsto nf$ defines a linear map, and hence the operations $\gamma\colon P(k)*P(n)\to P(kn)$ are also linear. Direct verification shows that the associativity and distributivity relations hold. □

In [**33**] Milgram constructed maps $j(n)\colon E^1(n)\to P(n)$. It is easy to check that these maps preserve the bioperad structure and hence define the bioperad map $j\colon E^1\to P$.

In subsequent chapters we shall present other examples of bioperads related to classical topological groups.

CHAPTER 2

Simplicial Objects and Homotopy Theory

2.1. Simplicial and cosimplicial objects

The notions of simplicial and cosimplicial objects are among the central concepts of today's algebraic topology and homotopy theory; this is due not only to the fact that the category of topological spaces can be replaced by the category of simplicial sets, but also to the fact that simplicial methods are used in various topological and algebraic constructions.

Let us recall the main definitions. For clearness of the presentation, we shall start with the definition of a geometric simplicial complex [**54**].

Let $x^0, \dots, x^n$ be a set of linearly independent points in $\mathbb{R}^m$, $m \geq n$, i.e., of points that do not belong to an $(n-1)$-dimensional hyperplane in $\mathbb{R}^m$. Then the convex hull x_n of these points is called an *n-dimensional simplex* in the space $\mathbb{R}^m$, and the points are called the *vertices* of the simplex. We shall also say that the simplex is spanned by its vertices. The simplices spanned by the vertices forming a subset of the initial set of vertices are called the *faces* of the initial simplex.

A *geometric simplicial complex* in $\mathbb{R}^m$ is a finite set K of simplices in $\mathbb{R}^m$ satisfying the following conditions:

1. If $x_n \in K$, then any face x_i of the simplex x_n also belongs to K.
2. Distinct simplices in K either are disjoint or have a common face as their intersection.

If K is a geometric simplicial complex in $\mathbb{R}^m$, then the union of all its simplices is a subset of $\mathbb{R}^m$; this union is denoted by $|K|$ and is called a *polyhedron*.

By abstracting from the space in which a geometric simplicial complex is considered, we can give a definition of an abstract simplicial complex. Namely, suppose that a finite set of objects, which are called vertices, is given, and in this set a family K of subsets, which are called simplices, is distinguished so that the above conditions 1 and 2 hold. Then this family K is called an *abstract simplicial complex*.

A *map* $f\colon K \to L$ *of abstract simplicial complexes* is a map of their vertices under which the simplices $x \in K$ are mapped onto simplices $y \in L$. Two abstract simplicial complexes K and L are said to be *isomorphic* if there exist mutually inverse maps $f\colon K \to L$ and $g\colon L \to K$.

A *realization of an abstract simplicial complex* K is a geometric simplicial complex L isomorphic to K.

It is readily seen that every abstract simplicial complex has a realization. Indeed, let $x^0, \dots, x^n$ be vertices of an abstract simplicial complex K, and let $y^0, \dots, y^n$ be vertices of a simplex in $\mathbb{R}^n$. Consider the transformation of vertices that sends each x^i to y^i. Let L be the geometric simplicial complex in $\mathbb{R}^n$ consisting of the simplices spanned by the vertices y^i for which the corresponding vertices x^i

are the vertices of a simplex in K. Then the above transformation of vertices induces a map $f: K \to L$ of the abstract simplicial complex K onto the geometric simplicial complex L, and this map is an isomorphism.

Although simplicial complexes make it possible to reduce many topological problems to purely combinatorial ones, they are not in fact very convenient. Thus, for example, it is rather difficult to define the product of simplicial complexes. The concept of a simplicial object defined for an arbitrary category turns out to be much more suitable. We shall recall the main definitions [**28**].

Let $\mathcal{X}$ be an arbitrary category. A *simplicial object in* $\mathcal{X}$ is a family $X_* = \{X_n\}_{n\geq 0}$ of objects X_n of the category $\mathcal{X}$ together with a family of morphisms

$$d_i: X_n \to X_{n-1}, \quad s_i: X_n \to X_{n+1}, \quad 0 \leq i \leq n,$$

which are called the *face* and *degeneracy operators*, respectively, and which satisfy the relations

$$\begin{aligned} d_i d_j &= d_{j-1} d_i, \quad i < j; \\ s_i s_j &= s_j s_{i-1}, \quad i > j; \\ d_i s_j &= \begin{cases} s_{j-1} d_i, & i < j, \\ \mathrm{Id}, & i = j,\ i = j+1, \\ s_j d_{i-1}, & i > j+1. \end{cases} \end{aligned}$$

A simplicial object X_* is said to be *augmented* if an object X_{-1} and a morphism $d_0: X_0 \to X_{-1}$ satisfying the relation $d_0 d_0 = d_0 d_1$ are additionally given (the morphism d_0 is called an *augmentation*).

A *map of simplicial objects* $f_*: X_* \to Y_*$ is a family $f_* = \{f_n\}_{n\geq 0}$ of morphisms $f_n: X_n \to Y_n$ in $\mathcal{X}$ that are permutable with the face and degeneracy operators, i.e., that satisfy the following relations:

$$d_i f_n = f_{n-1} d_i, \quad s_i f_n = f_{n+1} s_i, \quad 0 \leq i \leq n.$$

Dually, a *cosimplicial object in* $\mathcal{X}$ is a family $X^* = \{X^n\}_{n\geq 0}$ of objects of $\mathcal{X}$ together with morphisms

$$\delta^i: X^{n-1} \to X^n, \quad \sigma^i: X^{n+1} \to X^n, \quad 0 \leq i \leq n,$$

which are called the *coface* and *codegeneracy operators*, respectively, and which satisfy the relations

$$\begin{aligned} \delta^j \delta^i &= \delta^i \delta^{j-1}, \quad i < j; \\ \sigma^j \sigma^i &= \sigma^{i-1} \sigma^j, \quad i > j; \\ \sigma^j \delta^i &= \begin{cases} \delta^i \sigma^{j-1}, & i < j, \\ \mathrm{Id}, & i = j,\ i = j+1, \\ \delta^{i-1} \sigma^j, & i > j+1. \end{cases} \end{aligned}$$

A cosimplicial object X^* is said to be *augmented* if an object X^{-1} and a morphism $\delta^0: X^{-1} \to X^0$ such that $\delta^0 \delta^0 = \delta^1 \delta^0$ are additionally given (the morphism δ^0 is called an *augmentation*).

A *map of cosimplicial objects* $f^*: X^* \to Y^*$ is a family $f = \{f^n\}_{n \geq 0}$ of morphisms $f^n: X^n \to Y^n$ in $\mathcal{X}$ that are permutable with the coface and codegeneracy operators, i.e., that satisfy the following relations:

$$\delta^i f_{n-1} = f^n \delta^i, \quad \sigma^i f^{n+1} = f^n \sigma^i, \quad 0 \leq i \leq n.$$

Denote the category of simplicial objects in $\mathcal{X}$ and of their maps by $\mathcal{X}_s$, and denote the category of cosimplicial objects in $\mathcal{X}$ and of their maps by $\mathcal{X}^s$. It is clear that if X_* is a simplicial object in $\mathcal{X}$ and if $F: \mathcal{X} \to \mathcal{Y}$ is a covariant (contravariant) functor, then $F(X_*)$ is a simplicial (cosimplicial) object in $\mathcal{Y}$.

Let Δ be the category whose objects are the sets $\{0, 1, \ldots, n\}$, $n \in \mathbb{N}$, and whose morphisms are nondecreasing maps. Then giving a simplicial (cosimplicial) object in $\mathcal{X}$ is equivalent to giving a contravariant (covariant) functor $F: \Delta \to \mathcal{X}$, and giving a map of simplicial (cosimplicial) objects is equivalent to giving a natural transformation of the functors.

Suppose that $\mathcal{X}$ is a category with multiplication $\times$. For any simplicial objects X'_* and X''_* in this category, we define the product $X'_* \times X''_*$, which is a simplicial object, by setting

$$\begin{aligned}
(X'_* \times X''_*)_n &= X'_n \times X''_n, \\
d_i = d'_i \times d''_i &: X'_n \times X''_n \to X'_{n-1} \times X''_{n-1}, \\
s_i = s'_i \times s''_i &: X'_n \times X''_n \to X'_{n+1} \times X''_{n+1}.
\end{aligned}$$

Dually, for any cosimplicial objects X'^* and X''^* in this category, the product $X'^* \times X''^*$, which is a cosimplicial object, is defined by setting

$$\begin{aligned}
(X'^* \times X''^*)^n &= X'^n \times X''^n, \\
\delta^i = \delta'^i \times \delta''^i &: X'^{n-1} \times X''^{n-1} \to X'^n \times X''^n, \\
\sigma^i = \sigma'^i \times \sigma''^i &: X'^{n+1} \times X''^{n+1} \to X'^n \times X''^n.
\end{aligned}$$

Simplicial objects in the category of sets *Ens* are called *simplicial sets.* Simplicial objects in the category of topological spaces are called *simplicial spaces,* etc.

An example of a simplicial set is the simplicial set K_* arising from an abstract simplicial complex K. Namely, let us order the vertices of K, and for the elements of K_n let us take the collections $(x^0, \ldots, x^n)$ of (not necessarily distinct) vertices of K, $x^0 \leq \cdots \leq x^n$, such that the set $\{x^0, \ldots, x^n\}$ is the set of vertices of a simplex in K. The face and degeneracy operators are defined by

$$\begin{aligned}
d_i(x^0, \ldots, x^n) &= (x^0, \ldots, x^{i-1}, x^{i+1}, \ldots, x^n); \\
s_i(x^0, \ldots, x^n) &= (x^0, \ldots, x^i, x^i, \ldots, x^n).
\end{aligned}$$

For each $m \geq 0$, the simplicial set Δ^m, called the *standard m-dimensional simplex,* is defined as follows:

$$\begin{aligned}
\Delta^m_n &= \{(m_0, \ldots, m_n) : m_i \in \mathbb{Z}, 0 \leq m_0 \leq \cdots \leq m_n \leq m\}; \\
d_i(m_0, \ldots, m_n) &= (m_0, \ldots, m_{i-1}, m_{i+1}, \ldots, m_n); \\
s_i(m_0, \ldots, m_n) &= (m_0, \ldots, m_i, m_i, \ldots, m_n).
\end{aligned}$$

The family $\Delta^* = \{\Delta^m\}_{m\geq 0}$ is a cosimplicial object in the category of simplicial sets; the coface and codegeneracy operators in it are defined by

$$\delta^i(m_0,\dots,m_n) = (\delta^i(m_0),\dots,\delta^i(m_n)),$$
$$\sigma^i(m_0,\dots,m_n) = (\sigma^i(m_0),\dots,\sigma^i(m_n)),$$

where

$$\delta^i(p) = \begin{cases} p, & p < i, \\ p+1, & p \geq i; \end{cases} \qquad \sigma^i(p) = \begin{cases} p, & p \leq i, \\ p-1, & p > i. \end{cases}$$

In the category of topological spaces *Top* we also have a cosimplicial object denoted by Δ^*; in this case

$$\Delta^n = \{(t_1,\dots,t_n) \in \mathbb{R}^n \mid 0 \leq t_1 \leq \dots \leq t_n \leq 1\}.$$

The coface and codegeneracy operators

$$\delta^i\colon \Delta^{n-1} \to \Delta^n, \quad \sigma^i\colon \Delta^{n+1} \to \Delta^n, \quad 0 \leq i \leq n,$$

are given by

$$\delta^i(t_1,\dots,t_{n-1}) = \begin{cases} (0,t_1,\dots,t_{n-1}), & i = 0, \\ (t_1,\dots,t_i,t_i,\dots,t_{n-1}), & 0 < i < n, \\ (t_1,\dots,t_{n-1},1), & i = n, \end{cases}$$
$$\sigma^i(t_1,\dots,t_{n+1}) = (t_1,\dots,t_i,t_{i+2},\dots,t_{n+1}).$$

Let X be a topological space. The *singular simplicial set* $S_*(X)$ is defined by setting $S_n(X) = \mathrm{Hom}_{Top}(\Delta^n, X)$ (the space of continuous maps from Δ^n to X) with the face and degeneracy operators being induced by the coface and codegeneracy operators in Δ^*, i.e.,

$$d_i(f) = f\delta^i, \quad s_i(f) = f\sigma^i, \quad f\colon \Delta^n \to X, \quad 0 \leq i \leq n.$$

Let G be an associative monoid with identity $1 \in G$ acting on a space X on the right and on a space Y on the left. We construct a simplicial object $B_*(X,G,Y)$ in the category of topological spaces as follows: let

$$B_n(X,G,Y) = X \times G^{\times n} \times Y,$$

and let the face and degeneracy operators be defined by

$$d_i(x_0,\dots,x_{n+1}) = (x_0,\dots,x_i x_{i+1},\dots,x_{n+1}),$$
$$s_i(x_0,\dots,x_{n+1}) = (x_0,\dots,x_i,1,x_{i+1},\dots,x_{n+1}).$$

For the cases in which $X = *$ or $Y = *$, the corresponding simplicial objects are denoted by $B_*(G,Y)$, $B_*(X,G)$, and $B_*(G)$. If $X = G$ or $Y = G$, then the corresponding simplicial objects $B_*(G,G,Y)$ and $B_*(X,G,G)$ are simplicial resolutions of the G-spaces Y and X.

In the most general case, the simplicial objects B_* are defined in the category of functors. Namely, let $A\colon \mathcal{X} \to \mathcal{X}$ be a monad acting on a functor $L\colon \mathcal{Y} \to \mathcal{X}$ on the right and on a functor $M\colon \mathcal{X} \to \mathcal{Z}$ on the left. Then, in a similar way as above, we can define the simplicial object $B_*(L,A,M)$ in the category of functors. Dually, for any comonad K and for any functors L and M endowed with left and right coactions of K, we define the cosimplicial object $F^*(L,K,M)$.

One of the most important concepts related to simplicial and cosimplicial objects is realization. We shall consider the realization of a simplicial object in the category of compactly generated Hausdorff topological spaces; in what follows, this category will be called simply the category of topological spaces.

Let X_* be a simplicial space, i.e., a simplicial object in the category of topological spaces. Its *realization* is the quotient $|X_*|$ of the space

$$\bigcup_n X_n \times \Delta^n$$

by the equivalence relation $\sim$ generated by the relations

$$\begin{aligned}(d_i x, u) &\sim (x, \delta^i u), \quad x \in X_n, u \in \Delta^{n-1},\\ (s_i x, u) &\sim (x, \sigma^i u), \quad x \in X_n, u \in \Delta^{n+1},\end{aligned}$$

where $\Delta^* = \{\Delta^n\}$ is the above-mentioned cosimplicial object in the category of topological spaces.

The equivalence class containing an element (x, u), where $x \in X_n$, $u \in \Delta^n$, will be denoted by $[x, u]$.

Consider the quotient space $F_m|X_*|$ of

$$\bigcup_{n \le m} X_n \times \Delta^n$$

by the equivalence relation $\sim$ defined above. The spaces $F_m|X_*|$ form a filtration of $|X_*|$ by closed subspaces. Let us introduce a new topology on $|X_*|$ in which a set is closed if its intersection with each $F_m|X_*|$ is closed.

It is readily seen that a simplicial map $f_*\colon X_* \to Y_*$ induces a continuous map $|f_*|\colon |X_*| \to |Y_*|$. Thus we obtain a functor $|\ |\colon \mathit{Top}_s \to \mathit{Top}$ from the category of simplicial spaces to the category of topological spaces.

Note that a simplicial set X_* can be regarded as a simplicial space with the discrete topology. Hence the realization $|X_*|$ of a simplicial set X_* is defined; it is a topological space (and even a CW-complex), and thus we obtain a functor $|\ |\colon \mathit{Ens}_s \to \mathit{Top}$ from the category of simplicial sets to the category of topological spaces.

It is readily seen that the realization of the simplicial set Δ^n is the space Δ^n.

Let X be a topological space, and let X_* be the simplicial space with $X_n = X$ and the identity map taken for the face and degeneracy operators; then we have $|X_*| = X$.

One of the main properties of realization is the following.

For two simplicial spaces X'_* and X''_*, there is a natural homeomorphism

$$|X'_* \times X''_*| \cong |X'_*| \times |X''_*|.$$

To construct it, we denote by $p'_*\colon X'_* \times X''_* \to X'_*$ and $p''_*\colon X'_* \times X''_* \to X''_*$ the projections onto the first and onto the second factor, respectively. These projections induce maps of the realizations

$$p'\colon |X'_* \times X''_*| \to |X'_*|, \quad p''\colon |X'_* \times X''_*| \to |X''_*|.$$

Hence we obtain the map

$$p' \times p''\colon |X'_* \times X''_*| \to |X'_*| \times |X''_*|.$$

This map is the required homeomorphism. The inverse map is constructed purely combinatorially by using the standard triangulation of the product $\Delta^n \times \Delta^m$.

Consider the functor $S_*\colon Top \to Ens_s$ from the category of topological spaces to the category of simplicial sets; this functor takes a space X to the singular simplicial set $S_*(X)$. Let us show that the functors $|\ |$ and S_* are adjoint. To do this, we must prove that for a simplicial set X_* and a topological space Y, there are natural bijective correspondences

$$\mathrm{Hom}_{Ens_s}(X_*, S_*(Y)) \leftrightarrow \mathrm{Hom}_{Top}(|X_*|, Y).$$

Indeed, let $f_*\colon X_* \to S_*(Y)$ be a simplicial map. We define the continuous map $f\colon |X_*| \to Y$ by setting

$$f([x,u]) = f_n(x)(u), \quad x \in X_n, \quad u \in \Delta^n.$$

It is easily verified that the map f so defined is continuous and independent of the choice of the representatives.

Conversely, let $f\colon |X_*| \to Y$ be a continuous map. We define a simplicial map $f_*\colon X_* \to S_*(Y)$ by setting $f_n(x)(u) = f([x,u])$. It is easily verified that the map so defined is simplicial.

It follows directly from these definitions that the correspondences $f_* \longmapsto f$ and $f \longmapsto f_*$ constructed above yield the required bijection.

Thus in the case when the topological space X is the realization of a simplicial set X_*, for any topological space Y, there is a bijection

$$\mathrm{Hom}_{Ens_s}(X_*, S_*(Y)) \leftrightarrow \mathrm{Hom}_{Top}(X, Y),$$

and hence the study of continuous maps of topological spaces is reduced to the study of maps of simplicial sets.

In fact, the notion of realization of a simplicial object can be defined for any category that has sums and products and in which a cosimplicial object Δ^* is given. For example, realization is defined for simplicial objects in the category of simplicial sets. In this case, for Δ^* we must take the cosimplicial object consisting of the simplicial sets Δ^n. Realization is also defined for simplicial objects in the categories of simplicial modules, chain complexes, etc.

Realization of a simplicial object allows us to construct classifying spaces. Namely, let G be a topological group. Consider the simplicial object $E_*G = B_*(G, G, *)$. Its realization is denoted by EG. The points of this space are the collections $(g_0, \ldots, g_n, u)$, $u \in \Delta^n$, $g_i \in G$, satisfying the following relations:

$$(g_0, \ldots, g_i g_{i+1}, \ldots, g_n, u) \sim (g_0, \ldots, g_n, \delta^i u), \quad u \in \Delta^{n-1}, g_i \in G;$$
$$(g_0, \ldots, g_i, 1, g_{i+1}, \ldots, g_n, u) \sim (g_0, \ldots, g_n, \sigma^i u), \quad u \in \Delta^{n+1}, g_i \in G.$$

In each equivalence class $[g_0, \ldots, g_n, u]$ we can choose a representative $(g_0, \ldots, g_n, u)$ such that u is an interior point of Δ^n and $g_i \neq 1$ for $1 \leq i \leq n$. Besides, this representative is uniquely defined.

The definition of realization directly implies that the group G acts freely on EG by the formula

$$g[g_0, \ldots, g_n, u] = [gg_0, \ldots, g_n, u].$$

The quotient space EG/G is the realization of the simplicial object $B_*(G) = B_*(*, G, *)$ and is denoted by BG. It is called the *classifying space of the group G*. Its points are the elements $[g_1, \ldots, g_n, u]$. By taking specific groups, we can obtain various classifying spaces. Thus, for the orthogonal group $O(n)$ of degree n, we obtain the classifying space $BO(n)$. For the unitary group $U(n)$ of degree n, we obtain the classifying space $BU(n)$. For the symplectic group $Sp(n)$ of degree n, we obtain the classifying space $BSp(n)$, etc.

Consider the group $O(n)$. It acts on the n-dimensional sphere S^n and hence the simplicial object $B_*(*, O(n), S^n)$ is defined, which is also denoted by $MO_*(n)$. The realization of this simplicial object yields the *Thom space* of the group $O(n)$; this space is denoted by $MO(n)$.

The maps $SS^n \to S^{n+1}$ and $O(n) \to O(n+1)$ induce maps $SMO(n) \to MO(n+1)$, thus making the sequence $MO(n)$ into a spectrum. The latter is denoted by MO and is referred to as the *Thom spectrum*.

Similarly, the group $U(n)$ acts on the $2n$-dimensional sphere S^{2n}, and hence the simplicial object $B_*(*, U(n), S^{2n})$ is defined, which is denoted by $MU_*(2n)$. The realization of this simplicial object yields the Thom space of the group $U(n)$; this space is denoted by $MU(2n)$.

Let $MU(2n+1) = SMU(2n)$. Then the sequence of spaces $MU(n)$ is made into a spectrum, which is denoted by MU and is referred to as the Thom spectrum.

The Thom spectra MSp, MSO, and MSU are constructed in a similar way.

Note that the maps $S^n \wedge S^m \to S^{n+m}$ and $O(n) \times O(m) \to O(n+m)$ induce maps $MO_*(n) \wedge MO_*(m) \to MO_*(n+m)$ and a map

$$\begin{aligned} MO(n) \wedge MO(m) &= |MO_*(n)| \wedge |MO_*(m)| \\ &\to |MO_*(n) \wedge MO_*(m)| \to |MO_*(n+m)| = MO(n+m), \end{aligned}$$

which make the spectrum MO into a ring spectrum.

Similarly, the other spectra considered above are ring spectra.

2.2. Simplicial sets

Let us study the category of simplicial sets in more detail. Suppose that $f_*, g_*: X_* \to Y_*$ are maps of simplicial sets. We say that f_* is *homotopic to* g_* (and we write $f_* \simeq g_*$) if there is a simplicial map $h_*: X_* \times \Delta^1 \to Y_*$ (called a *homotopy between* f_* *and* g_*) that satisfies the conditions

$$h_*|_{X_* \times 0_*} = f_*, \quad h_*|_{X_* \times 1_*} = g_*,$$

where Δ^1 is the simplicial set of the standard 1-dimensional simplex; 0_*, 1_* are the simplicial subsets in Δ^1 generated by 0-dimensional simplices $(0), (1)$.

Thus the above definition of the homotopy relation is quite similar to the corresponding definition for the category of topological spaces. However, in contrast to the case of continuous map, the homotopy relation on maps of simplicial sets is not an equivalence relation. It is readily seen that reflexivity holds, but in general neither symmetry nor transitivity holds.

To eliminate this drawback, it is conventional to consider only simplicial sets satisfying the Kan extension condition, rather than arbitrary ones. Namely, a simplicial set X_* is said to satisfy the *Kan extension condition* if for any collection of n-dimensional simplices $x^0, \dots, x^{k-1}, x^{k+1}, \dots, x^{n+1}$ satisfying the compatibility condition

$$d_i x^j = d_{j-1} x^i, \quad i < j, \quad i \neq k, \quad j \neq k,$$

there is an $(n+1)$-dimensional simplex x such that $d_i(x) = x^i$ for $i \neq k$.

For example, the singular simplicial set $S_*(X)$ of a topological space X satisfies the Kan extension condition. This follows from the fact that the union of $n+1$ faces of the simplex Δ^{n+1} is a retract of Δ^{n+1}. Hence any continuous map defined on this union can be extended to the whole simplex Δ^{n+1}. On the other hand, simplicial sets generated by geometric complexes do not satisfy the Kan extension condition.

It turns out that if Y_* is a simplicial set satisfying the Kan extension condition, then the homotopy relation of simplicial maps $f_*, g_*: X_* \to Y_*$ is an equivalence relation.

Let us reformulate the definition of being homotopic to make it applicable to maps of simplicial objects in an arbitrary category. We shall show that simplicial maps $f_*, g_*: X_* \to Y_*$ are homotopic if and only if there are maps $h_i: X_n \to Y_{n+1}$, $0 \le i \le n$, satisfying the relations

$$\begin{aligned} d_0 h_0 &= g_n, \\ d_{n+1} h_n &= f_n, \\ d_i h_j &= \begin{cases} h_{j-1} d_i, & i < j, \\ d_j h_{j-1}, & i = j > 0, \\ h_j d_{i-1}, & i > j+1, \end{cases} \\ s_i h_j &= \begin{cases} h_{j+1} s_i, & i \le j, \\ h_j s_{i-1}, & i > j. \end{cases} \end{aligned}$$

Indeed, if $h_*: X_* \times \Delta^1 \to Y_*$ is a homotopy between f_* and g_*, then the maps h_i are defined by the formulas $h_i(x) = h_*(s_i(x), (0, \dots, 0, 1, \dots, 1))$, where the first 1 stands in the $(i+1)$th place. Conversely, if the maps h_i are as above, then the homotopy h_* is given by

$$h_*(x, (0, \dots, 0)) = f_*(x), \quad h_*(x, (1, \dots, 1)) = g_*(x),$$
$$h_*(x, (0, \dots, 0, 1, \dots, 1)) = d_{i+1} h_i(x),$$

where the first 1 stands in the ith place. Direct verification shows that the required relations hold.

Let us prove that the realization functor $|\ |: \mathit{Ens}_s \to \mathit{Top}$ from the category of simplicial sets to the category of topological spaces takes homotopic simplicial maps $f_*, g_*: X_* \to Y_*$ to homotopic continuous maps $|f_*|, |g_*|: |X_*| \to |Y_*|$. Indeed, if $h_*: X_* \times \Delta^1 \to Y_*$ is a simplicial homotopy between f_* and g_*, then the corresponding homotopy h between the continuous maps $|f_*|$ and $|g_*|$ is defined as the composition

$$|X_*| \times I = |X_*| \times |\Delta^1| \to |X_* \times \Delta^1| \to |Y_*|,$$

where I is the unit segment.

Let us show that the functor of singular simplicial set S_* takes homotopic continuous maps of topological spaces to homotopic maps of simplicial sets. Indeed, let $h: I \times X \to Y$ be a homotopy between simplicial maps $f, g: X \to Y$. There is a natural simplicial map $\Delta^1 \to S_*(I)$. The required simplicial homotopy h_* between the maps $S_*(f), S_*(g): S_*(X) \to S_*(Y)$ is defined as the composition

$$S_*(X) \times \Delta^1 \to S_*(X) \times S_*(I) \to S_*(X \times I) \to S_*(Y).$$

Thus we see that for any simplicial set X_* and for any topological space Y, the above-mentioned bijection

$$\mathrm{Hom}_{Ens_s}(X_*, S_*(Y)) \longleftrightarrow \mathrm{Hom}_{Top}(|X_*|, Y)$$

induces a bijection of the sets of homotopy classes

$$[X_*, S_*(Y)] \longleftrightarrow [|X_*|, Y].$$

Let G be a topological group acting on a space X on the left. This action $\mu: G \times X \to X$ induces a map of simplicial objects $\mu_*: B_*(G, G, X) \to X_*$, where X_* is the simplicial object with all $X_n = X$ and the identity transformation taken for the face and degeneracy operators. Let us show that this map $B_*(G, G, X) \to X_*$ is a homotopy equivalence, i.e., the simplicial object $B_*(G, G, X)$ is a simplicial resolution of the space X. To do this, we define the inverse map $i_*: X_* \to B_*(G, G, X)$ by setting $i_n(x) = 1[1, \ldots, 1]x$. Clearly, we have $\mu_* \circ i_* = \mathrm{Id}: X_* \to X_*$. We define the homotopy between the maps $i_* \circ \mu_*$ and Id by setting

$$h_i(g_0[g_1, \ldots, g_n]x) = 1[1, \ldots, 1, g_0 \cdot \ldots \cdot g_i, g_{i+1}, \ldots, g_n]x), \quad 0 \le i \le n.$$

Straightforward computations show that the family

$$h_i: B_n(G, G, X) \to B_{n+1}(G, G, X), \quad 0 \le i \le n,$$

constitutes the required simplicial homotopy.

Taking into account the facts that realization preserves homotopy and that $|X_*| = X$, we see that the bar construction $B(G, G, X)$ is homotopy equivalent to X. In particular, for $X = *$, the space $EG = B(G, G, *)$ is contractible.

Now let us recall the definition of the homotopy groups of a simplicial set. Suppose that X_* is a simplicial set with base point $*$, i.e., with a distinguished zero-dimensional simplex $*$, whose degeneracies are also denoted by $*$. Consider the n-dimensional simplices $x \in X_n$ satisfying the conditions $d_i(x) = *$, $0 \le i \le n$. For these simplices, let us define the homotopy relation $\simeq$ by setting $x' \simeq x''$ if there is a simplex $y \in X_{n+1}$ such that $d_i(y) = *$ for $0 \le i < n$, $d_n(y) = x'$, and $d_{n+1}(y) = x''$. The simplex y is called a *homotopy between x' and x''*.

Let us show that if a simplicial set X_* satisfies the Kan extension condition, then this homotopy relation is an equivalence relation.

The reflexivity is obvious. For the homotopy between x and x we can take the simplex $y = s_n(x)$.

Suppose that $x \simeq x'$, where y' is the corresponding homotopy; $x \simeq x''$, where y'' is the corresponding homotopy. Let us show that $x' \simeq x''$. To do this, consider the $(n+2)$-tuple of $(n+1)$-dimensional simplices $(*, \ldots, *, y', y'', -)$. This collection satisfies the compatibility condition, and hence there exists an $(n+2)$-dimensional simplex z such that $d_i(z) = *$ for $0 \le i < n$, $d_n(z) = y'$, $d_{n+1}(z) = y''$. Consider the simplex $y = d_{n+2}(z)$. It satisfies the relations

$$\begin{aligned} d_i(y) &= d_i d_{n+2}(z) = 0 \ \text{ if } \ 0 \le i < n, \\ d_n(y) &= d_n d_{n+2}(z) = d_{n+1} d_n(z) = d_{n+1}(y') = x', \\ d_{n+1}(y) &= d_{n+1} d_{n+2}(z) = d_{n+1} d_{n+1}(z) = d_{n+1}(y'') = x''. \end{aligned}$$

The set of equivalence classes of n-dimensional simplices of a simplicial set X_* modulo the equivalence relation defined above is denoted by $\pi_n(X_*)$. For $n \ge 1$, this set is a group; it is called the *n-dimensional homotopy group* of the simplicial set X_*.

Let us define the group operation in $\pi_n(X_*)$. Suppose that $[x'], [x''] \in \pi_n(X_*)$ are homotopy classes containing simplices x' and x'', respectively. Consider the tuple $(*, \dots, *, x', -, x'')$. It satisfies the compatibility condition, and hence there is an element $z \in X_{n+1}$ such that $d_i(z) = 0$ for $0 \le i < n-1$, $d_{n-1}(z) = x'$, $d_{n+1}(z) = x''$. We set $[x'] + [x''] = [d_n(z)]$.

It can be shown that this operation is well-defined, i.e., the result depends neither on the choice of the representatives of the homotopy classes nor on the choice of the element z. For $n \ge 1$, this operation defines a group structure on $\pi_n(X_*)$. Moreover, for $n \ge 2$, this operation is commutative, i.e., it defines an abelian group structure.

It is readily seen that if $X_* = S_*(X)$ is the singular simplicial set of a topological space X with base point $* \in X$, then the homotopy groups $\pi_n(S_*(X))$ coincide with the homotopy groups $\pi_n(X)$ of the topological space X.

A map of simplicial sets $f_*: E_* \to B_*$ is called a *Kan fibration* if for any collection $x_0, \dots, x_{k-1}, x_{k+1}, \dots, x_{n+1}$ of n-dimension simplices in E_* that satisfy the compatibility condition, i.e.,

$$d_i(x_j) = d_{j-1}(x_i), \quad i < j, \quad i \neq k, \quad j \neq k,$$

and for any $(n+1)$-simplex $y \in B_*$ satisfying the conditions $d_i(y) = p(x_i)$, $i \neq k$, there is an $(n+1)$-simplex $x \in E_*$ such that $d_i(x) = x_i$, $i \neq k$, $p(x) = y$.

If $*$ is the base point in B_*, then the simplicial set $F_* = p_*^{-1}(*)$ is called the *fiber* of the fibration $p_*: E_* \to B_*$.

It is not hard to show that if $p_*: E_* \to B_*$ is a Kan fibration, then F_* satisfies the Kan extension condition. Besides, if B_* satisfies the Kan extension condition, so does E_*. In the case when p_* is surjective, if E_* satisfies the Kan extension condition, so does B_*.

It can be shown that a map $p_*: E_* \to B_*$ is a Kan fibration if and only if it satisfies the homotopy lifting property. Namely, for any simplicial set X_*, any map $\widetilde{f}_*: X_* \to E_*$, and a homotopy $h_*: X_* \times \Delta^1 \to B_*$ such that

$$h_*|_{0_*} = p_* \circ \widetilde{f}_*,$$

there is a homotopy $\widetilde{h}_*: X_* \times \Delta^1 \to E_*$ such that

$$p_* \circ \widetilde{h}_* = h_*, \quad \widetilde{h}_*|_{0_*} = \widetilde{f}_*.$$

Let E_*, B_* be simplicial sets satisfying the Kan extension condition, and let $p_*: E_* \to B_*$ be a Kan fibration. Then there is a long exact sequence of homotopy groups

$$\dots \to \pi_{n+1}(B_*) \xrightarrow{d_{n+1}} \pi_n(F_*) \xrightarrow{i_n} \pi_n(E_*) \xrightarrow{p_n} \pi_n(B_*) \to \dots$$

The connecting homomorphism $d_{n+1}: \pi_{n+1}(B_*) \to \pi_n(F_*)$ is defined as follows. Let $[y] \in \pi_{n+1}(B_*)$, and let $y \in B_{n+1}$ be its representative. Consider the tuple $(*, \dots, *)$ of n-simplices in E_* and the simplex $y \in B_{n+1}$. Then there is a simplex $x \in E_{n+1}$ such that $d_i(x) = *$, $i \le n$, and $p_*(x) = y$. The simplex $d_{n+1}(x) \in E_n$ satisfies the conditions $d_i(d_{n+1}(x)) = *$, $0 \le i \le n$, and $p_*(d_{n+1}(x)) = *$, i.e., $d_{n+1}(x) \in F_*$. Let $d_{n+1}[y] = [d_{n+1}(x)]$. It can be shown that this definition depends neither on the representative y nor on the simplex x, and that d_{n+1} is a homomorphism.

We now pass to simplicial groups, i.e., simplicial objects in the category of groups. It can be shown that any simplicial group satisfies the Kan extension condition, and an arbitrary epimorphism of simplicial groups is a Kan fibration.

Let us show that if G_* is a simplicial group or even a simplicial monoid, then the group structure in the homotopy groups is determined by the multiplication in G_*.

Indeed, suppose that $x, y \in G_n$, $d_i(x) = d_i(y) = e_{n-1}$, $0 \leq i \leq n$. Let us show that $[x] + [y] = [xy]$. Let $z = s_{n-1}(x)s_n(y)$. Then $d_i(z) = e_{n-1}$ for $0 \leq i < n-1$, $d_{n-1}(z) = x$, $d_n(z) = xy$, $d_{n+1}(z) = y$. By definition, this means that the required relation holds.

Let us also show that in the case when G_* is a simplicial group or even a simplicial monoid, the homotopy groups are abelian.

Indeed, suppose that $x, y \in G_n$, $d_i(x) = d_i(y) = e_{n-1}$, $0 \leq i \leq n$. Let $w = s_n(x)s_{n-1}(y)$. Then $d_i(w) = e_{n-1}$ for $0 \leq i < n-1$, $d_{n-1}(w) = y$, $d_n(w) = xy$, $d_{n+1}(w) = x$. By definition, this implies the relation $[y] + [x] = [xy]$. Since, by the above, we have $[x] + [y] = [xy]$, the required relation $[x] + [y] = [y] + [x]$ holds.

Let X_* be a simplicial set, and let G_* be a simplicial group. A *twisting function* $\phi: X_* \to G_*$ is a family $\phi = \{\phi_n\}_{n \geq 1}$ of maps $\phi_n: X_n \to G_{n-1}$ satisfying the conditions

$$\begin{aligned} d_0\phi(x) &= (\phi(d_0x))^{-1}\phi(d_1x), \\ d_i\phi(x) &= \phi(d_{i+1}x), \ i > 0, \\ s_i\phi(x) &= \phi(s_{i+1}x), \ i \geq 0, \\ \phi(s_0x) &= e. \end{aligned}$$

Suppose that X_* and F_* are simplicial sets, G_* is a simplicial group acting on F_* on the left, and $\phi: X_* \to G_*$ is a twisting function. Then the *twisted Cartesian product with base X_*, fiber F_*, and structure group G_** is the simplicial set $X_* \times_\phi F_*$ which is defined as follows:

$$(X_* \times_\phi F_*)_n = X_n \times F_n,$$

and the face and degeneracy operators are given by

$$\begin{aligned} d_i(x, y) &= (d_ix, d_iy), \quad i > 0, \\ d_0(x, y) &= (d_0x, \phi(x)d_0y), \\ s_i(x, y) &= (s_ix, s_iy), \quad i \geq 0. \end{aligned}$$

It is easily checked that the conditions imposed on twisting functions imply the simplicial relations for the face and degeneracy operators.

Let us show that if F_* is a simplicial set satisfying the Kan extension property, then the projection $p_*: X_* \times_\phi F_* \to X_*$ is a Kan fibration. Indeed, let

$$(x_0, y_0), \ldots, (x_{k-1}, y_{k-1}), (x_{k+1}, y_{k+1}), \ldots, (x_{n+1}, y_{n+1})$$

be a compatible family of n-dimensional simplices in $X_* \times_\phi F_*$, and let x be an $(n+1)$-dimensional simplex in X_* such that $d_i(x) = x_i$, $i \neq k$.

If $k = 0$, then the family $y_1, \ldots, y_{n+1}$ is compatible, and hence there exists an $(n+1)$-simplex $y \in F_*$ such that $d_i(y) = y_i$, $i > 0$. In this case, the simplex (x, y) is the required $(n+1)$-simplex in $X_* \times_\phi F_*$.

If $k > 0$, then the family $\phi(x)^{-1}y_0, y_1, \ldots, y_{k-1}, y_{k+1}, \ldots, y_{n+1}$ is compatible, and hence there exists an $(n+1)$-simplex $y \in F_*$ such that $d_i(y) = y_i$, $i \neq k$. In this case, the simplex (x, y) is the required $(n+1)$-simplex in $X_* \times_\phi F_*$.

Let us define a simplicial analog of a loop space. Suppose that X_* is a simplicial set with a single zero-dimensional simplex e. Such simplicial sets are said to be *reduced.* Denote by GX_* the simplicial group in which GX_n is the quotient of the free group generated by the simplices in X_{n+1} modulo the relations $s_0x_n = e_n$. For any $x \in X_{n+1}$, we denote the equivalence class of x in GX_n by $\alpha(x)$ and define the face and degeneracy operators by setting their values on the generators as follows:

$$\begin{aligned} d_0\alpha(x) &= \alpha(d_0x)^{-1}\alpha(d_1x), \\ d_i\alpha(x) &= \alpha(d_{i+1}x), \ i > 0, \\ s_i\alpha(x) &= \alpha(s_{i+1}x), \ i \geq 0. \end{aligned}$$

The operators d_i and s_i so defined can be uniquely extended to group homomorphisms. Besides, $\alpha\colon X_* \to GX_*$ is a twisting function. This twisting function is universal in the sense that an arbitrary twisting function $\phi\colon X_* \to G_*$ generates a unique map of simplicial groups $f_*\colon GX_* \to G_*$ such that $f_*\alpha = \phi$. The simplicial group GX_* is the required simplicial analog of a loop space.

It can be shown that the simplicial set $E_* = X_* \times_\alpha GX_*$ is contractible. This, combined with the exactness of the homotopy sequence of a fibration, implies that there is a natural isomorphism of the homotopy groups $\pi_{n+1}(X_*) \cong \pi_n(GX_*)$.

Note that since a simplicial group satisfies the Kan extension condition, this isomorphism allows us to define the homotopy groups for arbitrary simplicial sets, and not just for those satisfying the Kan extension condition.

2.3. A_∞-simplicial objects and A_∞-topological groups

Here we define the notion of A_∞-topological group and prove that if a space $\widetilde{G}$ is obtained by a deformation retraction of a topological group G, then it has the structure of A_∞-topological group and there is an A_∞-homotopy equivalence $\widetilde{G} \simeq G$. Thus Stasheff's results concerning topological monoids [**45**] are extended to topological groups.

We shall start with some simplicial constructions.

DEFINITION. A family $X_* = \{X_n\}_{n\geq 0}$ of objects of a given category will be called a *presimplicial object* if there are given face and degeneracy operators

$$\begin{aligned} d_i\colon X_n &\to X_{n-1}, \quad 1 \leq i \leq n, \\ s_i\colon X_n &\to X_{n+1}, \quad 0 \leq i \leq n, \end{aligned}$$

that satisfy the following relations:

$$\begin{aligned} d_id_j &= d_{j-1}d_i, \quad i < j, \\ s_is_j &= s_{j+1}s_i, \quad i \leq j, \\ d_is_j &= \begin{cases} s_{j-1}d_i, & i < j, \\ \mathrm{Id}, & i = j, \ i = j+1, \\ s_jd_{i-1}, & i > j+1. \end{cases} \end{aligned}$$

Thus a presimplicial object differs from a simplicial one only by the face operator d_0. A simplicial object possesses this operator, while a presimplicial object does not.

DEFINITION. A *map* $f\colon X_* \to Y_*$ *of presimplicial objects* is a family $f_* = \{f_n\}_{n\geq 0}$ of maps $f_n\colon X_n \to Y_n$ commuting with the face and degeneracy operators, i.e., such that

$$\begin{aligned} d_i f_n &= f_{n-1} d_i, \\ s_i f_n &= f_{n+1} s_i. \end{aligned}$$

Dually, we give the following definition.

DEFINITION. A family $X^* = \{X^n\}_{n\geq 0}$ of objects of a given category will be called a *precosimplicial object* if there are given coface and codegeneracy operators

$$\begin{aligned} \delta^i\colon X^n &\to X^{n+1}, \quad 1 \leq i \leq n+1; \\ \sigma^i\colon X^n &\to X^{n-1}, \quad 0 \leq i \leq n-1, \end{aligned}$$

that satisfy the following relations

$$\begin{aligned} \delta^j\delta^i &= \delta^i\delta^{j-1}, \quad i<j, \\ \sigma^j\sigma^i &= \sigma^i\sigma^{j+1}, \quad i \leq j, \\ \sigma^j\delta^i &= \begin{cases} \delta^i\sigma^{j-1}, & i<j, \\ \mathrm{Id}, & i=j,\ i=j+1, \\ \delta^{i-1}\sigma^j, & i>j+1. \end{cases} \end{aligned}$$

Thus a precosimplicial object differs from a cosimplicial one only by the coface operator δ^0. A cosimplicial object possesses this operator, while a precosimplicial object does not.

DEFINITION. A *map* $f^*\colon X \to Y$ *of precosimplicial objects* is a family $f^* = \{f^n\}_{n\geq 0}$ of maps $f^n\colon X^n \to Y^n$ commuting with the coface and codegeneracy operators, i.e., such that

$$\begin{aligned} \delta^i f_n &= f^{n+1}\delta^i, \\ \sigma^i f^n &= f_{n-1}\sigma^i. \end{aligned}$$

In the categories of presimplicial and precosimplicial topological spaces, the notions of homotopic maps, homotopy equivalence of objects, etc. can be defined in a usual way.

It is clear that simplicial objects are examples of presimplicial objects. Let us give another example of a presimplicial object.

Suppose that G is a topological space with base point $* \in G$. Let $W_n(G) = G^{\times n}$, $W_0(G) = *$. We define maps $d_i\colon W_n(G) \to W_{n-1}(G)$ by setting

$$\begin{aligned} d_i(g_1,\dots,g_n) &= (g_1,\dots,g_{i-1},g_{i+1},\dots,g_n), \\ s_i(g_1,\dots,g_n) &= (g_1,\dots,g_i,g_i,\dots,g_n), \\ s_0(g_1,\dots,g_n) &= (*,g_1,\dots,g_n). \end{aligned}$$

It is readily seen that all necessary presimplicial relations hold, and hence $W_*(G) = \{W_n(G)\}$ is a presimplicial object in the category of topological spaces.

Now let us define the notion of A_∞-simplicial object in the category of topological spaces (or simply A_∞-simplicial space).

Let I^n be the unit cube, $I^n = \{(t_1, \ldots, t_n) |\ 0 \le t_i \le 1\}$. By

$$u_i^\epsilon: I^n \to I^{n+1}, \quad 1 \le i \le n+1, \quad \epsilon = 0, 1,$$
$$v_i: I^n \to I^{n-1}, \quad 0 \le i \le n,$$

denote the maps defined by

$$u_i^\epsilon(t_1, \ldots, t_n) = (t_1, \ldots t_{i-1}, \epsilon, t_i, \ldots, t_n);$$
$$v_i(t_1, \ldots, t_n) = \begin{cases} (t_2, \ldots, t_n), & i = 0, \\ (t_1, \ldots, t_i * t_{i+1}, \ldots, t_n), & 1 \le i \le n-1, \\ (t_1, \ldots, t_{n-1}), & i = n, \end{cases}$$

where $t_i * t_{i+1} = t_i + t_{i+1} - t_i \cdot t_{i+1}$.

DEFINITION. A presimplicial object $X_* = \{X_n\}$ in the category of topological spaces will be called an *A_∞-simplicial object* or just an *A_∞-simplicial space* if there are given face operators

$$d_0^m: X_n \times I^m \to X_{n-m-1}, \quad 0 \le m \le n-1,$$

that satisfy the following conditions:

$$d_0^m(1 \times u_i^0) = d_0^{m-1}(d_i \times 1), \quad 1 \le i \le m;$$
$$d_0^m(1 \times u_i^1) = d_0^{m-i}(d_0^{i-1} \times 1), \quad 1 \le i \le m;$$
$$d_0^m(d_i \times 1) = d_{i-m-1} d_0^m, \quad i > m+1,$$
$$d_0^m(s_i \times 1) = \begin{cases} d_0^{m-1}(1 \times v_i), & i \le m,\ m \ge 1, \\ \mathrm{Id}, & i = m = 0, \\ s_{i-m-1} d_0^m, & i > m. \end{cases}$$

It is clear that a simplicial object $X_* = \{X_n\}$ in the category of topological spaces can be regarded as an A_∞-simplicial object with trivial operators $d_0^m: X_n \times I^m \to X_{n-m-1}$ for $m \ge 1$. Namely, $d_0^m = d_0 \ldots d_0 p$, where $p: X_n \times I^m \to X_n$ is the projection.

A_∞-objects form a category whose morphisms are maps $f_*: X_* \to Y_*$ preserving the A_∞-structure.

Further, let us define A_∞-maps.

DEFINITION. Suppose that $X_* = \{X_n\}$ is an A_∞-simplicial space and $Y_* = \{Y_n\}$ is a simplicial space. An *A_∞-map from X_* to Y_** is a map $f_* = \{f_n\}$, $f_n: X_n \to Y_n$ of presimplicial objects together with a family of maps

$$f_n^m: X_n \times I^m \to Y_{n-m}, \quad 0 \le m \le n,$$

satisfying the following conditions:

$$\begin{aligned}
f_n^0 &= f_n;\\
f_n^m(1\times u_i^0) &= f_n^{m-1}(d_i\times 1), \quad 1\le i\le m;\\
f_n^m(1\times u_i^1) &= f_{n-i}^{m-i}(d_0^{i-1}\times 1), \quad 1\le i\le m;\\
f_{n-1}^m(d_i\times 1) &= d_{i-m}f_n^m, \quad i>m;\\
f_{n+1}^m(s_i\times 1) &= \begin{cases} f_n^{m-1}(1\times v_i), & i<m,\\ s_{i-m}f_n^m, & i\ge m.\end{cases}
\end{aligned}$$

We say that an A_∞-map is an A_∞*-homotopy equivalence* if the corresponding maps f_n are homotopy equivalences.

Dually, we give the following definition.

DEFINITION. A precosimplicial object $X^* = \{X^n\}$ in the category of topological spaces will be called an A_∞*-cosimplicial object* or just an A_∞*-cosimplicial space* if there are given coface operators

$$\delta_m^0 \colon X^n\times I^m \to X^{n+m+1}, \quad 0\le m\le n-1,$$

that satisfy the following relations:

$$\begin{aligned}
\sigma^0\delta_0^0 &= \mathrm{Id};\\
\delta_m^0(1\times u_i^0) &= \delta^i\delta_{m-1}^0, \quad 1\le i\le m;\\
\delta_m^0(1\times u_i^1) &= \delta_{i-1}^0(\delta_{m-1}^0\times 1), \quad 1\le i\le m;\\
\delta_{m-1}^0(1\times v_i) &= \sigma^i\delta_m^0, \quad 0\le i\le m, \quad m\ge 1;\\
\delta_m^0(\delta^i\times 1) &= \delta^{i+m+1}\delta_m^0, \quad i\ge 1;\\
\delta_m^0(\sigma^i\times 1) &= \sigma^{i+m+1}\delta_m^0, \quad i\ge 0.
\end{aligned}$$

It is clear that a cosimplicial object $X^* = \{X^n\}$ in the category of topological spaces can be regarded as an A_∞-cosimplicial object with trivial operators $\delta_m^0\colon X^n\times I^m \to X^{n+m+1}$ for $m\ge 1$.

A_∞-cosimplicial objects form a category whose morphisms are maps $f^*\colon X^*\to Y^*$ preserving the A_∞-structure.

Further, let us define A_∞-maps.

DEFINITION. Suppose that $X^*=\{X^n\}$ is an A_∞-cosimplicial space and $Y^* = \{Y^n\}$ is a cosimplicial space. We say that an A_∞*-map from* X^* *to* Y^* is a map $f^* = \{f^n\}$, $f^n\colon X^n\to Y^n$, of precosimplicial objects together with a family of maps

$$f_m^n\colon X^n\times I^m\to Y^{n+m}, \quad 0\le m\le n,$$

satisfying the following relations:

$$\begin{aligned}
f_0^n &= f^n;\\
f_m^n(1\times u_i^0) &= \delta^{i-1}f_{m-1}^n, \quad 1\le i\le m;\\
f_m^n(1\times u_i^1) &= f_i^{n+m-i-1}(\delta_{m-i}^0\times 1), \quad 1\le i\le m;\\
f_{n-1}^m(d_i\times 1) &= d_{i-m}f_n^m, \quad i>m;\\
f_{m-1}^n(1\times v_i) &= \sigma^i f_m^n, \quad 0\le i<m;\\
f_m^n(\delta^i\times 1) &= \delta^{i+m}f_m^{n-1}, \quad i\ge 1;
\end{aligned}$$

$$f^n_{m-1}(\sigma^i \times 1) = \sigma^{i+m} f^m_n, \quad i \geq 0.$$

An A_∞-map will be called an *A_∞-homotopy equivalence* if the corresponding maps f_n are homotopy equivalences.

Let $X_* = \{X_n\}$ be an A_∞-simplicial object in the category of topological spaces. We define its *realization* $\|X_*\|$ by setting

$$\|X_*\| = \bigcup_n X_n \times I^n/_\sim,$$

where the equivalence $\sim$ is generated by the relations

$$\begin{aligned} x_n \times (t_1, \dots, t_{i-1}, 0, t_{i+1}, \dots, t_n) &\sim d_i(x_n) \times (t_1, \dots, t_{i-1}, t_{i+1}, \dots, t_n); \\ x_n \times (t_1, \dots, t_{i-1}, 1, t_{i+1}, \dots, t_n) &\sim d_0^{i-1}(x_n \times (t_1, \dots, t_{i-1})) \times (t_{i+1}, \dots, t_n); \\ s_i(x_n) \times (t_1, \dots, t_{n+1}) &\sim x_n \times v_i(t_1, \dots, t_{n+1}). \end{aligned}$$

THEOREM. *If a simplicial object X_* in the category of topological spaces is regarded as an A_∞-simplicial object with trivial higher operators d_0^m, $m \geq 1$, then there is a natural homeomorphism $\|X_*\| \cong |X_*|$.*

PROOF. We define the map $f\colon \|X_*\| \to |X_*|$ by setting

$$f(x_n \times (t_1, \dots, t_n)) = x_n \times (t'_1, \dots, t'_n),$$

where $t'_i = t_1 * \cdots * t_i$.

The inverse map $g\colon |X_*| \to \|X_*\|$ is defined by

$$g(x_n \times (t'_1, \dots, t'_n)) = x_n \times (t_1, \dots, t_n),$$

where $t_1 = t'_1$, $t_i = \frac{t'_i - t'_{i-1}}{1 - t'_{i-1}}$ for $i > 1$ (by convention $\frac{0}{0} = 1$); see [**7**]. $\square$

THEOREM. *The realization functor $\|\ \|$ from the category of A_∞-simplicial topological spaces to the category of topological spaces is adjoint to the singular simplicial space functor S_*.*

PROOF. The natural transformation $\Phi\colon \|S_*(X)\| \to X$ is defined by

$$\Phi(x_n \times (t_1, \dots, t_n)) = x_n(t'_1, \dots, t'_n).$$

The natural transformation

$$\Psi\colon X_* \to S_*\|X_*\|, \quad \Psi = \{\Psi^m_n\}, \quad \Psi^m_n\colon X_n \times I^m \to S_{n-m}\|X_*\|,$$

is defined by

$$\Psi(x_n \times (t_1, \dots, t_m))(t'_{m+1}, \dots, t'_n) = x_n \times (t_1, \dots, t_n). \quad \square$$

DEFINITION. We say that a topological space $\widetilde{X}$ is *S-homotopy equivalent* to a topological space X if we are given maps $\xi: \widetilde{X} \to X$, $\eta: X \to \widetilde{X}$ and a homotopy $h: X \times I \to X$ that satisfy the following relations:

$$\eta\xi = \mathrm{Id}; \quad h(x,0) = x; \quad h(x,1) = \xi\eta(x);$$
$$\eta h(x,t) = \eta(x); \quad h(\xi(x),t) = \xi(x); \quad h(h(x,t_1),t_2) = h(x, t_1 * t_2).$$

Note that the usual contracting homotopy of the real line $\mathbb{R}$ by the formula $h(x,t) = x(1-t)$ satisfies these relations. Hence the theorem on deformation retracts can be strengthened and reformulated as follows.

THEOREM. *If a CW-complex A is a deformation retract of a CW-complex X, then A is S-homotopy equivalent to X.*

THEOREM. *Suppose that $X_* = \{X_n\}$ is a simplicial object in the category of topological spaces, and $\widetilde{X}_* = \{\widetilde{X}_n\}$ is a presimplicial object which is S-homotopy equivalent to X_* in the category of presimplicial objects. Then $\widetilde{X}_*$ has the structure of an A_∞-simplicial object, and there is an A_∞-simplicial homotopy equivalence between X_* and $\widetilde{X}_*$.*

PROOF. Let $\xi_*: \widetilde{X}_* \to X_*$, $\eta_*: X_* \to \widetilde{X}_*$, and $h_*: X_* \times I \to X_*$ be the corresponding maps. We define operators $d_0^m: \widetilde{X}_n \times I^m \to \widetilde{X}_{n-m-1}$, $0 \le m \le n-1$, by setting

$$d_0^m = \eta d_0 h_{n-m} \dots d_0 h_{n-1} d_0 \xi.$$

Straightforward computations show that the required relations hold.

Further, define maps $\xi_n^m: \widetilde{X}_n \times I^m \to X_{n-m}$ by setting

$$\xi_n^m = h_{n-m} d_0 \dots h_{n-1} d_0 \xi.$$

These maps yield an A_∞-simplicial homotopy equivalence between $\widetilde{X}_*$ and X_*. □

Dually, the following theorem holds.

THEOREM. *Suppose that $X^* = \{X^n\}$ is a cosimplicial object in the category of topological spaces, and $\widetilde{X}^* = \{\widetilde{X}^n\}$ is a precosimplicial object which is S-homotopy equivalent to X^* in the category of precosimplicial objects. Then $\widetilde{X}^*$ has the structure of an A_∞-cosimplicial object, and there is an A_∞-cosimplicial homotopy equivalence between X^* and $\widetilde{X}^*$.*

Of course, the notions of A_∞-simplicial and A_∞-cosimplicial objects can be defined not only for the category of topological spaces, but for some other categories as well. For example, it can be done for the categories of simplicial sets, chain complexes, etc. To do this, it is necessary to use an analog of the unit n-dimensional cube I^n in each of these categories.

Now let us pass to the definition of A_∞-topological group. First, we shall reformulate the definition of the notion of topological group.

Let G be a space with base point $* \in G$. Consider the presimplicial object $W_*(G) = \{W_n(G)\}$. If G is a topological group, then $W_*(G)$ can be made into a simplicial object by defining the face operator d_0 by

$$d_0(g_1, \dots, g_n) = (g_2 - g_1, \dots, g_n - g_1).$$

Conversely, let us show that any face operator d_0 in $W_*(G)$ satisfying the simplicial relations generates a group structure on G. Indeed, let us define the operation $y - x$ in G by setting $y - x = d_0(x, y)$.

The simplicial relation $d_0 s_0(x) = (x)$ implies that $x - * = x$. The relation $d_0 s_1(x) = (*)$ implies that $x - x = *$. Finally, the relation $d_0 d_1(x, y, z) = d_0 d_0(x, y, z)$ implies that $(z - x) - (y - x) = z - y$.

Thus the operation $y - x$ satisfies the following relations:

1. $x - * = x$,
2. $x - x = *$,
3. $(z - x) - (y - x) = z - y$.

It is readily seen that giving such an operation is equivalent to giving a group structure. Hence the definition of topological group can be reformulated in terms of simplicial objects. To define a group structure on G is the same as to define a face operator d_0 making $W_*(G)$ a simplicial object. Its realization $|W_*(G)|$ is denoted by $W(G)$. One can show that it is homeomorphic to the classifying space $B(G)$ of the topological group G.

Our aim is to define an A_∞-topological group structure on a pointed space G so that giving it is equivalent to giving the structure of an A_∞-simplicial object on $W_*(G)$.

The simplicial relations imply that to define the operators

$$d_0^m \colon W_n(G) \times I^m \to W_{n-m-1}(G),$$

it suffices to define the operators $d_0^m \colon W_{m+2}(G) \times I^m \to W_1(G) = G$. Denote them by $\phi^m \colon G^{\times m+2} \times I^m \to G$. Then the operators $d_0^m \colon W_n(G) \times I^m \to W_{n-m-1}(G)$ are given by

$$d_0^m((g_1, \dots, g_n), t) = (\phi^m((g_1, \dots, g_{m+1}, g_{m+2}), t) \dots, \phi^m((g_1, \dots, g_{m+1}, g_n), t)).$$

Let us find the relations that should hold for the operators ϕ^m so that the corresponding operators d_0^m define an A_∞-simplicial structure on $W_*(G)$. Note that the relations $d_0^m(d_i \times 1) = d_{i-m-1} d_0^m$ for $i > m + 1$ have already been used in the reduction of d_0^m to ϕ^m.

We define the $\cup$-product $\phi^j \cup \phi^i \colon G^{i+j+3} \times I^{i+j} \to G$ of the operations ϕ^i and ϕ^j by setting

$$\begin{aligned}(\phi^j \cup \phi^i)&((g_1, \dots, g_{i+j+3}), (t_1, \dots, t_{i+j})) \\ &= \phi^j((\phi^i((g_1, \dots, g_{i+1}, g_{i+2}), (t_1, \dots, t_i)), \\ &\qquad \dots, \phi^i((g_1, \dots, g_{i+1}, g_{i+j+3}), (t_1, \dots, t_i))), (t_{i+1}, \dots, t_{i+j})).\end{aligned}$$

It is readily seen that the relations of A_∞-simplicial structure on $W_*(G)$ can be rewritten in the form

$$\begin{aligned}\phi^m(1 \times u_i^0) &= \phi^{m-1}(d_i \times 1), \quad 1 \le i \le m; \\ \phi^m(1 \times u_i^1) &= \phi^{m-i} \cup \phi^{i-1}, \quad 1 \le i \le m; \\ \phi^m(s_i \times 1) &= \phi^{m-1}(1 \times v_i), \quad i \le m, \quad m \ge 1; \\ \phi^0(s_0 \times 1) &= \mathrm{Id}; \\ \phi^m(s_{m+1} \times 1) &= *.\end{aligned}$$

DEFINITION. A family of operations $\phi^m: G^{m+2} \times I^m \to G$ satisfying the above relations will be called an *A_∞-group structure on G.*

The above arguments imply the following assertion.

THEOREM. *Giving an A_∞-topological group structure on a topological space G with base point $* \in G$ is equivalent to giving an A_∞-simplicial structure on the presimplicial object $W_*(G)$.*

The realization of the A_∞-simplicial object $W_*(G)$ will be denoted by $W(G)$ and will be called the *classifying space of the A_∞-topological group G.*

Suppose that G is an A_∞-topological group, and H is simply a topological group. The simplicial relations imply that to define an A_∞-map from $W_*(G)$ to $W_*(H)$, it suffices to define maps $f^m: G^{\times m+1} \times I^m \to H$ satisfying the following relations:

$$\begin{aligned} f^m(1 \times u_i^0) &= f^{m-1}(d_i \times 1), \quad 1 \le i \le m; \\ f^m(1 \times u_i^1) &= f^{m-i} \cup \phi^{i-1}, \quad 1 \le i \le m; \\ f^m(s_i \times 1) &= f^{m-1}(1 \times v_i), \quad i < m; \\ f^m(s_m \times 1) &= *. \end{aligned}$$

DEFINITION. A family of maps $f^m: G^{m+1} \times I^m \to H$ satisfying the above relations will be called an *A_∞-map.*

The above discussion implies the following assertion.

THEOREM. *Giving an A_∞-map from an A_∞-topological group G to a topological group H is equivalent to giving an A_∞-map from the A_∞-simplicial object $W_*(G)$ to the simplicial object $W_*(H)$.*

THEOREM. *Suppose that G is a topological group and $\widetilde{G}$ is a pointed topological space which is S-homotopy equivalent to G. Then $\widetilde{G}$ possesses an A_∞-topological group structure, and there is an A_∞-homotopy equivalence $\widetilde{G} \simeq G$.*

PROOF. It is readily seen that in this situation the presimplicial object $W_*(\widetilde{G})$ is S-homotopy equivalent to the simplicial object $W_*(G)$, and hence we can apply the theorem proved above. □

THEOREM. *For any A_∞-topological group G, there are a topological group H and an A_∞-homotopy equivalence $G \simeq H$.*

PROOF. Consider the classifying space $W(G) = \|W_*(G)\|$. Let $S_*W(G)$ be the singular simplicial object, and let the simplicial group $F_*S_*W(G)$ be the analog of the loop space. Then the realization $|F_*S_*W(G)|$ is the required topological group H. □

2.4. Homotopy type and A_∞-group structure

Elucidation of the algebraic structure of homotopy groups and homotopy classification of continuous maps of topological spaces are among the oldest problems in algebraic topology. Here we shall define and study algebraic operations closely related to the group structure on the homotopy groups of a topological space.

We shall start with some simplicial constructions related to the standard simplices Δ^n. Let n and m be positive integers. Denote by $\overline{\Delta}^n_m$ the set of increasing maps

$$\mu\colon \{1,2,\dots,m\} \to \{1,2,\dots,n\}$$

assigning to each positive integer i, $1 \le i \le m$, a positive integer $\mu(i) = n_i$, $1 \le n_i \le n$, so that $1 \le n_1 \le \dots \le n_m \le n$.

If there is at least one positive integer i such that $n_i = n_{i+1}$, then the corresponding element μ will be called *degenerate.*

Note that we have the composition operation

$$\gamma\colon \overline{\Delta}^n_m \times \overline{\Delta}^m_k \to \overline{\Delta}^n_k$$

assigning to maps $\mu \in \overline{\Delta}^n_m$, $\nu \in \overline{\Delta}^m_k$ their composition $\mu \circ \nu \in \overline{\Delta}^n_k$.

Denote by $\delta^i \in \overline{\Delta}^n_{n-1}$, $1 \le i \le n$, the map that does not take the value i, that is,

$$\delta^i(j) = \begin{cases} j, & j < i, \\ j+1, & j \ge i. \end{cases}$$

The restriction of the composition operation $\gamma\colon \overline{\Delta}^n_{n-1} \times \overline{\Delta}^{n-1}_m \to \overline{\Delta}^n_m$ to the element $\delta^i \in \overline{\Delta}^n_{n-1}$ yields a map which will be denoted by

$$\delta^i\colon \overline{\Delta}^{n-1}_m \to \overline{\Delta}^n_m, \quad 1 \le i \le n.$$

Similarly, by $\sigma^i \in \overline{\Delta}^n_{n+1}$, $1 \le i \le n$, denote the map that takes the value i twice, i.e.,

$$\sigma^i(j) = \begin{cases} j, & j \le i, \\ j-1, & j > i. \end{cases}$$

The restriction of the composition operation $\gamma\colon \overline{\Delta}^n_{n+1} \times \overline{\Delta}^{n+1}_m \to \overline{\Delta}^n_m$ to the element $\sigma^i \in \overline{\Delta}^n_{n+1}$ yields a map which will be denoted by

$$\sigma^i\colon \overline{\Delta}^{n+1}_m \to \overline{\Delta}^n_m, \quad 1 \le i \le n.$$

The family $\{\overline{\Delta}^n_m\}_{m \ge 0}$ will be denoted by $\overline{\Delta}^n$, and the corresponding families of maps by

$$\delta^i\colon \overline{\Delta}^{n-1} \to \overline{\Delta}^n, \quad \sigma^i\colon \overline{\Delta}^{n+1} \to \overline{\Delta}^n, \quad 1 \le i \le n.$$

It is clear that the following relations hold for the composition operation:

$$\delta^i \gamma = \gamma(\delta^i \times 1); \quad \sigma^i \gamma = \gamma(\sigma^i \times 1).$$

Now let $G = \{G_m\}_{m \ge 1}$ be a family of sets G_m with base points $* \in G_m$. Consider the families $x\colon \overline{\Delta}^n \to G$ of maps $x^m\colon \overline{\Delta}^n_m \to G_m$ such that x^m takes degenerate elements of $\overline{\Delta}^n_m$ to the base point $* \in G_m$. These maps $x^m\colon \overline{\Delta}^n_m \to G_m$ will also be denoted by

$$x_n(n_1, \dots, n_m), \quad 1 \le n_1 < \dots < n_m \le n.$$

Denote the set of these maps by $W^m_n(G)$. Note that $W^m_n(G)$ can be thought of as the product of G_m by itself $\binom{n}{m}$ times. For example,

$$W^m_m(G) = G_m, \quad W^m_{m+1}(G) = \underbrace{G_m \times \dots \times G_m}_{m+1},$$

and so on.

Denote the family of maps $x\colon \overline{\Delta}^n \to G$ by $W_n(G) = \{W_n^m(G)\}_{m\geq 1}$. The maps δ^i and σ^i induce maps

$$d_i\colon W_n(G) \to W_{n-1}(G), \quad s_i\colon W_n(G) \to W_{n+1}(G), \quad 1 \leq i \leq n,$$

which are called the face and degeneracy operators. In the above notation, they are given by the following formulas:

$$(d_i x)(n_1, \dots, n_m) = x(n_1, \dots, n_k, n_{k+1}+1, \dots, n_m+1),$$

where $n_k < i$, $n_{k+1} \geq i$, $1 \leq i \leq n$;

$$(s_i x)(n_1, \dots, n_m) = x(n_1, \dots, n_k, n_{k+1}-1, \dots, n_m-1),$$

where $n_k \leq i$, $n_{k+1} > i$, $1 \leq i \leq n$.

We define one more degeneracy operator $s_0\colon W_n(G) \to W_{n+1}(G)$ by setting

$$(s_0 x)(n_1, \dots, n_m) = \begin{cases} x(n_1-1, \dots, n_m-1), & n_1 > 1, \\ *, & n_1 = 1. \end{cases}$$

The operators

$$d_i\colon W_n(G) \to W_{n-1}(G), \quad 1 \leq i \leq n,$$

and

$$s_i\colon W_n(G) \to W_{n+1}(G), \quad 0 \leq i \leq n,$$

satisfy the simplicial relations

$$\begin{aligned} d_j d_i &= d_{i-1} d_j, \quad i > j; \\ s_j s_i &= s_i s_{j-1}, \quad i < j; \\ d_j s_i &= \begin{cases} s_i d_{j-1}, & j > i+1, \\ \mathrm{Id}, & j = i, i+1, \\ s_{i-1} d_j, & j < i. \end{cases} \end{aligned}$$

Note that to complete the definition of a simplicial structure on $W(G) = \{W_n(G)\}$, we need one more face operator $d_0\colon W_n(G) \to W_{n-1}(G)$.

If $G = \{G_m\}$ is a family of groups (for $m \geq 2$, we shall assume the groups G_m to be abelian), then the operator d_0 is defined in a standard way. Namely, denote for brevity the element $x_m(1, \dots, i-1, i+1, \dots m+1)$ by x^i, and let

$$d_0(x^1, \dots, x^{m+1}) = x^1 - x^2 + \dots + (-1)^m x^{m+1}.$$

On the remaining elements, the operator d_0 is given as follows:

$$\begin{aligned} (d_0 x)(n_1, \dots, n_m) = x_n(n_1+1, \dots, n_m+1) - x_n(1, n_2+1, \dots, n_m+1) \\ + \dots + (-1)^m x_n(1, n_1+1, \dots, n_{m-1}+1). \end{aligned}$$

Straightforward computations show that the required relations hold.

Note that in this case the simplicial sets $W^m(G)$ are the Eilenberg–MacLane spaces $K(G_m, m)$. In particular, if the family G consists of a single group G_1, then

$W_n(G)$ is the product $G_1^{\times n}$, whose elements are denoted by $[x_1, \dots, x_n]$, and the face and degeneracy operators are defined by

$$\begin{aligned} d_i[x_1, \dots, x_n] &= [x_1, \dots, x_{i-1}, x_{i+1}, \dots, x_n], \quad 1 \le i \le n, \\ s_i[x_1, \dots, x_n] &= [x_1, \dots, x_i, x_i, \dots, x_n], \quad 1 \le i \le n, \\ s_0[x_1, \dots, x_n] &= [*, x_1, \dots, x_n], \\ d_0[x_1, \dots, x_n] &= [x_2 - x_1, \dots, x_n - x_1]. \end{aligned}$$

Let us consider arbitrary simplicial operators

$$d_0 \colon W_n(G) \to W_{n-1}(G)$$

which, taken together with the rest of the face and degeneracy operators, make $W(G)$ into a simplicial set.

The simplicial relations imply that to define these operators it suffices to set their values in $W_{n-1}^{n-1}(G) = G_{n-1}$. Indeed, let

$$x \in W_n(G), \quad d_0(x) = \{y_{n-1}(n_1, \dots, n_m)\}, \quad 1 \le n_1 < \cdots < n_m \le n-1.$$

Assume that no element among $n_1, \dots, n_m$ in $y_{n-1}(n_1, \dots, n_m)$ is equal to i. By n_k denote the greatest element less than i. Then we have the relation

$$(d_i y_{n-1})(n_1, \dots, n_k, n_{k+1} - 1, \dots, n_m - 1) = y_{n-1}(n_1, \dots, n_m).$$

The simplicial relation $d_i d_0 = d_0 d_{i+1}$ implies that this value $y_{n-1}(n_1, \dots, n_m)$ is determined by the value of the operator d_0 on the element $d_{i+1}(x)$. Thus, by reducing the dimension, we shall arrive at the situation in which all the elements from 1 to $n-1$ occur among $n_1, \dots, n_m$. In this case, we have

$$y_{n-1}(n_1, \dots, n_m) = y_{n-1}(1, \dots, n-1) \in G_{n-1}.$$

Let $p_n \colon W_n(G) \to G_n$ be the projection onto G_n. We set

$$\phi_n = p_{n-1} d_0 \colon W_n(G) \to G_{n-1}.$$

The operators ϕ_n can be thought of as certain operations on the family $G = \{G_n\}$. Let us find out the relations which must hold for the family of operations $\phi = \{\phi_n\}$ so that the corresponding operator d_0 will satisfy the required simplicial relations.

Note that the relations $d_0 d_i = d_{i-1} d_0$, $i \ge 2$, have already been used when we reduced defining the operator d_0 to defining the operations ϕ_n.

The relations $d_0 s_0(x) = x$ and $d_0 s_i(x) = s_{i-1} d_0(x)$, $i \ge 1$, for the operations ϕ_n mean that $\phi_n s_0(x) = p_{n-1}(x)$ and $\phi_n s_i(x) = *$ for $x \in W^{n-1}(G)$ and $i \ge 1$.

To elucidate the last relation $d_0 d_0 = d_0 d_1$, consider the correspondence

$$G = \{G_m\} \longmapsto W(G) = \{W^n(G)\},$$

which defines a functor on the category of pointed sets.

The composition operations $\gamma \colon \overline{\Delta}_m^n \circ \overline{\Delta}_k^m \to \overline{\Delta}_k^n$ induce maps

$$\nabla \colon W_n(G) \to W_n(W(G)).$$

Denote by $\sigma \colon \overline{\Delta}_{m-1}^{n-1} \to \overline{\Delta}_m^n$ the maps defined by the correspondences

$$(n_1, \dots, n_{m-1}) \longmapsto (1, n_1 + 1, \dots, n_{m-1} + 1).$$

These maps induce maps $\sigma: W_n(G) \to W_{n-1}(G)$. The definition of the operations $\phi = \{\phi_n\}$ implies the relation

$$d_0 = (\sigma\phi)\nabla: W_n(G) \to W_{n-1}(G).$$

Let us define the $\cup$-product $\phi \cup \phi$ by setting $\phi \cup \phi = \phi(\sigma\phi)\nabla$. Then the fact that the relation $d_0 d_0 = d_0 d_1$ holds for the operations ϕ means that the relation $\phi \cup \phi = \phi d_1$ holds.

Now let us state the main definition.

DEFINITION. An *A_∞-group structure on a family $G = \{G_m\}$ of pointed sets* is a family $\phi = \{\phi_n\}$ of operations $\phi_n: W_n(G) \to G_{n-1}$ satisfying the following relations:

1. $\phi_n s_0(x) = p_{n-1}(x)$ and $\phi_n s_i(x) = *$ for $x \in W_{n-1}(G)$ and $i \geq 1$;
2. $\phi d_1 = \phi \cup \phi$.

The above arguments imply the following assertion.

THEOREM. *Giving an A_∞-group structure $\phi = \{\phi_n\}$ on a family $G = \{G_m\}$ of pointed sets is equivalent to giving a simplicial operator $d_0: W_n(G) \to W_{n-1}(G)$ on $W(G) = \{W_n(G)\}$ which, together with the other face and degeneracy operators, endows $W(G)$ with the structure of a simplicial set (denoted by $W_\phi(G)$).*

Suppose that two A_∞-group families (G', ϕ'), (G'', ϕ'') are given, and $W_{\phi'}G'$, $W_{\phi''}G''$ are the corresponding simplicial sets. Consider a simplicial map $f: W_{\phi'}G' \to W_{\phi''}G''$. As above, it can be shown that to define such a map it suffices to set its values in $W_n^n G'' = G''_n$. Denote the corresponding maps by

$$\psi = \{\psi_n\}, \quad \psi_n: W_n G' \to G''_n.$$

The map f can be uniquely reconstructed by ψ. Namely, $f = (1 \times \psi)\nabla$. Besides, for the map f to be simplicial, it is necessary and sufficient that the maps ψ satisfy the relations

$$\psi s_i = *, \quad \phi' \cup \psi = \psi \cup \phi'',$$

where $\phi' \cup \psi = \phi'(1 \times \psi)\nabla$, $\psi \cup \phi'' = \psi(\sigma \times \phi'')\nabla$.

DEFINITION. We shall say that a *map* $\psi: (G'\phi') \to (G'', \phi'')$ is a family $\psi = \{\psi_n\}$ of maps $\psi_n: W_n(G') \to G''_n$ satisfying the following relations:

1. $\psi s_i = *$,
2. $\phi' \cup \psi = \psi \cup \phi''$.

The above arguments imply the following assertion.

THEOREM. *Giving a map $\psi: (G', \phi') \to (G'', \phi'')$ is equivalent to giving a simplicial map $f: W_{\phi'}G' \to W_{\phi''}G''$.*

DEFINITION. Two A_∞-group families (G', ϕ'), (G'', ϕ'') are said to be *equivalent* (and we write $(G', \phi') \sim (G'', \phi'')$) if there is a map $\psi: (G', \phi') \to (G'', \phi'')$ such that the restrictions of the maps $\psi_n: W_n G' \to G''_n$ to G'_n are isomorphisms.

Let us prove the following result.

THEOREM. *The relation $(G', \phi') \sim (G'', \phi'')$ introduced above is an equivalence relation.*

Indeed, the reflexivity is obvious, since for ψ we can take the map with $\psi_0 = \mathrm{Id}$ and $\psi_k = 0$ for $k > 0$.

Let us prove the transitivity. Suppose that $(G', \phi') \sim (G'', \phi'')$, $(G'', \phi'') \sim (G''', \phi''')$ and ψ', ψ'' are the corresponding maps. Direct verification shows that the map $\psi = \psi'' \cup \psi'$ yields the required equivalence

$$(G', \phi') \sim (G''', \phi''').$$

Let us prove the symmetry. Suppose that $(G', \phi') \sim (G'', \phi'')$ and ψ is the corresponding map. We define the map ψ' that yields the equivalence $(G'', \phi'') \sim (G', \phi')$ by setting $\psi_0' = \psi_0^{-1}$ and defining the remaining ψ_k' (for $k \geq 1$) recursively by the relation

$$\sum_{i=0}^{k} \psi_i' \cup \psi_{k-i} = 0.$$

Direct verification shows that the map ψ' so defined is the required one.

THEOREM. *Let X be a topological space or a simplicial set with base point. Then the family $\pi_* = \{\pi_m(X)\}$ of its homotopy groups has a unique, up to equivalence, A_∞-group structure, which determines the weak homotopy type of X.*

PROOF. Let X be a pointed simplicial set. For each element $x \in \pi_m(X)$, let us choose a representative $\xi(x) \in X_m$. We obtain a map

$$\xi\colon W_m^m(\pi_*) = \pi_m(X) \to X_m.$$

Let us define the map

$$\xi\colon W_{m+1}^m(\pi_*) \to X_{m+1}$$

which, together with the previous one, will be permutable with the face and degeneracy operators. Let

$$x \in W_{m+1}^m(\pi_*), \quad x\colon \Delta_m^{m+1} \to \pi_m(X).$$

This map can be thought of as a collection of elements $x_1, \dots, x_{m+1}$ in $\pi_m(X)$. The representatives $\xi(x_1), \dots, \xi(x_{m+1})$ define compatible maps of the faces of the $(m+1)$-simplex into X, with the exception of the single face which is opposite to the zero vertex.

Let us define the map of this face by setting the corresponding value equal to $\xi(x_1 - \dots + (-1)^m x_{m+1})$. We define the face operator

$$d_0^0\colon W_{m+1}^m(\pi_*) \to W_m^m(\pi_*) = \pi_m(X)$$

by setting $d_0^0(x) = x_1 - \dots + (-1)^m x_{m+1}$.

Thus we obtain a map of the boundary of the $(m+1)$-simplex. Besides, since the maps

$$\xi(x_1 - \dots + (-1)^m x_{m+1}), \quad \xi(x_1) - \dots + (-1)^m \xi(x_{m+1})$$

are homotopic, this map can be extended to a map of the entire $(m+1)$-simplex $y\colon \Delta^{m+1} \to X$. Let $\xi(x) = y$.

Assume that the face operators

$$d_0^0\colon W_{m+k+1}^m(\pi_*) \to W_{m+k}^m(\pi_*), \dots, d_0^{k-1}\colon W_{m+k+1}^m(\pi_*) \to W_{m+k}^{m+k-1}(\pi_*)$$

and the maps $\xi\colon W_{m+i}^m(\pi_*) \to X$ for all $i \leq k$ are defined. Let us define the operator

$$d_0^k\colon W_{m+k+1}^m(\pi_*) \to W_{m+k}^{m+k}(\pi_*) = \pi_{m+k}(X)$$

and the map $\xi\colon W_{m+k+1}^m(\pi_*) \to X$.

Let $x \in W^m_{m+k+1}(\pi_*)$. The elements

$$\xi(d_0^0 x + \dots + d_0^{k-1} x), \xi(d_1 x), \dots, \xi(d_{m+k+1} x)$$

constitute a compatible family of maps of the faces of the $(m+k+1)$-simplex $\dot{\Delta}^{m+k+1} \to X$, whose homotopy class defines an element $z \in \pi_{m+k}(X)$. We set $d_0^k(x) = z$. Then the elements

$$\xi(d_0^0 x + \dots + d_0^{k-1} x + d_0^k x), \quad \xi(d_1 x), \dots, \xi(d_{m+k+1} x)$$

constitute a compatible family of maps of the faces of the $(m+1)$-simplex, and this family can be extended to a map of the $(m+1)$-simplex $y\colon \Delta^{m+k+1} \to X$. We set $\xi(x) = y$. This completes the inductive construction of the operators $d_0\colon W_n(\pi_*) \to W_{n-1}(\pi_*)$ and the simplicial map $\xi\colon W_\phi \pi_* \to X$.

This map induces an isomorphism in homotopy groups, and hence it is a weak homotopy equivalence. Moreover, as can be easily checked, $W_\phi \pi_*$ is a minimal complex, and its construction is a refinement of the construction of a minimal complex for a simplicial set [**28**]. Therefore, if we take a minimal complex K for X, then the map $\xi\colon W_\phi \pi_* \to K$ is an isomorphism.

The construction of the operator d_0 and the map ξ is not uniquely defined. If d_0', d_0'' are two face operators of this kind, and

$$\xi'\colon W_{\phi'}\pi_* \to K, \quad \xi''\colon W_{\phi''}\pi_* \to K$$

are the corresponding isomorphisms with the minimal complex K, then the isomorphism

$$(\xi'')^{-1} \circ \xi'\colon W_{\phi'}\pi_* \to W_{\phi''}\pi_*$$

yields an equivalence of the corresponding A_∞-group structures on the family of groups π_*. □

This result refines the theory of the Postnikov k-invariants [**36**] and makes it possible to reduce the problem of describing the homotopy type of topological spaces to the purely algebraic problem of investigating A_∞-group structures on their homotopy groups.

CHAPTER 3

Algebraic Structures on Chain Complexes

3.1. Chain complexes

Let us recall the necessary facts concerning chain complexes. Suppose that R is a commutative ring with identity. A *chain complex* or, alternatively, a *differential graded module* is an R-module X endowed with a grading by submodules X_n, $n \in \mathbb{Z}$, such that $X = \sum_n X_n$, and with a homomorphism $d: X \to X$ (called a *differential*) that lowers the degree by 1, i.e., $d(X_n) \subset X_{n-1}$, and satisfies the relation $d \circ d = 0$.

A chain complex X is said to be *augmented* if $X_n = 0$ for $n < 0$ and an epimorphism $a: X_0 \to R$ satisfying the relation $a \circ d = 0$ is given (this epimorphism is called an *augmentation*).

The *homology* of a chain complex X is the graded module

$$H_*(X) = \sum_n H_n(X), \quad H_n(X) = \frac{\operatorname{Ker}\{d: X_n \to X_{n-1}\}}{\operatorname{Im}\{d: X_{n+1} \to X_n\}}.$$

A homomorphism $f: X \to Y$ of chain complexes is called a *chain map* if it preserves the degree, i.e., $f(X_n) \subset Y_n$, and commutes with the differential, i.e., $d \circ f = f \circ d$. If the complexes X and Y are augmented, then the homomorphism $f: X \to Y$ is additionally required to commute with the augmentation, i.e., $a_Y \circ f = a_X$.

It is clear that a chain map $f: X \to Y$ induces a map in homology

$$H_*(f): H_*(X) \to H_*(Y), \quad H_*(f) = \sum_n H_n(f).$$

Two chain maps $f, g: X \to Y$ are said to be *homotopic* (and we write $f \simeq g$) if there exists a map $h: X \to Y$ (a *homotopy*) raising degree by 1 and satisfying the relation $d \circ h + h \circ d = g - f$.

By I we denote the chain complex with a single one-dimensional generator (01) and two zero-dimensional generators (0) and (1). Let the differential be defined by the formula $d(01) = (1) - (0)$. Then giving a chain homotopy $h: X \to Y$ between chain maps $f, g: X \to Y$ is equivalent to giving a chain map $H: X \otimes I \to Y$ that satisfies the relations

$$H(x \otimes (0)) = f(x), \quad H(x \otimes (1)) = g(x).$$

The relationship between the maps H and h is given by the formula $H(x \otimes (01)) = h(x)$.[1]

It can be shown easily that homotopic chain maps $f \simeq g: X \to Y$ induce the same map in homology:

$$H_*(f) = H_*(g): H_*(X) \to H_*(Y).$$

---

[1]For the definition of the product of two chain complexes, see page 47.

Two chain complexes X and Y are said to be *homotopy equivalent* (written $X \simeq Y$) if there exist chain maps $f: X \to Y$ and $g: Y \to X$ such that $g \circ f \simeq \mathrm{Id}: X \to X$ and $f \circ g \simeq \mathrm{Id}: Y \to Y$. It is readily seen that homotopy equivalent complexes have isomorphic homology.

Direct verification shows that the homotopy relation on chain maps and the homotopy equivalence on chain complexes are equivalence relations.

A basic example of an augmented chain complex is provided by the *singular chain complex*

$$C_*(X;R) = \sum_{n\geq 0} C_n(X;R)$$

of a topological space X with coefficients in R. To define it, consider the free simplicial R-module $S_*(X;R)$ generated by the singular simplicial set $S_*(X)$. We put $C_n(X;R) = S_n(X;R)$ and define the differentials $d: C_n(X;R) \to C_{n-1}(X;R)$ by

$$d = \sum_{i=0}^{n} (-1)^i d_i.$$

The augmentation $a: C_0(X;R) \to R$ on the zero-dimensional simplices is set to $1 \in R$. Straightforward computations show that the relations $d \circ d = 0$, $a \circ d = 0$ hold; hence $C_*(X;R)$ is an augmented chain complex.

The *normalized chain complex $N_*(X;R)$ of a topological space X with coefficients in R* is the quotient complex of $C_*(X;R)$ by the submodule generated by the degenerate simplices, i.e.,

$$N_n(X;R) = \frac{C_n(X;R)}{\sum_i \mathrm{Im}\{s_i: C_{n-1}(X;R) \to C_n(X;R)\}}.$$

The differential in $N_*(X;R)$ is induced by the differential in $C_*(X;R)$.

We have the chain equivalence $N_*(X;R) \simeq C_*(X;R)$.

The homology of the chain complex $C_*(X;R)$ is referred to as the *homology of the space X with coefficients in R* and is denoted by $H_*(X;R)$.

Clearly, by associating to any topological space X the singular chain complex $C_*(X;R)$, we define a functor from the category of topological spaces to the category of chain complexes.

Similarly, for an arbitrary simplicial set X_*, the chain complex $C_*(X_*;R)$ is defined, where $C_n(X_*;R)$ is the free R-module generated by X_n and the differential $d: C_n(X_*;R) \to C_{n-1}(X_*;R)$ is induced by the face operators in X_*. The corresponding normalized complex is denoted by $N_*(X_*;R)$.

Along with the singular chain complex $C_*(X;R)$ of the topological space X, we consider the *cochain complex* $C^*(X;R)$ dual to $C_*(X;R)$, i.e.,

$$C^n(X;R) = \mathrm{Hom}(C_n(X;R);R).$$

The homology of this complex is referred to as the *cohomology of the topological space X with coefficients in R* and is denoted by $H^*(X;R)$.

For any chain complex X, by SX we denote the *suspension of* X, i.e., the complex with $(SX)_n = X_{n-1}$. The elements of SX will be denoted by $[x]$, where $x \in X$. The differential is defined by the formula $d[x] = -[dx]$. The desuspension $S^{-1}X$ is defined similarly.

Let X and Y be chain complexes. We define the chain complexes $X \otimes Y$ and $\mathrm{Hom}(X, Y)$ by setting

$$X \otimes Y = \sum_n (X \otimes Y)_n, \quad \mathrm{Hom}(X,Y) = \sum_n \mathrm{Hom}_n(X,Y),$$
$$(X \otimes Y)_n = \sum_{p+q=n} X_p \otimes Y_q, \quad \mathrm{Hom}_n(X,Y) = \prod_i \mathrm{Hom}(X_i, Y_{n+i}).$$

For two maps $f' \in \mathrm{Hom}(X', Y')$ and $f'' \in \mathrm{Hom}(X'', Y'')$, we define the tensor product $f' \otimes f'' \in \mathrm{Hom}(X' \otimes X'', Y' \otimes Y'')$ by setting

$$f' \otimes f''(x' \otimes x'') = (-1)^\epsilon f'(x') \otimes f''(x''),$$

where $\epsilon = \deg(f'') \cdot \deg(x')$.

Note that the sign ϵ appears as a result of the transposition of f'' and x', being the product of the dimensions of the corresponding elements. This rule is applied not only in this case, but in other cases as well.

Now let us define the differential d in $X \otimes Y$ by setting $d = d \otimes 1 + 1 \otimes d$. By applying this formula to the elements $x_p \otimes y_q \in X_p \otimes Y_q$, we obtain

$$d(x_p \otimes y_q) = d(x_p) \otimes y_q + (-1)^p x_p \otimes d(y_q).$$

If X and Y are augmented chain complexes, then their tensor product $X \otimes Y$ is also an augmented chain complex with the augmentation defined by the formula $a(x \otimes y) = a(x) \cdot a(y)$.

Similarly, we define the differential in $\mathrm{Hom}(X, Y)$ by setting

$$d(f_n) = d \circ f_n - (-1)^n f_n \circ d,$$

where $f_n \in \mathrm{Hom}_n(X, Y)$.

Now let us consider the notion of realization for simplicial and cosimplicial objects in the category of chain complexes. Denote by Δ^n the normalized chain complex of the standard n-dimensional simplex. The family $\Delta^* = \{\Delta^n\}$ forms a cosimplicial object in the category of chain complexes. The coface operators $\delta^i \colon \Delta^n \to \Delta^{n+1}$ and the codegeneracy operators $\sigma^i \colon \Delta^n \to \Delta^{n-1}$ are induced by the coface and codegeneracy operators for the standard simplices. It is clear that $\Delta^1 = I$.

Let $X_* = \{X_n\}$ be a simplicial object in the category of chain complexes. Then its *realization* is the complex

$$|X_*| = \sum_n X_n \otimes \Delta^n / \sim,$$

where the equivalence relation is generated by the relations

$$d_i(x) \otimes u \sim x \otimes \delta^i(u), \quad x \in X_n, \ u \in \Delta^{n-1};$$
$$s_i(x) \otimes u \sim x \otimes \sigma^i(u), \quad x \in X_{n-1}, \ u \in \Delta^n.$$

Dually, the *realization of a cosimplicial object* X^* in the category of chain complexes is the complex

$$|X^*| = \mathrm{Hom}(\Delta^*, X^*),$$

where Hom is taken in the category of cosimplicial objects, i.e., it consists of the families $f^* = \{f^n\}$ of maps $f^n \colon \Delta^n \to X^n$ commuting with the coface and codegeneracy operators.

A special case of the realization of a simplicial object in the category of chain complexes is the realization of a simplicial object in the category of modules, which are regarded as chain complexes with zero grading and zero differential. Let X be a topological space; the realization $|S_*(X;R)|$ of a simplicial module $S_*(X;R)$ is a chain complex, which, as can be easily verified, is isomorphic to the normalized chain complex $N_*(X;R)$.

In general, the realization $|Y_*|$ of a simplicial module Y_* is a chain complex isomorphic to the normalized chain complex $N(Y_*)$, where

$$N_n(Y_*) = Y_n / \sum_i \{\mathrm{Im}\ s_i \colon Y_{n-1} \to Y_n\}.$$

The differentials $d\colon N_n(Y_*) \to N_{n-1}(Y_*)$ are induced by the maps

$$\sum_{i=0}^{n} (-1)^i d_i \colon Y_n \to Y_{n-1}.$$

Conversely, for any chain complex X, the family

$$\mathrm{Hom}(\Delta^*, X) = \{\mathrm{Hom}(\Delta^n, X)\},$$

where Hom is taken in the category of chain complexes, is a simplicial module. The face and degeneracy operators are induced by the corresponding face and degeneracy operators in Δ^*.

The correspondence $X \longmapsto \mathrm{Hom}(\Delta^*, X)$ defines a functor $N^{-1}\colon Chain \to Mod_s$, which is adjoint to the functor $N\colon Mod_s \to Chain$. Moreover, for any chain complex X, there is a natural isomorphism $N(N^{-1}(X)) \cong X$.

For two simplicial modules Y_*' and Y_*'', we have the Eilenberg–Zilber chain map

$$\psi\colon N(Y_*' \otimes Y_*'') \to N(Y_*') \otimes N(Y_*'')$$

defined on the generators by

$$\psi(y_n' \otimes y_n'') = \sum_{i=0}^{n} d_{i+1} \dots d_n(y_n') \otimes d_0^i(y_n'').$$

On the other hand, we have the Alexander–Whitney chain map

$$\phi\colon N(Y_*') \otimes N(Y_*'') \to N(Y_*' \otimes Y_*'')$$

defined on the generators by

$$\phi(y_p' \otimes y_q'') = \sum_{(\mu,\nu)} (-1)^{\epsilon} s_{\nu_q} \dots s_{\nu_1}(y_p') \otimes s_{\mu_p} \dots s_{\mu_1}(y_q''),$$

where the summation is over all (p,q)-shuffles (μ,ν), and ϵ is the sign of the shuffle.

In particular, if X' and X'' are topological spaces, then we have the Eilenberg–Zilber map

$$\psi\colon C_*(X' \times X''; R) \to C_*(X'; R) \otimes C_*(X''; R)$$

and the Alexander–Whitney map

$$\phi\colon C_*(X'; R) \otimes C_*(X''; R) \to C_*(X' \times X''; R).$$

By using the method of acyclic models, it can be shown that the Eilenberg–Zilber and Alexander–Whitney maps are mutually inverse chain equivalences.

If $f, g: X \to Y$ are homotopic continuous maps of topological spaces, and if $h: X \times I \to Y$ is the corresponding homotopy, then the induced chain maps $C_*(f), C_*(g): C_*(X; R) \to C_*(Y; R)$ are chain homotopic. The corresponding homotopy $H: C_*(X; R) \otimes I \to C_*(Y; R)$ is the composite map

$$C_*(X; R) \otimes I \xrightarrow{\phi} C_*(X \times I; R) \xrightarrow{C_*(h)} C_*(Y; R).$$

Similarly, if $f_*, g_*: X_* \to Y_*$ are homotopic maps of simplicial modules, then the induced maps of chain complexes are chain homotopic. Moreover, the functors N and N^{-1} induce a homotopy equivalence between the category of simplicial modules and the category of chain complexes.

3.2. Algebras and modules over an algebra

A *differential graded algebra* or simply an *algebra* is a chain complex A together with a chain map $\pi: A \otimes A \to A$ (called a *multiplication*) satisfying the *associativity condition*

$$\pi(\pi \otimes 1) = \pi(1 \otimes \pi),$$

which can also be represented as the commutative diagram

$$\begin{array}{ccc} A \otimes A \otimes A & \xrightarrow{\pi \otimes 1} & A \otimes A \\ {\scriptstyle 1 \otimes \pi}\downarrow & & \downarrow{\scriptstyle \pi} \\ A \otimes A & \xrightarrow{\pi} & A \end{array}$$

The multiplication in an algebra will also be denoted by a dot, i.e., instead of $\pi(x \otimes y)$ we shall write $x \cdot y$.

An *identity element* of an algebra A is an element $e \in A_0$ such that $e \cdot x = x \cdot e = x$ for any $x \in A$.

An algebra A is said to be *augmented* if there is an augmentation $a: A \to R$ of the complex A that preserves the multiplication, i.e., $a(x \cdot y) = a(x) \cdot a(y)$. If A is an algebra with identity, then the augmentation is additionally required to take the identity element of the algebra to the identity element of the ring R. An algebra A is said to be *commutative* if the relation $\pi = \pi \circ T$ holds, where $T: A \otimes A \to A \otimes A$ is the permutation operator defined by

$$T(x \otimes y) = (-1)^{\epsilon} y \otimes x, \quad \epsilon = \deg(x) \cdot \deg(y).$$

Let A be an algebra. A *left module over* A is a chain complex M together with ma map $\mu: A \otimes M \to M$ satisfying the condition

$$\mu(\pi \otimes 1) = \mu(1 \otimes \mu),$$

which can also be represented as the commutative diagram

$$\begin{array}{ccc} A \otimes A \otimes M & \xrightarrow{\pi \otimes 1} & A \otimes M \\ {\scriptstyle 1 \otimes \mu}\downarrow & & \downarrow{\scriptstyle \mu} \\ A \otimes M & \xrightarrow{\mu} & A \end{array}$$

The notion of right module over an algebra is defined similarly.

An example of an algebra is given by the tensor algebra $TX = \sum_n X^{\otimes n}$ generated by a chain complex X. The multiplication in TX is defined by

$$(x_1 \otimes \cdots \otimes x_n) \cdot (x_{n+1} \otimes \cdots \otimes x_{n+m}) = x_1 \otimes \cdots \otimes x_{n+m}.$$

The tensor algebra TX possesses the following universal property.

For any algebra A and any map $f: X \to A$, there is a unique algebra map $\widetilde{f}: TX \to A$ such that the following diagram is commutative:

$$\begin{array}{ccc} X & \xrightarrow{i} & TX \\ & \searrow^{f} & \downarrow \tilde{f} \\ & & A \end{array}$$

An example of a commutative algebra is the symmetric algebra T_sX obtained from the tensor algebra TX as the quotient of TX by the actions of the symmetric groups Σ_n on the tensor products $X^{\otimes n}$. The symmetric algebra T_sX possesses the following universal property.

For any commutative algebra A and any chain map $f: X \to A$, there is a unique algebra map $\widetilde{f}: T_sX \to A$ such that the following diagram is commutative:

$$\begin{array}{ccc} X & \xrightarrow{i} & T_sX \\ & \searrow^{f} & \downarrow \tilde{f} \\ & & A \end{array}$$

Dually, a chain complex K is called a *differential graded coalgebra* or simply a *coalgebra* if there is given a chain map $\nabla: K \to K \otimes K$ (called a *comultiplication*) that satisfies the associativity condition

$$(\nabla \otimes 1)\nabla = (1 \otimes \nabla)\nabla,$$

which can also be expressed as the commutative diagram

$$\begin{array}{ccc} K & \xrightarrow{\nabla} & K \otimes K \\ {\scriptstyle\nabla}\downarrow & & \downarrow{\scriptstyle\nabla\otimes 1} \\ K \otimes K & \xrightarrow{1\otimes\nabla} & K \otimes K \otimes K \end{array}$$

A coalgebra K is said to be *augmented* if an augmentation of the complex K preserves the comultiplication. A coalgebra K is said to be *commutative* if the relation $T \circ \nabla = \nabla$ holds.

Let K be a coalgebra. A *left comodule over* K is a chain complex L together with a map $\tau: L \to K \otimes L$ satisfying the condition

$$(\nabla \otimes 1)\tau = (1 \otimes \tau)\tau,$$

which can be expressed as the commutative diagram

$$\begin{array}{ccc} L & \xrightarrow{\tau} & K \otimes L \\ {\scriptstyle\tau}\downarrow & & \downarrow{\scriptstyle\nabla\otimes 1} \\ K \otimes L & \xrightarrow{1\otimes\tau} & K \otimes K \otimes L \end{array}$$

The notion of right comodule over a coalgebra is defined similarly.

It is readily seen that if K is a coalgebra, then the dual complex $A = \mathrm{Hom}(K, R)$ is an algebra with respect to the multiplication induced by the comultiplication in K.

An example of a coalgebra is provided by the tensor coalgebra $TX = \sum_n X^{\otimes n}$ generated by a chain complex X. The comultiplication in TX is defined by

$$\nabla(x_1 \otimes \cdots \otimes x_n) = \sum_i (x_1 \otimes \cdots \otimes x_i) \otimes (x_{i+1} \otimes \cdots \otimes x_n).$$

In general, the tensor coalgebra TX does not possess any universal property. For a universal property to hold, let us consider coalgebras of finite type. Namely, we say that K is a *coalgebra of finite type* if for any $x \in K$ there exists a positive integer n such that $\nabla(n)(x) = 0$, where $\nabla(n) = (\nabla \otimes \cdots \otimes 1) \ldots (\nabla \otimes 1)\nabla$ is the iteration of the comultiplication ∇.

In the category of coalgebras of finite type, the tensor coalgebra possesses the following universal property.

For any coalgebra K of finite type and for any chain map $f\colon K \to X$, there is a unique coalgebra map $\widetilde{f}\colon K \to TX$ such that the following diagram is commutative:

$$\begin{array}{ccc} X & \xleftarrow{p} & TX \\ & \nwarrow^{f} & \uparrow \widetilde{f} \\ & & K \end{array}$$

An example of a commutative coalgebra is provided by the coalgebra T_cX which is the subcomplex of TX consisting of all its elements that are invariant with respect to the actions of the symmetric groups Σ_n on the tensor products $X^{\otimes n}$. The coalgebra T_cX possesses the following universal property.

For any commutative coalgebra K of finite type and for any chain map $f\colon K \to X$, there is a unique coalgebra map $\widetilde{f}\colon K \to T_cX$ such that the following diagram is commutative:

$$\begin{array}{ccc} X & \xleftarrow{p} & T_cX \\ & \nwarrow^{f} & \uparrow \widetilde{f} \\ & & K \end{array}$$

One of the main examples of coalgebras is the singular chain complex $C_*(X; R)$ of a topological space X. The comultiplication

$$\nabla\colon C_*(X; R) \to C_*(X; R) \otimes C_*(X; R)$$

is the composition

$$C_*(X; R) \to C_*(X \times X; R) \xrightarrow{\psi} C_*(X; R) \otimes C_*(X; R)$$

of the map induced by the diagonal $X \to X \times X$ and the Eilenberg–Zilber map.

This comultiplication acts on singular simplices $x_n \in C_n(X; R)$ by the formula

$$\nabla(x_n) = \sum_{i=0}^{m} d_{i+1} \ldots d_n(x_n) \otimes d_0^i(x_n).$$

The comultiplication on chains induces a multiplication on cochains

$$\cup\colon C^*(X; R) \otimes C^*(X; R) \to C^*(X; R),$$

which endows the singular cochain complex of a topological space with an algebra structure.

This multiplication has the form

$$(f \cup g)(x) = \sum_{i=0}^{m} f(d_{q+1} \dots d_{p+q}(x_n)) \otimes g(d_0^p(x)),$$

where $f \in C^p(X; R)$, $g \in C^q(X; R)$, $x \in C^{p+q}(X; R)$.

An algebra A is called a *Hopf algebra* if in addition to the multiplication $\pi\colon A \otimes A \to A$, a comultiplication $\nabla\colon A \to A \otimes A$ is given that makes A a coalgebra and satisfies the *Hopf relation*

$$\nabla \circ \pi = (\pi \otimes \pi) \circ (1 \otimes T \otimes 1) \circ (\nabla \otimes \nabla),$$

which can be expressed as the commutative diagram

$$\begin{array}{ccccc} A \otimes A & \xrightarrow{\nabla \otimes \nabla} & A \otimes A \otimes A \otimes A & \xrightarrow{1 \otimes T \otimes 1} & A \otimes A \otimes A \otimes A \\ \pi \big\downarrow & & & & \big\downarrow \pi \otimes \pi \\ A & \xrightarrow{\nabla} & A \otimes A & \xrightarrow{=} & A \otimes A \end{array}$$

where $T\colon A \otimes A \to A \otimes A$ is the transposition map.

An example of a Hopf algebra is provided by the tensor algebra TX with the comultiplication defined on the generators by

$$\nabla(x) = x \otimes 1 + 1 \otimes x, \quad x \in X.$$

By using the Hopf relation, it is readily shown that in general the comultiplication is given by

$$\nabla(x_1 \otimes \dots \otimes x_n) = \sum_{p+q=n} (x_1 \otimes \dots \otimes x_p) \otimes (x_{p+1} \otimes \dots \otimes x_{p+q})\sigma(p, q),$$

where the summation is over all (p, q)-shuffles $\sigma(p, q)$ of $1, \dots, n$.

By a similar formula, one can define a Hopf algebra structure on the algebra $T_s X$.

Dually, regarding TX as a coalgebra, we can endow it with a multiplication making it a Hopf algebra, by setting

$$(x_1 \otimes \dots \otimes x_p) \cdot (x_{p+1} \otimes \dots \otimes x_{p+q}) = \sum (x_1 \otimes \dots \otimes x_{p+q})\sigma(p, q).$$

A similar formula defines a Hopf algebra structure on the coalgebra $T_c X$.

Another example of a Hopf algebra is given by the singular chain complex $C_*(G; R)$ of a topological space G with an associative multiplication $\pi\colon G \times G \to G$. The multiplication in $C_*(G; R)$ is the composition

$$C_*(G; R) \otimes C_*(G; R) \xrightarrow{\phi} C_*(G \times G; R) \to C_*(G; R)$$

of the Alexander–Whitney map and the map induced by the multiplication in G. In particular, the singular chain complex $C_*(G; R)$ of a topological group G and the singular chain complex $C_*(\Omega_l X; R)$ of the loop space $\Omega_l X$ of a space X are Hopf algebras.

Let us consider the question as to what structure exists on the homology of an algebra. Suppose that A is a differential graded algebra with multiplication

$$\pi\colon A \otimes A \to A.$$

This multiplication induces a multiplication in homology,

$$\pi_*: H_*(A) \otimes H_*(A) \to H_*(A),$$

and hence the homology of a differential graded algebra is a graded algebra. Along with this multiplication, there are other operations on the homology of a differential algebra, i.e., the Massey products, which are partial and multivalued operations of the form

$$m_n: H_*(A)^{\otimes n+2} \to H_*(A), \quad n \geq 1,$$

and which raise the dimension by n. Let us recall the definitions.

The first of them,

$$m_1: H_*(A) \otimes H_*(A) \otimes H_*(A) \to H_*(A),$$

is defined for the elements $x_1 \otimes x_2 \otimes x_3 \in A \otimes A \otimes A$ such that $x_1 \cdot x_2 = 0$, $x_2 \cdot x_3 = 0$. By $\overline{x}_1, \overline{x}_2, \overline{x}_3$ we denote representatives of the homology classes x_1, x_2, x_3. Then there are elements $y_1, y_2 \in A$ such that

$$d(y_1) = x_1 \cdot x_2, \quad d(y_2) = x_2 \cdot x_3.$$

It is readily seen that the element $y_1 \cdot x_3 + (-1)^{n_1+1} x_1 \cdot y_2$, where $n_1 = \dim(x_1)$, is a cycle and hence determines an element of the homology $H_*(A)$. By definition, this element is regarded as the value of the operation $m_1(x_1 \otimes x_2 \otimes x_3)$.

Of course, this definition depends on the choice of the representatives of the homology classes. Various representatives provide not one value, but a set of values of the operation $m_1(x_1 \otimes x_2 \otimes x_3)$. The value of the operation is thought of as being zero if the set $m_1(x_1 \otimes x_2 \otimes x_3)$ contains zero.

Similarly, the operation

$$m_2: H_*(A)^{\otimes 4} \to H_*(A)$$

is defined for the elements $x_1 \otimes x_2 \otimes x_3 \otimes x_4 \in A^{\otimes 4}$ such that

$$x_1 \cdot x_2 = 0, \quad x_2 \cdot x_3 = 0, \quad x_3 \cdot x_4 = 0;$$
$$m_1(x_1 \otimes x_2 \otimes x_3) = 0, \quad m_1(x_2 \otimes x_3, \otimes x_4) = 0.$$

By $\overline{x}_1, \overline{x}_2, \overline{x}_3, \overline{x}_4$ we denote representatives of the homology classes x_1, x_2, x_3, x_4. Then there exist elements $y_1, y_2, y_3 \in A$ and $z_1, z_2 \in A$ such that

$$d(y_1) = x_1 \cdot x_2, \quad d(y_2) = x_2 \cdot x_3 \quad d(y_3) = x_3 \cdot x_4;$$
$$d(z_1) = y_1 \cdot x_3 + (-1)^{n_1+1} x_1 \cdot y_2, \quad d(z_2) = y_2 \cdot x_4 + (-1)^{n_2+1} x_x \cdot y_3,$$

where $n_1 = \dim(x_1)$, $n_2 = \dim(x_2)$.

It is readily seen that the element $z_1 \cdot x_4 + (-1)^{n_1} x_1 \cdot z_2 + (-1)^{n_1+n_2} y_1 \cdot y_3$ is a cycle and hence determines an element of the homology $H_*(A)$. By definition, this element is regarded as the value of the operation $m_2(x_1 \otimes x_2 \otimes x_3 \otimes x_4)$.

The higher operations m_n are defined in a similar way.

The Massey products play an important role in the computation of the homology of various differential algebras, since they make it possible to obtain new elements from already known ones by applying the Massey products. Their fundamental disadvantage is that they are partial and multivalued operations. In what follows, we shall eliminate this defect by considering the notion of an A_∞-algebra.

3.3. Twisted tensor products and the functor $\mathcal{D}$

Twisted tensor products were introduced by E. Brown [**12**] to describe the chain complex of a fibration. Let us recall the main definitions.

Let K be a coalgebra, and let A be an algebra. A *twisting cochain* $\phi: K \to A$ is a homomorphism that lowers the dimension by 1 and satisfies the condition

$$d(\phi) + \phi \cup \phi = 0,$$

where $\phi \cup \phi = \pi(\phi \otimes \phi)\nabla: K \to A$; see [**12**].

If A is an augmented algebra, then the twisting cochain $\phi: K \to A$ is additionally required to satisfy the relation $a \circ \phi = 0$.

If L is a right K-comodule, M is a left A-module, and $\phi: K \to A$ is a twisting cochain, then the *twisted tensor product* $L \otimes_\phi M$ is the chain complex that coincides as a graded module with the ordinary tensor product, and in which the differential d_ϕ is defined by

$$d_\phi = d \otimes 1 + 1 \otimes d + \phi\cap,$$

where $\phi\cap = (1 \otimes \mu)(1 \otimes \phi \otimes 1)(\tau \otimes 1): L \otimes M \to L \otimes M$; see [**12**].

In particular, the following twisted tensor products are defined:

$$K \otimes_\phi M, \quad L \otimes_\phi A, \quad K \otimes_\phi A.$$

If L is a left K-comodule, M is a left A-module, and $\phi: K \to A$ is a twisting cochain, then the *twisted module of homomorphisms* $\mathrm{Hom}_\phi(L, M)$ is the differential graded module that coincides as a graded module with the ordinary module of homomorphisms, and in which the differential is defined by

$$d_\phi(f) = d(f) + \phi \cup f, \quad f \in \mathrm{Hom}(L, M),$$

where $\phi \cup f = \mu(\phi \otimes f)\tau$.

The importance of twisted tensor products depends upon the fact that for any Serre fibration $p: E \to B$ with arcwise connected base B and fiber F, there is a chain equivalence

$$C_*(E; R) \simeq C_*(B; R) \otimes_\phi C_*(F; R),$$

where $\phi: C_*(B; R) \to C_*(\Omega_l B; R)$ is a certain twisting cochain; see [**12**].

In the case of a fiber bundle with structure group G, there is a twisting cochain $\phi: C_*(B; R) \to C_*(G; R)$, and we have a similar chain equivalence.

In the case of a twisted Cartesian product $X \times_\xi F$ with twisting function $\xi: X \to G$, we also have a twisting cochain $\phi: C_*(X; R) \to C_*(G; R)$ and a chain equivalence

$$C_*(X \times_\xi F; R) \simeq C_*(X; R) \otimes_\phi C_*(F; R).$$

These chain equivalences make it possible to reduce the computation of the homology of fiber spaces or twisted Cartesian products to the computation of the homology of the corresponding twisted tensor products.

Let $L \otimes_\phi M$ be a twisted tensor product, and let $\phi: K \to A$ be the corresponding twisting cochain. Consider the filtration by the skeletons of the complex L, i.e.,

$$F^n = \sum_{i \leq n} L_i \otimes M \subset L \otimes_\phi M.$$

The differential d_ϕ preserves this filtration, and hence we have a spectral sequence for the homology of the twisted tensor product. If L is a free R-module, then the first term E^1 of this spectral sequence is isomorphic to the tensor product $L \otimes M_*$, where $M_* = H_*(M)$. The map $\phi_1: K_1 \to A_0$ and the action $\mu: A_0 \otimes M \to M$ induce a map $\phi_{1*}: K_1 \to \operatorname{Hom}(M_*; M_*)$. We define the differential $d_{\phi_{1*}}$ in $L \otimes M_*$ by setting

$$d_{\phi_{1*}} = d \otimes 1 + \phi_{1*} \cap .$$

Denote the corresponding chain complex by $L \otimes_{\phi_{1*}} M_*$. It is readily seen that its homology is isomorphic to the term E^2 of the spectral sequence.

A stronger assertion connecting the complexes $L \otimes_\phi M$ and $L \otimes M_*$ was proved by E. Brown.

THEOREM (Brown). *If R is a field, then for any twisted tensor product $L \otimes_\phi M$, where L is a comodule of finite type over a coalgebra K, there exist a twisting cochain $\phi_*: K \to \operatorname{Hom}(M_*; M_*)$ and a chain equivalence $L \otimes_\phi M \simeq L \otimes_{\phi_*} M_*$.*

PROOF. For any twisting cochain $\phi: K \to A$, by ϕ we also denote the composite map $K \otimes M \xrightarrow{\phi\otimes 1} A \otimes M \xrightarrow{\mu} M$. Let $\xi: M_* \to M$ be a representative map for the homology classes. Since the ground ring R is a field, this map can be assumed to be a homomorphism. Further, let $\eta: M \to M_*$ be the inverse map, $\eta \circ \xi = \mathrm{Id}$, and let $h: M \to M$ be the chain homotopy, $d(h) = \xi \circ \eta - \mathrm{Id}$. We define the map $\phi_*: K \otimes M_* \to M_*$ by setting

$$\phi_* = \eta\phi(1 \otimes \xi) + \eta\phi(1 \otimes h)(1 \otimes \phi)(\nabla \otimes 1)(1 \otimes \xi) + \dots .$$

Straightforward computations show that ϕ_* is a twisting cochain.

We define the maps $\widetilde{\xi}: L \otimes M_* \to L \otimes M$ and $\widetilde{\eta}: L \otimes M \to L \otimes M_*$ by setting

$$\widetilde{\xi} = 1 \otimes \xi + (1 \otimes h)(1 \otimes \phi)(\tau \otimes 1)(1 \otimes \xi) + \dots;$$
$$\widetilde{\eta} = 1 \otimes \eta + (1 \otimes \eta)(1 \otimes \phi)(1 \otimes 1 \otimes h)(\tau \otimes 1) + \dots .$$

Straightforward computations show that these maps commute with the differentials and that $\widetilde{\eta} \circ \widetilde{\xi} = \mathrm{Id}$.

We define the homotopy $\widetilde{h}: L \otimes_\phi M \to L \otimes_\phi M$ by setting

$$\widetilde{h} = 1 \otimes h + (1 \otimes h)(1 \otimes \phi)(1 \otimes 1 \otimes h)(\tau \otimes 1) + \dots .$$

Straightforward computations show that $d(\widetilde{h}) = \widetilde{\xi} \circ \widetilde{\eta}$. This completes the proof of the theorem. □

Note that if the coalgebra K itself is taken for the K-comodule L, then the twisted tensor products $K \otimes_\phi M$ and $K \otimes_{\phi_*} M_*$ have, in addition, the structure of K-comodules. The maps $\widetilde{\xi}$, $\widetilde{\eta}$ and the homotopy $\widetilde{h}$ preserve this structure and hence provide a chain equivalence of K-comodules.

We shall state and prove an analog of the Brown theorem which allows us to replace the coalgebra K by the cohomology K_* in the twisted tensor product $K \otimes_\phi M$. To do this, we slightly extend the notion of twisting cochain. Namely, let A be an algebra, and let L be a chain complex. We say that a *twisting cochain* $\phi: L \to L \otimes A$ is a map that lowers the dimension by 1 and satisfies the relation $d(\phi) + \phi \cup \phi = 0$, where $\phi \cup \phi = (1 \otimes \pi)(\phi \otimes 1)\phi$.

THEOREM. *If R is a field and if certain finiteness conditions hold (for example, if the algebra A is connected), then for any twisted tensor product $L \otimes_\phi M$ there exist a twisting cochain $\phi_*: L_* \to L_* \otimes A$ and a chain equivalence*

$$L \otimes_\phi M \simeq L_* \otimes_{\phi_*} M.$$

PROOF. Suppose that $\xi: L_* \to L$ is a representative map for the homology classes, $\eta: L \to L_*$ is the inverse map, $\eta \circ \xi = \mathrm{Id}$, and $h: L \to L$ is a chain homotopy, $d(h) = \xi \circ \eta - \mathrm{Id}$. We define the map $\phi_*: L_* \to L_* \otimes A$ by setting

$$\phi_* = (\eta \otimes 1)\phi\xi + (\eta \otimes 1)(1 \otimes \pi)(\phi \otimes 1)(h \otimes 1)\phi\xi + \dots$$

The finiteness conditions are imposed so that this sum is finite. Straightforward computations show that ϕ_* is a twisting cochain.

We define the maps $\widetilde{\xi}: L_* \otimes M \to L \otimes M$, $\widetilde{\eta}: L \otimes M \to L_* \otimes M$ by setting

$$\widetilde{\xi} = \xi \otimes 1 + (1 \otimes \mu)(h \otimes 1 \otimes 1)(\phi \otimes 1)(\xi \otimes 1) + \dots;$$
$$\widetilde{\eta} = \eta \otimes 1 + (\eta \otimes 1)(1 \otimes \mu)(\phi \otimes 1)(h \otimes 1) + \dots.$$

Straightforward computations show that these maps are permutable with the differentials and $\widetilde{\eta} \circ \widetilde{\xi} = \mathrm{Id}$.

We define the homotopy $\widetilde{h}: L \otimes_\phi M \to L \otimes_\phi M$ by setting

$$\widetilde{h} = h \otimes 1 + (1 \otimes \mu)(h \otimes 1 \otimes 1)(\phi \otimes 1)(h \otimes 1) + \dots.$$

Straightforward computations show that $d(\widetilde{h}) = \widetilde{\xi} \circ \widetilde{\eta}$. This completes the proof of the theorem. □

Note that if the algebra A itself is taken for the A-module M, then the twisted tensor products $L \otimes_\phi A$ and $L_* \otimes_{\phi_*} A$ have, in addition, the structure of A-modules. The maps $\widetilde{\xi}$, $\widetilde{\eta}$, and the homotopy $\widetilde{h}$ preserve this structure and hence provide a chain equivalence of A-modules.

The twisted tensor products are closely related to the functor $\mathcal{D}$, which was introduced by N. A. Berikashvili [**6**] to describe the differentials in the spectral sequence of a fibration.

Let K be a coalgebra and A an algebra. Denote by $M(K, A)$ the set of all twisting cochains $\phi: K \to A$. For an arbitrary map $c: K \to A$ preserving the grading and satisfying the relation $c_0 = 0$, we define the transformation $F_c: M(K, A) \to M(K, A)$ by setting $F_c(\phi) = \phi'$, where $\phi': K \to A$ satisfies the relations

$$\phi'_n = \phi_n + \sum_{i=1}^{n-1} c_i \cup \phi_{n-i} - \sum_{i=1}^{n-1} \phi'_i \cup c_{n-i} - dc_n + c_{n-1}d,$$

i.e., ϕ' satisfies the relation

$$\phi' = \phi + c \cup \phi - \phi' \cup c - d(c).$$

Straightforward computations show that the map $\phi' = F_c(\phi)$ so defined is a twisting cochain. Besides, we have

$$F_{c'+c+c'\cup c} = F_{c'} \circ F_c.$$

We introduce the equivalence relation $\sim$ on the set of twisting cochains $M(K, A)$ by setting $\phi \sim \phi'$ if there is a map $c: K \to A$ such that $F_c(\phi) = \phi'$. The reflexivity of this relation is obvious. The transitivity follows from the formula $F_{c'+c+c'\cup c} = F_{c'} \circ F_c$. Let us prove the symmetry. Suppose that $F_c(\phi) = \phi'$. Consider the map c' defined inductively by

$$c'_n = -c_n - \sum_{i=1}^{n-1} c'_i \cup c_{n-i}.$$

Then we have $c' + c + c' \cup c = 0$, and hence

$$F_{c'}(\phi') = F_{c'} \circ F_c(\phi) = F_{c'+c+c'\cup c}(\phi) = \phi.$$

THEOREM. *Twisted tensor products corresponding to equivalent twisting cochains are isomorphic.*

PROOF. Suppose that $\phi, \phi' \colon K \to A$ are equivalent twisting cochains, $c \colon K \to A$, $F_c(\phi) = \phi'$, L is a K-comodule, M is an A-module. Let us show that the twisted tensor products $L \otimes_\phi M$ and $L \otimes_{\phi'} M$ are isomorphic. Consider a map $c' \colon K \to A$ such that $c' + c + c' \cup c = 0$, and let the maps

$$\psi_c \colon L \otimes_\phi M \to L \otimes_{\phi'} M, \quad \psi_{c'} \colon L \otimes_{\phi'} M \to L \otimes_\phi M$$

be defined by the formulas $\psi_c(x) = x + c \cap x$, $\psi_{c'}(y) = y + c' \cap y$. Straightforward computations show that these maps are permutable with the differential and are mutually inverse. □

Denote the quotient set of $M(K, A)$ modulo the equivalence relation defined above by $\mathcal{D}(K, A)$. The correspondence $(K, A) \longmapsto \mathcal{D}(K, A)$ defines a functor $\mathcal{D}$ into the category of pointed sets; it is contravariant with respect to the first argument and covariant with respect to the second argument.

The functor $\mathcal{D}$ can be defined not only for twisting cochains, but in some other situations as well. For example, this can be done for the perturbation theory of differentials [**21**]. Namely, let X be a chain complex and let $d \colon X \to X$ be its differential. Consider a perturbation of this differential in the form $\widetilde{d} = d + \phi$, where $\phi \colon X \to X$ is a homomorphism that lowers the dimension by one. The relation $\widetilde{d} \circ \widetilde{d} = 0$ is equivalent to the relation $d(\phi) + \phi \circ \phi = 0$. Hence ϕ can be regarded as a twisting cochain defining the perturbation of the differential.

We say that two perturbations $\widetilde{d}'$, $\widetilde{d}''$ of the differential d are *equivalent* (and we write $\widetilde{d}' \sim \widetilde{d}''$) if the corresponding twisting cochains ϕ', ϕ'' are equivalent, i.e., if there exists a homomorphism $c \colon X \to X$ such that

$$d(c) = \phi' - \phi'' + c \circ \phi' - \phi'' \circ c.$$

Denote the set of equivalence classes of perturbations of the differential modulo the equivalence relation defined above by $\mathcal{D}(X)$.

Let us consider a more general situation in which the functor $\mathcal{D}$ arises. Let A be a differential bigraded algebra,

$$A = \sum_{n,m} A^n_m,$$

where the differential preserves the upper grading and decreases the lower grading by one. A family $\phi = \{\phi_n\}_{n \geq 1}$ of elements $\phi_n \in A^n_{-1}$ will be called a *twisting family* if $d(\phi) + \phi \cup \phi = 0$, where $\cup$ denotes the multiplication in A.

We say that two twisting families ϕ', ϕ'' are *equivalent* (and we write $\phi' \sim \phi''$) if there exists a family $c = \{c_n\}_{n \geq 0}$ of elements $c_n \in A^n_0$ satisfying the relations $c_0 = 1$ and $d(c) = c \cup \phi' - \phi'' \cup c$.

Denote the set of equivalence classes of twisting families modulo the equivalence relation defined above by $\mathcal{D}(A)$.

The functor $\mathcal{D}$ can be computed by using the Massey products. Namely, let A be a differential bigraded algebra. Consider twisting families $\phi = \{\phi_n\}_{n\geq 1}$, $\phi_n \in A^n_{-1}$. We rewrite the relation $d(\phi) + \phi \cup \phi = 0$ as the system of equations

$$\begin{gathered}
d(\phi_1) = 0; \\
d(\phi_2) + \phi_1 \cup \phi_1 = 0; \\
d(\phi_3) + \phi_1 \cup \phi_2 + \phi_2 \cup \phi_1 = 0; \\
\dots \\
d(\phi_n) + \sum_{i=1}^{n-1} \phi_i \cup \phi_{n-i} = 0; \\
\dots
\end{gathered}$$

The first equation implies that ϕ_1 is a cycle and hence defines an element $h_1 \in H_{-1}(A^1)$. The second equation implies that the product $h_1 \cup h_1$ is zero. The third equation implies that the Massey product $m_1(h_1 \otimes h_1 \otimes h_1)$ is zero, and so on.

Thus, for the element $h_1 \in H_{-1}(A^1)$, all Massey products are zero, i.e., for each $n \geq 0$ we have

$$m_n(h_1 \otimes \cdots \otimes h_1) = 0.$$

By dimension arguments, we see that this relations can be rewritten as

$$\sum_n m_n(h_1 \otimes \cdots \otimes h_1) = 0.$$

Denote the set of such elements h_1 by $J_1 H_{-1}(A)$.

Suppose that ϕ and ϕ' are equivalent twisting families. This means that there is a family $c = \{c_n\}_{n\geq 0}$, $c_n \in A^n_0$, satisfying the relations $c_0 = 1$ and

$$d(c) = c \cup \phi - \phi' \cup c, \quad c \in A_0, \ c = 1 + c_1 + c_2 + \dots .$$

The last relation can be rewritten as the system of equations

$$\begin{gathered}
d(c_1) = \phi_1 - \phi_1'; \\
d(c_2) = \phi_2 - \phi_2' + c_1 \cup \phi_1 - \phi_1' \cup c_1; \\
\dots \\
d(c_n) = \phi_n - \phi_n' + \sum_{i=1}^{n-1} c_i \cup \phi_{n-i} - \sum_{i=1}^{n-1} \phi_i' \cup c_{n-i}; \\
\dots
\end{gathered}$$

The first equation implies that equivalent twisting families ϕ and ϕ' define the same element h_1. Thus we have a map

$$p_1 \colon \mathcal{D}(A) \to J_1 H_{-1}(A).$$

Note that if the module A^1_{-1} is finitely generated, then we can chose representatives $\psi^1_n \in A^n_{-1}$ for the element h_1 so that the relation of a twisting family holds.

Besides, the twisting family $\psi^1 = \{\psi^1_n\}$ will satisfy the relation $d(c_1) = \phi_1 - \psi^1_1$. Therefore, the twisting family $F_{c_1}(\phi)$, which is equivalent to ϕ, satisfies the relation $F_{c_1}(\phi)_1 = \psi_1$.

Let us improve the twisting cochain ϕ by replacing it with $F_{c_1}(\phi)$. Consider the family $\phi^1 = \phi - \psi^1$. It satisfies the relations

$$d(\phi^1) + \psi^1 \cup \phi^1 + \phi^1 \cup \psi^1 + \phi^1 \cup \phi^1 = 0, \quad \phi^1_1 = 0,$$

which follow from the fact that ψ^1 and $\phi^1 + \psi^1$ are twisting families.

The first of these relations can be rewritten as the system of equations

$$\begin{gathered} d(\phi^1_2) = 0; \\ d(\phi^1_3) + \psi^1_1 \cup \phi^1_2 + \phi^1_2 \cup \psi^1_1 = 0; \\ d(\phi^1_4) + \psi^1_1 \cup \phi^1_3 + \psi^1_2 \cup \phi^1_2 + \phi^1_2 \cup \psi^1_2 + \phi^1_3 \cup \psi^1_1 + \phi^1_2 \cup \phi^1_2 = 0; \\ \dots \\ d(\phi^1_n) + \sum_{i=1}^{n-2} \psi^1_i \cup \phi^1_{n-i} + \sum_{i=2}^{n-1} \psi^1_i \cup \phi^1_{n-i} + \sum_{i=2}^{n-2} \phi^1_i \cup \phi^1_{n-i} = 0; \\ \dots \end{gathered}$$

The first equation implies that the element ϕ^1_2 is a cycle and hence defines an element $h_2 \in H_{-1}(A^2)$. The second equation implies that the product $h_1 \cup h_2 + h_2 \cup h_1$ is zero. The third equation implies that the Massey product

$$m_1(h_1 \otimes h_1 \otimes h_2 + h_1 \otimes h_2 \otimes h_1 + h_2 \otimes h_1 \otimes h_1) + h_2 \cup h_2$$

is zero, and so on.

Thus the element $h_2 \in H_{-1}(A^2)$ satisfies the relations

$$\sum m_{k-2}(h_{i_1} \otimes \dots \otimes h_{i_k}) = 0$$

for all $n \geq 2$, where the summation is over all $k \geq 2$ and all $i_1, \dots, i_k$ such that

$$i_1 + \dots + i_k = n, \quad \max\{i_1, \dots, i_k\} = 2.$$

Let $h^1 = h_1$, $h^2 = h_1 + h_2$. By dimensional arguments, it follows that these relations together with the relations for h_1 can be rewritten in the form

$$\sum_n m_n(h^1 \otimes \dots \otimes h^1) = 0; \quad \sum_n m_n(h^2 \otimes \dots \otimes h^2) = 0.$$

Let ϕ and ϕ' be equivalent twisting families. Then the corresponding improved twisting families are also equivalent, i.e., there is a family $c = \{c_n\}_{n \geq 1}$ satisfying the relation $d(c) = c \cup \phi - \phi' \cup c$. We rewrite this relation as the system

$$\begin{gathered} d(c_1) = 0; \\ d(c_2) = \phi_2 - \phi'_2 + c_1 \cup \phi_1 - \phi_1 \cup c_1; \\ \dots \\ d(c_n) = \phi_n - \phi'_n + \sum_{i=1}^{n-1} c_i \cup \phi_{n-i} - \sum_{i=1}^{n-1} \phi'_i \cup c_{n-i}; \\ \dots \end{gathered}$$

The first equation of this system implies that c_1 is a cycle and hence defines an element $g_1 \in H_0(A^1)$. The second equation implies that the elements h_2, h'_2 corresponding to the twisting families ϕ, ϕ' satisfy the relation

$$h_2 - h'_2 = g_1 \cup h_1 - h_1 \cup g_1,$$

and so on.

We define the equivalence relation $\sim$ for elements of the form $h^2 = h_1 + h_2$ by setting $h^2 \sim h'^2$ if $h_1 = h'_1$ and there exists an element $g^1 = 1 + g_1$ satisfying the relations

$$\sum_n m_n(h'^2 \otimes \cdots \otimes g^1 \otimes \cdots \otimes h^2) = 0,$$

where the summation is also over all possible positions (indexed by i, $1 \le i \le n+2$) of the factor g^1 in the tensor product. Denote the set of equivalence classes of these elements h^2 by $J_2H_{-1}(A)$. As a result, we obtain a map

$$p_2 \colon \mathcal{D}(A) \to J_2H_{-1}(A).$$

If the module A^2_{-1} is finitely generated, then we can choose representatives $\psi^2 = \{\psi^2_n\}_{n \ge 2}$ of h_2 that satisfy the relations

$$\phi^1_2 - \psi^2_2 = d(c_2);$$
$$d(\psi^2) + \psi^1 \cup \psi^2 + \psi^2 \cup \psi^1 + \psi^2 \cup \psi^2 = 0.$$

Let us improve the twisting family ϕ by replacing it by $F_{c_2}(\phi)$. Consider the family $\phi^2 = \phi^1 - \psi^2$. Repeating the above construction for the family ϕ^2, we obtain the element $h_3 \in H_{-1}(A^3)$, and so on.

Consider the set of sequences $h = \{h_i\}_{i \ge 1}$ of elements $h_i \in H_{-1}(A^i)$ such that the elements $h^n = h_1 + \cdots + h_n$ satisfy the relations

$$\sum_k m_k(h^n \otimes \cdots \otimes h^n) = 0.$$

By dimensional arguments, it is easy to check that these relations hold if and only if the relations

$$\sum m_{k-2}(h_{i_1} \otimes \cdots \otimes h_{i_k}) = 0$$

hold for any $i \ge 1$ and $n > i$, where the summation is over all $k \ge 2$ and $i_1, \dots, i_k$ such that

$$i_1 + \cdots + i_k = n, \quad \max\{i_1, \dots, i_k\} = i.$$

We define the equivalence relation for these sequences by setting $h \sim h'$ if there exists a sequence $g = \{g_i\}_{i \ge 0}$ of elements $g_i \in H_0(A^i)$, $g_0 = 1$, such that the elements $g^n = 1 + g_1 + \cdots + g_n$ satisfy the relations

$$\sum_k m_k(h'^n \otimes \cdots \otimes g^n \otimes \cdots \otimes h^n) = 0.$$

Denote the corresponding quotient set by $JH_{-1}(A)$. The above process provides us with a map

$$p \colon \mathcal{D}(A) \to JH_{-1}(A).$$

Straightforward verification shows that it is a bijection.

Thus we have proved the following theorem.

THEOREM. *For any differential bigraded algebra A with finitely generated modules A^n_{-1}, there is a bijection $\mathcal{D}(A) \longleftrightarrow JH_{-1}(A)$.*

The above map $p_1: \mathcal{D}(A) \to J_1H_{-1}(A)$ induces a short exact sequence of pointed sets

$$* \to \mathcal{D}^1(A) \xrightarrow{i^1} \mathcal{D}(A) \xrightarrow{p_1} J_1H_{-1}(A) \to *,$$

where $\mathcal{D}^1(A)$ is the subset of $\mathcal{D}(A)$ generated by the twisting families ϕ such that $\phi_1 = 0$.

Let ϕ be a twisting family satisfying the relation $\phi_1 = 0$. The relation $d(\phi) + \phi \cup \phi$ can be rewritten as the following system of equations:

$$\begin{aligned} d(\phi_2) &= 0; \\ d(\phi_3) &= 0; \\ d(\phi_4) + \phi_2 \cup \phi_2 &= 0; \\ &\dots \\ d(\phi_n) + \sum_{i=2}^{n-2} \phi_i \cup \phi_{n-i} &= 0; \\ &\dots \end{aligned}$$

The first equation implies that the element ϕ_2 is a cycle and hence defines an element $h_2 \in H_{-1}(A^2)$. The second equation implies that ϕ_3 is a cycle and hence defines an element $h_3 \in H_{-1}(A^3)$. The third equation means that the product $h_2 \cup h_2$ is zero, and so on.

Thus for any $m \geq 4$, the elements h_2, h_3 satisfy the relations

$$\sum_{k \geq 2} m_{k-2}(h_{i_1} \otimes \cdots \otimes h_{i_k}) = 0,$$

where the summation is over all $i_1, \dots, i_k = 2, 3$.

Denote by $J^2H_{-1}(A)$ the set of all elements $h_2 \in H_{-1}(A^2)$ such that there is an element $h_3 \in H_{-1}(A^3)$ satisfying the above relations.

The correspondence $\phi \longmapsto h_2$ defines a map

$$p^2: \mathcal{D}^1(A) \to J^2H_{-1}(A)$$

and a short exact sequence

$$* \to \mathcal{D}^2(A) \xrightarrow{i^2} \mathcal{D}^1(A) \xrightarrow{p^2} J^2H_{-1}(A) \to *,$$

where $\mathcal{D}^2(A)$ is the subset of $\mathcal{D}^1(A)$ generated by the twisting families ϕ such that $\phi_1 = 0$ and $\phi_2 = 0$.

In general, for $n \geq 1$, by $J^nH_{-1}(A)$ we denote the set of the elements $h_n \in H_{-1}(A^n)$ such that there exist elements $h_i \in H_{-1}(A^i)$, $n < i < 2n$, satisfying the relations

$$\sum_{k \geq 2} m_{k-2}(h_{i_1} \otimes \cdots \otimes h_{i_k}) = 0$$

for all $m \geq 2n$, where the summation is over all $n \leq i_1, \dots, i_k \leq 2n-1$.

The correspondence $\phi \longmapsto h_n$ defines a map

$$p^n: \mathcal{D}^n(A) \to J^nH_{-1}(A)$$

and a short exact sequence

$$* \to \mathcal{D}^{n+1}(A) \xrightarrow{i^{n+1}} \mathcal{D}^n(A) \xrightarrow{p^n} J^n H_{-1}(A) \to *,$$

where $\mathcal{D}^n(A)$ is the subset of $\mathcal{D}(A)$ generated by the twisting families ϕ such that $\phi_1 = 0, \dots, \phi_{n-1} = 0$.

Thus we obtain the following theorem.

THEOREM. *If A is a differential bigraded algebra with finitely generated modules A^n_{-1}, then for any $n \geq 1$ there is a short exact sequence*

$$* \to \mathcal{D}^{n+1}(A) \xrightarrow{i^{n+1}} \mathcal{D}^n(A) \xrightarrow{p^n} J^n H_{-1}(A) \to *.$$

In particular, $\mathcal{D}(A) = 0$ if and only if $J^n H_{-1}(A) = 0$ for all $n \geq 1$.

3.4. Bar constructions and cobar constructions

Let A be an algebra; we shall define a coalgebra BA, which is called the *bar construction on* A, and a twisting cochain $\beta: BA \to A$.

As a graded coalgebra, BA coincides with the tensor coalgebra TSA over the suspension SA of A. The generators of BA are denoted by $[x_1, \dots, x_n]$, $x_i \in A$; the dimension of such a generator equals the sum of the dimensions of the x_i plus n. The comultiplication $\nabla: BA \to BA \otimes BA$ is given by

$$\nabla[x_1, \dots, x_n] = \sum_{i=1}^{n-1} [x_1, \dots, x_i] \otimes [x_{i+1}, \dots, x_n].$$

The twisting cochain $\beta: BA \to A$ is defined by

$$\beta[x_1, \dots, x_n] = \begin{cases} x_1, & n = 1, \\ 0, & n > 1. \end{cases}$$

The differential in the coalgebra BA is determined by the twisting cochain condition for β and has the form

$$\begin{aligned} d[x_1, \dots, x_n] &= \sum_{i=1}^{n} (-1)^{\epsilon_{i-1}+1} [x_1, \dots, d(x_i), \dots, x_n] \\ &\quad + \sum_{i=1}^{n-1} (-1)^{\epsilon_i} [x_1, \dots, \pi(x_i \otimes x_{i+1}), \dots, x_n], \end{aligned}$$

where $\epsilon_i = \dim(x_1) + \dots + \dim(x_i) + i$.

Let us pass to the dual situation. Suppose that K is a coalgebra. We define an algebra FK, which is called the *cobar construction on* K, and a twisting cochain $\alpha: K \to FK$; see [**1,12**]. As a graded algebra, FK coincides with the tensor algebra $TS^{-1}K$ over the desuspension $S^{-1}K$ of K. The generators in FK are denoted by $[x_1, \dots, x_n]$, $x_i \in K$, and are of dimension equal to the sum of the dimensions of the x_i minus n. The multiplication $\pi: FK \otimes FK \to FK$ is defined by

$$\pi([x_1, \dots, x_n] \otimes [x_{n+1}, \dots, x_{n+m}]) = [x_1, \dots, x_{n+m}].$$

The twisting cochain $\alpha: K \to FK$ is defined by the formula $\alpha(x) = [x]$.

The differential in the algebra FK is determined by the twisting cochain condition for α and has the form

$$d[x_1,\dots,x_n]=\sum_{i=1}^{n}(-1)^{\epsilon_{i-1}+1}[x_1,\dots,d(x_i),\dots,x_n] + \sum_{i=1}^{n}(-1)^{\epsilon_i}[x_1,\dots,x_i',x_i'',\dots,x_n],$$

where $\nabla(x_i)=\sum x_i'\otimes x_i''$.

Each twisting cochain $\phi\colon K\to A$ induces an algebra map $F(\phi)\colon FK\to A$,

$$F(\phi)[x_1,\dots,x_n]=\pi(\phi(x_1)\otimes\cdots\otimes\phi(x_n)).$$

As for the coalgebra map $B(\phi)\colon K\to BA$, the situation is somewhat more difficult. Namely, this map should be defined by

$$B(\phi)(x)=[\phi(x)]+\sum[\phi(x'),\phi(x'')]+\sum[\phi(x'),\phi(x''),\phi(x''')]+\dots,$$

where $\sum x'\otimes x''=\nabla(x)$, $\sum x'\otimes x''\otimes x'''=(\nabla\otimes 1)\nabla(x)$, etc. However, as a result we possibly obtain a sum of infinitely many terms. This can be avoided if we assume that K is a coalgebra of finite type.

In this case we obtain a coalgebra map $B(\phi)\colon K\to BA$. Otherwise, if K is not of finite type, let us consider the filtration $K^1\subset K^2\subset\cdots\subset K$, where

$$K^1=\{x\in K\mid\nabla(x)=0\},\quad\dots,\quad K^n=\left\{x\in K\mid\nabla(x)\in\sum_{i=1}^{n-1}K^i\otimes K^{n-i}\right\},\dots$$

Let $\check K=\varinjlim K^n$. It is clear that $\check K$ is a subcoalgebra of finite type in K, and there is a coalgebra map $B(\phi)\colon\check K\to BA$.

In particular, the twisting cochain $\alpha\colon K\to FK$ induces a coalgebra map $B(\alpha)\colon\check K\to BFK$, and the twisting cochain $\beta\colon BA\to A$ induces an algebra map $F(\beta)\colon FBA\to A$.

THEOREM. *For any algebra A, the map $F(\beta)\colon FBA\to A$ is a chain equivalence.*

PROOF. We denote the map $F(\beta)$ by η for brevity, and we define the map $\xi\colon A\to FBA$ by setting $\xi(x)=[[x]]$. It is clear that ξ is a chain map (but not an algebra map) and $\eta\circ\xi=\mathrm{Id}$.

Let us show that $\xi\circ\eta\simeq\mathrm{Id}\colon FBA\to FBA$. To this end, we define the chain homotopy $h\colon FBA\to FBA$ by setting

$$h[[x_1],[x_2,\dots,x_n],\dots,[y_1,\dots,y_m]]=[[x_1,\dots,x_n],\dots,[y_1,\dots,y_m]],$$

and $h(x)=0$ otherwise.

Straightforward computations show that the required relation holds, i.e., h is the sought-for chain homotopy. □

COROLLARY. *For an arbitrary chain complex X, there is a chain equivalence $FTX\simeq X$.*

Let A be an algebra, and $\hat A$ its completion with respect to the standard filtration

$$A=A^1\supset A^2=A\cdot A\supset\dots,\qquad\hat A=\varprojlim A/A^n.$$

There is a natural algebra map $A \to \hat{A}$. The algebra A is said to be *complete* if this map is an isomorphism.

THEOREM. *If A is a complete algebra, then there is a chain equivalence of algebras $\hat{F}BA \simeq A$.*

PROOF. Indeed, the algebra map $\eta: FBA \to A$ induces a map of completions $\hat{\eta}: \hat{F}BA \to A$. Besides, there is an inclusion $\hat{\xi}: A \to \hat{F}BA$, $\hat{\eta} \circ \hat{\xi} = \mathrm{Id}$. The above-mentioned homotopy $h: FBA \to FBA$ preserves the filtration and hence induces a homotopy $\hat{h}: \hat{F}BA \to \hat{F}BA$, $d(\hat{h}) = \hat{\xi} \circ \hat{\eta} - \mathrm{Id}$. □

THEOREM. *If K is an arbitrary coalgebra of finite type, then the map $B(\alpha): K \to BFK$ is a chain equivalence.*

PROOF. We denote the map $B(\alpha)$ by ξ for brevity, and we define the map $\eta: BFK \to K$ by setting $\eta[[x]] = x$ and $\eta(x) = 0$ otherwise. Clearly, η is a chain map (but not a coalgebra map) and $\eta \circ \xi = \mathrm{Id}$. Let us show that $\xi \circ \eta \simeq \mathrm{Id}$. To this end, we define the chain homotopy $h: BFK \to BFK$ by setting

$$h[[x_1, \ldots, x_n], \ldots, [y_1, \ldots, y_m]] = [[x_1], [x_2, \ldots, x_n], \ldots, [y_1, \ldots, y_m]], \quad n \geq 2,$$

and 0 if $n = 1$.

Straightforward computations show that the required relation holds, i.e., h is the desired chain homotopy. □

COROLLARY. *For any chain complex X, there is a chain equivalence $BTX \simeq X$.*

Now let us consider the case in which K is not a coalgebra of finite type. The standard filtration on the cobar construction FK induces a filtration on the bar construction BFK. Denote by $\hat{B}FK$ its completion with respect to this filtration. The twisting cochain $\alpha: K \to FK$ induces a coalgebra map $\xi = \hat{B}(\alpha): K \to \hat{B}FK$. Besides, there is a projection $\eta: \hat{B}FK \to K$. Since the above homotopy $h: BFK \to BFK$ preserves the filtration, it induces a homotopy $h: \hat{B}FK \to \hat{B}FK$, $d(h) = \xi \circ \eta - \mathrm{Id}$. Therefore, we obtain the following theorem.

THEOREM. *For any coalgebra K, the map $\hat{B}(\alpha): K \to \hat{B}FK$ is a chain equivalence.*

Let M be a left A-module; we shall define a left comodule $B(A, M)$ over BA, which is called the *left-sided bar construction on M*.

As a graded module, $B(A, M)$ coincides with the tensor product $B(A) \otimes M$. The generators in $B(A, M)$ are denoted by $[x_1, \ldots, x_n]x$, $x_i \in A$, $x \in M$, and are of dimension equal to the sum of the dimensions of the x_i and x plus n.

The differential is defined on these elements by

$$\begin{aligned} d([x_1, \ldots, x_n]x) = &\sum_{i=1}^{n} (-1)^{\epsilon_{i-1}+1} [x_1, \ldots, d(x_i), \ldots, x_n]x \\ &+ (-1)^{\epsilon_n+1} [x_1, \ldots, x_n] d(x) \\ &+ \sum_{i=1}^{n-1} (-1)^{\epsilon_i} [x_1, \ldots, x_i \cdot x_{i+1}, \ldots, x_n]x \\ &+ (-1)^{\epsilon_n} [x_1, \ldots, x_{n-1}] x_n \cdot x. \end{aligned}$$

Similarly, for any right A-module M, we define the *right-sided bar construction* $B(M, A)$, which is a right BA-comodule.

Suppose that M' is a right A-module and M'' is a left A-module. We define the *two-sided bar construction* $B(M', A, M'')$ as follows.

As a graded module, $B(M', A, M'')$ coincides with the tensor product $M' \otimes B(A) \otimes M''$. The generators in $B(M', A, M'')$ are denoted by $x'[x_1, \dots, x_n]x''$, $x' \in M'$, $x_i \in A$, $x'' \in M''$, and are of dimension equal to the sum of the dimensions of x', the x_i, and x'' plus n. The differential is defined on these elements by

$$\begin{aligned} d(x'[x_1, \dots, x_n]x'') = d(x')[x_1, \dots, x_n]x'' & \\ + \sum_{i=1}^{n} (-1)^{\epsilon_{i-1}+1} x'[x_1, \dots, d(x_i), \dots, x_n]x'' & \\ + (-1)^{\epsilon_n+1} x'[x_1, \dots, x_n]d(x'') & \\ + x' \cdot x_1[x_2, \dots, x_n]x'' & \\ + \sum_{i=1}^{n-1} (-1)^{\epsilon_i} x'[x_1, \dots, x_i \cdot x_{i+1}, \dots, x_n]x'' & \\ + (-1)^{\epsilon_n} x'[x_1, \dots, x_{n-1}]x_n \cdot x''. & \end{aligned}$$

Suppose that A is an algebra with unity. Then the two-sided bar construction $B(M', A, M'')$ can be thought of as the realization of a simplicial object

$$B_*(M', A, M'') = \{B_n(M', A, M'')\},$$

where

$$B_n(M', A, M'') = M' \otimes A^{\otimes n} \otimes M'',$$

and the face and degeneracy operators are defined by

$$\begin{aligned} d_0(x' \otimes x_1 \otimes \cdots \otimes x_n \otimes x'') &= x' \cdot x_1 \otimes x_2 \otimes \cdots \otimes x_n \otimes x'', \\ d_i(x' \otimes x_1 \otimes \cdots \otimes x_n \otimes x'') &= x' \otimes \cdots \otimes x_i \cdot x_{i+1} \otimes \cdots \otimes x'', \quad 1 \le i \le n-1, \\ d_n(x' \otimes x_1 \otimes \cdots \otimes x_n \otimes x'') &= x' \otimes x_1 \otimes \cdots \otimes x_{n-1} \otimes x_n \cdot x'', \\ s_i(x' \otimes x_1 \otimes \cdots \otimes x_n \otimes x'') &= x' \otimes \cdots \otimes x_i \otimes e \otimes x_{i+1} \otimes \cdots \otimes x_n \otimes x'', \quad 0 \le i \le n. \end{aligned}$$

Since the simplicial objects $B_*(A, A, M'')$ and $B_*(M', A, A)$ are, respectively, left-sided and right-sided simplicial resolutions of the modules M' and M'', there are chain equivalences

$$B(M', A, A) \simeq M', \quad B(A, A, M'') \simeq M''.$$

Dually, for any left K-comodule L, we define the *left-sided cobar construction* $F(K, L)$, which is a left module over the algebra FK.

The generators in $F(K, L)$ are denoted by $[x_1, \dots, x_n]x$, $x_i \in K$, $x \in L$, and are of dimension equal to the sum of the dimensions of the x_i and x minus n.

The differential on these elements is defined by

$$\begin{aligned} d([x_1,\dots,x_n]x) &= \sum_{i=1}^{n}(-1)^{\epsilon_{i-1}+1}[x_1,\dots,d(x_i),\dots,x_n]x \\ &+ (-1)^{\epsilon_n+1}[x_1,\dots,x_n]d(x) \\ &+ \sum_{i=1}^{n}(-1)^{\epsilon_i}[x_1,\dots,x_i',x_i'',\dots,x_n]x \\ &+ (-1)^{\epsilon_{n+1}}[x_1,\dots,x_n,x']x'', \end{aligned}$$

where $\sum x_i'\otimes x_i'' = \nabla(x_i)$, $\sum x'\otimes x'' = \tau(x)$.

Similarly, for any right K-comodule L, we define the *right-sided cobar construction* $F(L,K)$, which is an FK-module.

For any right K-comodule L' and any left K-comodule L'', the *two-sided cobar construction* $F(L',K,L'')$ is defined.

In the case of an augmented coalgebra K, the two-sided cobar construction $F(L',K,L'')$ can be thought of as the realization of the cosimplicial object $F^*(L',K,L'') = \{F^n(L',K,L'')\}$.

Since the cosimplicial objects $F^*(K,K,L'')$ and $F^*(L',K,K)$ are, respectively, left-sided and right-sided cosimplicial resolutions of the comodules L' and L'', there are chain equivalences

$$F(L',K,K)\simeq L', \quad F(K,K,L'')\simeq L''.$$

The following theorems are proved in the same way as their analogs for algebras and coalgebras.

THEOREM. *For any left module M over an algebra A, there is a chain equivalence $F(BA,B(A,M))\simeq M$.*

THEOREM. *For any left comodule L over a coalgebra K of finite type, there is a chain equivalence $B(FK,F(K,L))\simeq L$.*

THEOREM. *For any left comodule L over a coalgebra K, there is a chain equivalence $\hat{B}(FK,F(K,L))\simeq L$.*

Similar assertions hold for right modules and comodules.

Also note the following expressions for bar constructions in terms of twisted tensor products:

$$\begin{aligned} B(A,M)&\cong B(A)\otimes_\beta M, &\quad B(M,A,A)&\cong B(M,A)\otimes_\beta A, \\ F(L,K)&\cong L\otimes_\alpha F(K), &\quad F(K,K,L)&\cong K\otimes_\alpha F(K,L). \end{aligned}$$

In particular, the twisted tensor products $K\otimes_\alpha FK$ and $BA\otimes_\beta A$ are contractible.

3.5. Lie modules and comodules

Here we define the notions of Lie module and Lie comodule, which will subsequently be extended to the case of Lie algebras and Lie coalgebras over operads.

DEFINITION. Let X be a chain complex, and let K be a coalgebra. We say that X is a *Lie module over* K if there is given a map $\phi\colon K\otimes X\to X$ that lowers the dimension by one and satisfies the twisting cochain condition $d(\phi)+\phi\cup\phi=0$, where $\phi\cup\phi=\phi(1\otimes\phi)(\nabla\otimes 1)\colon K\otimes X\to X$.

Let M be a left module over an algebra A, and let $\psi\colon K \to A$ be a twisting cochain. Then M can be endowed with the structure of a Lie module over the coalgebra K, $\phi\colon K \otimes M \to M$, by setting $\phi = \mu(\psi \otimes 1)$.

Examples of Lie modules are provided by the bar and cobar constructions considered above. In particular, if M is a module over an algebra A, $\mu\colon A \otimes M \to M$, then the twisting cochain $\beta\colon BA \to A$ induces a map

$$BA \otimes M \xrightarrow{\beta\otimes 1} A \otimes M \xrightarrow{\mu} M,$$

which defines a Lie $B(A)$-module structure on M.

If L is a left comodule over a coalgebra K, then the cobar construction $X = F(K, L)$ is a left module over the algebra FK, and hence X is a Lie K-module.

Conversely, if X is a Lie module over a coalgebra K, then the twisting cochain $\phi\colon K \otimes X \to X$ induces an action $\mu\colon FK \otimes X \to X$ of the cobar construction FK on X, and hence X is an FK-module.

If X is a Lie module over a coalgebra K and L is a right K-comodule, then the twisted tensor product $L \otimes X$ can be endowed with a new differential

$$d_\phi = d \otimes 1 + 1 \otimes d + \phi \cap,$$

which turns $L \otimes X$ into a twisted tensor product, denoted by $L \otimes_\phi X$. In particular, for any coalgebra K, we have the twisted tensor product $K \otimes_\phi X$.

If X is a Lie module over a coalgebra K, and L is a left K-comodule, then the module of homomorphisms $\mathrm{Hom}(L, X)$ can be endowed with a new differential

$$d_\phi(f) = d(f) + \phi \cup f, \quad f \in \mathrm{Hom}(L, X),$$

which turns $\mathrm{Hom}(L, X)$ into a twisted module of homomorphisms, denoted by $\mathrm{Hom}_\phi(L, X)$.

Thus we have a functor $F = F(K, -)$ from the category of left K-comodules to the category of Lie K-modules (this functor takes each K-comodule L to the Lie K-module $F(K, L)$). Denote by G the functor from the category of Lie K-modules to the category of K-comodules that takes each Lie K-module X to the comodule $G(X) = K \otimes_\phi X$ over K. There is a natural transformation of functors $\alpha\colon \mathrm{Id} \to G \circ F$, $\alpha(L)\colon L \to K \otimes_\phi F(K, L)$, which is a chain equivalence.

It is similarly shown that there is a natural transformation of functors $\beta\colon F \circ G \to \mathrm{Id}$, $\beta(X)\colon F(K, K \otimes_\phi X) \to X$, which is a chain equivalence.

It is readily seen that the functors F and G are adjoint, and the following theorem holds.

THEOREM. *For any coalgebra K, the category of K-comodules is chain equivalent to the category of Lie K-modules.*

Restating the Brown theorem in the language of Lie modules yields the following assertion.

THEOREM. *If R is a field, then for any coalgebra K of finite type and any Lie K-module X, $\phi\colon K \otimes X \to X$, the homology $X_* = H_*(X)$ possesses a Lie K-module structure $\phi_*\colon K \otimes X_* \to X_*$, and there is a chain equivalence of K-comodules $K \otimes_\phi X \simeq K \otimes_{\phi_*} X_*$.*

Note that in this theorem we can take for X_* not only the homology of X, but also an arbitrary differential graded module which is chain equivalent to X in the sense that there are maps $\xi: X_* \to X$, $\eta: X \to X_*$ and a chain homotopy $h: X \to X$ satisfying the conditions listed in the proof of the Brown theorem.

COROLLARY. *Let R be a field, and let M be a module over an algebra A; then the homology $M_* = H_*(M)$ possesses a Lie BA-module structure, and there is a chain equivalence of BA-comodules $BA \otimes_\phi M \simeq BA \otimes_{\phi_*} M_*$.*

As in the theorem, for M_* we can take any differential graded module which is chain equivalent to M.

Now consider the situation in which the coalgebra K is not of finite type. In this case it suffices to consider Lie K-modules X with a filtration $X = X^1 \supset X^2 \supset \dots \supset X^n \supset \dots$ such that $\phi(K \otimes X^n) \subset X^{n+1}$. By $\hat{X}$ we denote the completion of X with respect to this filtration.

We shall use the following version of the perturbation theory of differential graded modules [**21**].

THEOREM (Gugenheim–Lambe–Stasheff). *Suppose that X is a differential graded module with a filtration $X = X^1 \supset X^2 \supset \dots$ and the differential d in it is equal to the sum $d' + d''$, where the differential d' preserves filtration, and the differential d'' increases filtration by one. Let $\hat{X}'_* = H_*(\hat{X}, d')$. Then it is possible to define on $\hat{X}'_*$ a new differential d''_*, maps $\xi: \hat{X}'_* \to \hat{X}$, $\eta: \hat{X} \to \hat{X}'_*$, and a homotopy $h: \hat{X} \to \hat{X}$ satisfying the conditions listed above.*

PROOF. Indeed, let $\xi': \hat{X}'_* \to \hat{X}$ be a representative map for the homology classes, and let the maps $\eta': \hat{X} \to \hat{X}'_*$, $h': \hat{X} \to \hat{X}$ satisfy the relations $\eta' \circ \xi' = \mathrm{Id}$, $d'(h') = \xi' \circ \eta' - \mathrm{Id}$, $\eta' \circ h' = 0$, $h' \circ \xi' = 0$, $h' \circ h' = 0$. Then the required differential d''_*, the maps $\xi: \hat{X}_* \to \hat{X}$, $\eta: \hat{X} \to \hat{X}_*$, and the homotopy $h: \hat{X} \to \hat{X}$ are given by

$$\begin{aligned} d''_* &= \eta' d'' \xi' + \eta' d'' h' d'' \xi' + \dots, \\ \xi &= \xi' + h' d'' \xi' + h' d'' h' d'' \xi' + \dots, \\ \eta &= \eta' + \eta' d'' h' + \eta' d'' h' d'' h' + \dots, \\ h &= h' + h' d'' h' + h' d'' h' d'' h' + \dots. \quad \square \end{aligned}$$

Combined with the Brown theorem, this implies the following assertion.

THEOREM. *Let K be a coalgebra, and let X be a Lie K-module with a filtration and with a differential $d = d' + d''$, where d' is a filtration-preserving differential, and d'' raises filtration by one. Then $\hat{X}'_* = H_*(\hat{X}, d')$ can be endowed with a differential d''_* and with a Lie K-module structure $\phi_*: K \otimes \hat{X}'_* \to \hat{X}'_*$ yielding a chain equivalence of K-comodules $K \otimes_\phi \hat{X} \simeq K \otimes_{\phi_*} \hat{X}'_*$.*

Dually, we define the notion of Lie comodule over an algebra.

DEFINITION. A chain complex Y is called a *Lie comodule over the algebra A* if there is given a map $\psi: Y \to Y \otimes A$ that lowers the dimension by one and satisfies the twisting cochain condition $d(\psi) + \psi \cup \psi = 0$.

The above properties of Lie modules are easily dualized to the case of Lie comodules. In particular, the following assertion holds.

THEOREM. *If R is a field, and if certain finiteness conditions hold, then for any Lie A-comodule Y, $\phi: Y \to Y \otimes A$, the homology $Y_* = H_*(Y)$ possesses a Lie A-comodule structure $\phi_*: Y_* \to Y_* \otimes A$, and there is a chain equivalence of A-modules $Y \otimes_\phi A \simeq Y_* \otimes_{\phi_*} A$.*

3.6. $\vee$-algebras and $\wedge$-coalgebras

Suppose that X is a graded module; by $X \wedge X$ we denote the quotient module of the tensor product $X \otimes X$ by the relation $x \otimes y + T(x \otimes y) \sim 0$, where $T: K \otimes K \to K \otimes K$ is the permutation operator,

$$T(x \otimes y) = (-1)^{|x|\cdot|y|} y \otimes x;$$

$|x|$, $|y|$ are the dimensions of x and y, respectively.

Similarly, by $X^{\wedge n} = X \wedge \cdots \wedge X$ we denote the quotient module of $X^{\otimes n} = X \otimes \cdots \otimes X$ by these relations.

DEFINITION. A chain complex X is called a $\wedge$-*coalgebra* if there is given a map $\nabla: X \to X \wedge X$ satisfying the relation

$$(\nabla \wedge 1 - 1 \wedge \nabla)\nabla = 0: X \to X \wedge X \wedge X.$$

Dually, for any module Y, by $Y \vee Y$ we denote the submodule of the tensor product $Y \otimes Y$ consisting of the elements $z \in Y \otimes Y$ such that $T(z) + z = 0$. The submodule $Y^{\vee n}$ of the tensor product $Y^{\otimes n}$ is defined similarly.

DEFINITION. A chain complex Y is called a $\vee$-*algebra* if there is given a map $\pi: Y \vee Y \to Y$ satisfying the condition

$$\pi(\pi \vee 1 - 1 \vee \pi) = 0: Y \vee Y \vee Y \to Y.$$

If K is an ordinary coalgebra, then the comultiplication $\nabla: K \to K \otimes K$ induces a comultiplication $\nabla: K \to K \wedge K$, thus making K into a $\wedge$-coalgebra.

If A is an ordinary algebra, then the multiplication $\pi: A \otimes A \to A$ induces a multiplication $\pi: Y \vee Y \to Y$, thus making Y into a $\vee$-algebra.

If A is a Hopf algebra with multiplication π and comultiplication ∇, then the module $QA = A/\operatorname{Im}\pi$ of its indecomposable elements is a $\wedge$-coalgebra with respect to the comultiplication $Q\nabla: QA \to QA \wedge QA$ induced by ∇. Similarly, the module $PA = \operatorname{Ker}\nabla$ of its primitive elements is a $\vee$-algebra with respect to the multiplication $P\pi: PA \vee PA \to PA$ induced by π.

If R is a field of characteristic not equal to 2, then for any module Y there is a map

$$\psi: Y \otimes Y \to Y \otimes Y, \quad \psi(x \otimes y) = \frac{1}{2}\left(x \otimes y - (-1)^{|x||y|} y \otimes x\right).$$

If Y is a $\vee$-algebra, then the composition $\pi \circ \psi$ defines a multiplication

$$[\ ,\ ]: Y \otimes Y \to Y,$$

which satisfies the relations

$$[x, y] + (-1)^{|x||y|}[y, x] = 0,$$
$$[[x, y], z] - (-1)^{|y||z|}[[x, z], y] + (-1)^{|x||y|+|x||z|}[[y, z], x] = 0,$$

i.e., makes Y a Lie algebra. Conversely, if Y is a Lie algebra with multiplication $[\ ,\ ]: Y \otimes Y \to Y$, then this multiplication induces a map $\pi: Y \vee Y \to Y$, which makes Y into a $\vee$-algebra.

Thus, for a field of characteristic not equal to 2, the notions of $\vee$-algebra and Lie algebra coincide. However, even in this case the language of $\vee$-algebras turns out to suit our purposes better.

Sometimes it is convenient to assume that the multiplication in a Lie algebra is an operation lowering the dimension by one. In this case the definition of a $\vee$-algebra is also modified. Namely, for $Y \vee Y$ we should take the submodule of $Y \otimes Y$ consisting of all elements that are fixed by the permutation operator. Similarly, in the definition of a $\wedge$-coalgebra X in which the comultiplication lowers the dimension by one, we should take for $X \wedge X$ the quotient module of $X \otimes X$ by the action of the symmetric group Σ_2.

It is clear that if Y is a $\vee$-algebra with dimension-preserving multiplication, then the suspension SY is a $\vee$-algebra in which multiplication lowers the dimension by one, and vice versa. Similarly, if X is a $\wedge$-coalgebra with dimension-preserving comultiplication, then the desuspension $S^{-1}X$ is a $\wedge$-coalgebra in which comultiplication lowers the dimension by one, and vice versa.

Usually, in what follows we consider $\vee$-algebras and $\wedge$-coalgebras with dimension-preserving structures.

For a $\vee$-algebra Y, as well as for a Lie algebra, we can define the *universal enveloping algebra* $U(Y)$ as the quotient of the tensor algebra TY by the ideal generated by the elements $i(z) - \pi(z)$, where $z \in Y \vee Y$, and $i: Y \vee Y \to Y \otimes Y$ is the inclusion map. The universality means that for any algebra A and any $\vee$-algebra map $f: Y \to A$, there is a unique algebra map $\widetilde{f}: U(Y) \to A$ such that $\widetilde{f} \circ i = f$.

If the multiplication in a $\vee$-algebra Y is trivial, then the universal enveloping algebra $U(Y)$ is the quotient of the tensor algebra by the ideal generated by the elements $i(z)$, $z \in Y \vee Y$, i.e., the elements $z \in Y \otimes Y$ such that $T(z) = -z$. This quotient algebra will be denoted by $T_s^- Y$.

For the universal enveloping algebra $U(Y)$ of a $\vee$-algebra Y, we define in a standard way (see [**43**]) the filtration

$$U_0(Y) = R \subset U_1(Y) \subset \cdots \subset U_n(Y) \subset \dots$$

and the associative algebra

$$\operatorname{gr} U(Y) = \sum_{n \geq 0} \operatorname{gr}_n U(Y), \quad \operatorname{gr}_n U(Y) = U_n(Y)/U_{n-1}(Y),$$

where the multiplication $\operatorname{gr}_p U(Y) \otimes \operatorname{gr}_q U(Y) \to \operatorname{gr}_{p+q} U(Y)$ is induced by the multiplication in $U(Y)$.

As in the case of ordinary Lie algebras [**43**], the following theorem can be proved.

THEOREM (Poincaré–Birkhoff–Witt). *The canonical map of modules $Y \to \operatorname{gr} U(Y)$ can be extended to an algebra map $T_s^-(Y) \to \operatorname{gr} U(Y)$, which is an isomorphism.*

Dually, we can define the *universal enveloping coalgebra* $V(X)$ of a $\wedge$-coalgebra X and establish the corresponding properties.

Let M be a module; we define the $\wedge$-coalgebra

$$T_\wedge M = \sum_{n\geq 1} T^n_\wedge M$$

as follows. We set $T^1_\wedge M = M$, $T^2_\wedge M = M \wedge M$, and we define $\nabla^2 : T^2_\wedge M \to T^1_\wedge M \wedge T^1_\wedge M$ to be the identity map.

Assume that the modules $T^1_\wedge M, \ldots, T^{n-1}_\wedge M$ and the maps $\nabla^1, \ldots, \nabla^{n-1}$ are already defined. Consider the map

$$\nabla \wedge 1 - 1 \wedge \nabla : \sum_{i=1}^{n-1} T^i_\wedge M \wedge T^{n-i}_\wedge M \to \sum_{i,j} T^i_\wedge M \wedge T^j_\wedge M \wedge T^{n-i-j}_\wedge M.$$

We define $T^n_\wedge M$ to be the kernel of this map, and the comultiplication ∇^n to be the natural inclusion map.

THEOREM. *The $\wedge$-coalgebra $T_\wedge M$ is a free $\wedge$-coalgebra over M, i.e., for any $\wedge$-coalgebra X of finite type and any map $f: X \to M$, there is a unique $\wedge$-coalgebra map $\widetilde{f}: X \to T_\wedge M$ such that $p \circ \widetilde{f} = f$.*

PROOF. We set $\widetilde{f}^1 = f: X \to M$. Assume that the maps $\widetilde{f}^1, \ldots, \widetilde{f}^{n-1}$ are already defined. To define the map $\widetilde{f}^n : X \to T^n_\wedge M$, consider the map

$$(\widetilde{f} \wedge \widetilde{f})\nabla : X \to \sum_{i=1}^{n-1} T^i_\wedge M \wedge T^{n-i}_\wedge M.$$

Straightforward computations show that the image of this map lies in the kernel of the map $\nabla \wedge 1 - 1 \wedge \nabla$. Hence this map can be lifted to a map $\widetilde{f}^n : X \to T^n_\wedge M$. $\square$

Dually, for any module M, we can define the $\vee$-algebra $T_\vee M$, which is a free $\vee$-algebra over M.

We define the comultiplication $\nabla: TM \to TM \otimes TM$ in the tensor algebra TM by setting $\nabla(x) = x \otimes 1 + 1 \otimes x$ for the generators $x \in M$; thus the tensor algebra becomes a cocommutative Hopf algebra. This Hopf algebra structure on TY induces a Hopf algebra structure on the universal enveloping algebra $U(Y)$ of the $\vee$-algebra Y. Dually, for any $\wedge$-coalgebra X, the universal enveloping coalgebra $V(X)$ has a commutative Hopf algebra structure.

THEOREM. *For any module M, there are Hopf algebra isomorphisms*

$$U(T_\vee M) \cong TM, \quad V(T_\wedge M) \cong TM.$$

PROOF. Indeed, the inclusion map $M \to TM$ induces a $\vee$-algebra map $T_\vee M \to TM$ and an algebra map $U(T_\vee M) \to TM$. On the other hand, the map $M \to U(T_\vee M)$ induces an algebra map $TM \to U(T_\vee M)$, which is the inverse of the map $U(T_\vee M) \to TM$. $\square$

As in the case of Lie algebras, it can be shown that for any $\vee$-algebra Y, the module of primitive elements $PU(Y)$ in the universal enveloping algebra $U(Y)$, which is regarded here as a Hopf algebra, is isomorphic to the $\vee$-algebra Y itself. Similarly, for the universal enveloping coalgebra $V(X)$ of a $\wedge$-coalgebra X, the module of indecomposable elements $QV(X)$ is isomorphic to the $\wedge$-coalgebra X itself. This implies the following assertion.

THEOREM. *For any module M, there are isomorphisms of $\vee$-algebras and $\wedge$-coalgebras $T_\vee M \cong PTM$ and $T_\wedge M \cong QTM$, where in the first case TM is regarded as a cocommutative Hopf algebra, and in the second case as a commutative Hopf algebra.*

If K is a commutative coalgebra, then the ordinary cobar construction FK is a Hopf algebra with comultiplication $\nabla\colon FK \to FK \otimes FK$ defined on the generators $[x] \in FK$ by the formula $\nabla([x]) = 1 \otimes [x] + [x] \otimes 1$. By the above, the module of primitive elements PFK in this Hopf algebra is a $\vee$-algebra; it is denoted by $F_\vee K$ and is referred to as the *$\vee$-cobar construction on the commutative coalgebra K*. The inclusion map $F_\vee K \to FK$ induces an algebra map $U(F_\vee K) \to FK$, which is an algebra isomorphism.

Dually, if A is a commutative algebra, the ordinary bar construction BA is a Hopf algebra with multiplication defined by

$$[x_1, \ldots, x_n] \cdot [x_{n+1}, \ldots, x_{n+m}] = \sum (-1)^\epsilon [x_{i_1}, \ldots, x_{i_{n+m}}],$$

where the summation is over all (n,m)-shuffles of $1, \ldots, n+m$, and ϵ is the sign of the shuffle.

The module of indecomposable elements QBA in this Hopf algebra is a $\wedge$-coalgebra; it is denoted by $B_\wedge A$ and is referred to as the *$\wedge$-bar construction on the commutative algebra A*. The projection $BA \to B_\wedge A$ induces a coalgebra map $BA \to V(B_\wedge A)$, which is a coalgebra isomorphism.

The notion of twisting cochain can be defined for $\vee$-algebras and $\wedge$-coalgebras as well as for ordinary algebras and coalgebras. Namely, let X be a $\wedge$-coalgebra, and let A be a commutative algebra. A *twisting cochain* is a map $\phi\colon X \to A$ that lowers the dimension by one and satisfies the relation $d(\phi) + \phi \cup \phi = 0$, where $\phi \cup \phi = \pi(\phi \wedge \phi)\nabla$.

For the $\wedge$-bar construction $B_\wedge A$ on a commutative algebra A, we have the standard twisting cochain $\beta\colon B_\wedge A \to A$.

Dually, let K be a commutative coalgebra, and let Y be a $\vee$-algebra; then a *twisting cochain* is a map $\phi\colon K \to Y$ that lowers the dimension by one and satisfies the relation $d(\phi) + \phi \cup \phi = 0$, where $\phi \cup \phi = \pi(\phi \vee \phi)\nabla$.

For the $\vee$-cobar construction $F_\vee K$ on a commutative coalgebra K, we have the standard twisting cochain $\alpha\colon K \to F_\vee K$.

Now we define bar constructions and cobar constructions on $\vee$-algebras and $\wedge$-coalgebras, respectively.

Let X be a $\wedge$-coalgebra with comultiplication $\nabla\colon X \to X \wedge X$. Consider the free commutative algebra $T_s S^{-1} X$ over the desuspension of X. To make it a commutative differential graded algebra, we define the differential by setting

$$d[x] = -[d(x)] - \sum (-1)^{|x'|}[x', x''],$$

where $\sum x' \wedge x'' = \nabla(x)$. On the remaining elements, the differential is defined by the formula for differentiating a product. The resulting commutative differential graded algebra is referred to as the *cobar construction on X* and is denoted by $F_s X$.

Dually, for any $\vee$-algebra Y, we define the *bar construction* $B_c Y$, which is a commutative differential graded coalgebra, and which coincides, as a graded coalgebra, with the free commutative coalgebra $T_c SY$ over the suspension of Y.

Thus we have a functor $F_\vee$ from the category of commutative coalgebras to the category of $\vee$-algebras and a functor B_c from the category of $\vee$-algebras to the category of commutative coalgebras.

For any commutative coalgebra K of finite type, the twisting cochain $\alpha: K \to F_\vee K$ induces a coalgebra map $B(\alpha): K \to B_c F_\vee K$ and a natural transformation of functors $\mathrm{Id} \to B_c \circ F_\vee$.

For any $\vee$-algebra Y, the twisting cochain $\beta: B_c Y \to Y$ induces a $\vee$-algebra map $F_\vee(\beta): F_\vee B_c Y \to Y$ and a natural transformation of functors $F_\vee \circ B_c \to \mathrm{Id}$.

It is readily seen that the functors $F_\vee$ and B_c are adjoint for the categories of commutative coalgebras of finite type and of $\vee$-algebras.

In the case of a field of characteristic zero, Quillen proved that the category of commutative coalgebras is homotopy equivalent to the category of Lie algebras [**40**]. In particular, the following theorem holds.

THEOREM (Quillen). *If the ground ring is a field of characteristic zero, then for any commutative coalgebra K of finite type and for any $\vee$-algebra Y, the maps*

$$B(\alpha): K \to B_c F_\vee K, \quad F_\vee(\beta): F_\vee B_c Y \to Y$$

are chain equivalences.

Dually, the functors $B_\wedge$ and F_s are adjoint for the categories of commutative algebras and of $\wedge$-coalgebras of finite type, and the following theorem holds.

THEOREM. *If the ground ring is a field of characteristic zero, then for any commutative algebra A and for any $\wedge$-coalgebra X of finite type, the maps*

$$F(\beta): F_s B_\wedge A \to A, \quad B(\alpha): X \to B_\wedge F_s X$$

are chain equivalences.

For example, let us show that for any commutative coalgebra K of finite type, there is a chain equivalence $B_c F_\vee K \simeq K$. For any chain complex Y, we have the inclusion map $i: T_c Y \to TY$, which is a coalgebra map. In the case when the ground ring is a field of characteristic zero, there is an inverse map $p: TY \to T_c Y$, which is defined on generators by

$$p(y_1 \otimes \cdots \otimes y_n) = \frac{1}{n!} \sum_{\sigma \in \Sigma_n} \sigma(y_1 \otimes \cdots \otimes y_n).$$

Similarly, we have the inclusion $i: T_\vee K \to TK$, which is a $\vee$-algebra map. In the case when the ground ring is a field of characteristic zero, there is an inverse map $p: TK \to T_\vee K$.

Therefore, for any commutative coalgebra K, there are an inclusion map $i: B_c F_\vee K \to BFK$ and a projection $p: BFK \to B_c F_\vee K$. As we know, for any coalgebra K of finite type there is a chain equivalence $BFK \simeq K$. Let $h: BFK \to BFK$ be the corresponding chain homotopy. We define the chain homotopy $\widetilde{h}: B_c F_\vee K \to B_c F_\vee K$ by setting $\widetilde{h} = p \circ h \circ i$. It is readily seen that this homotopy yields the required chain equivalence between $B_c F_\vee K$ and K.

Note that the Massey products and the functor $\mathcal{D}$ can be defined in the homology of $\vee$-algebras in much the same way as for the case of ordinary algebras.

CHAPTER 4

A_∞-Structures on Chain Complexes

4.1. A_∞-algebras and A_∞-coalgebras

Let us consider the notions of A_∞-algebra and A_∞-coalgebra, which were introduced by Stasheff in [**45**].

An A_∞*-algebra* is a chain complex A endowed with operations

$$\pi_n\colon A^{\otimes n+2} \to A, \quad n \geq 0,$$

that raise the dimension by n and satisfy the relations

$$d(\pi_{n+1}) = \sum_{i=0}^{n} (-1)^\epsilon \pi_i(1 \otimes \cdots \otimes \pi_{n-i} \otimes \cdots \otimes 1),$$

where the summation is also over all possible positions (indexed by k) of the factor π_{n-i} in the tensor product, and $\epsilon = nk + ik + n + k$.

In particular, if $n = 0$, then we obtain

$$d(\pi_1) = \pi_0(1 \otimes \pi_0 - \pi_0 \otimes 1),$$

which means that the operation π_0 is homotopy associative.

If $n = 1$, then we obtain

$$d(\pi_2) = \pi_1(\pi_0 \otimes 1 \otimes 1 - 1 \otimes \pi_0 \otimes 1 + 1 \otimes 1 \otimes \pi_0) - \pi_0(\pi_1 \otimes 1 + 1 \otimes \pi_1),$$

which signifies the homotopy associativity between the operations π_0 and π_1. And so on.

An ordinary differential algebra can also be regarded as an A_∞-algebra with trivial higher operations π_n, $n \geq 1$.

Another example of an A_∞-algebra is provided by the singular chain complex $C_*(X)$ of an A_∞-space X. In particular, the singular chain complex $C_*(\Omega X)$ of a loop space ΩX is an A_∞-algebra.

Dually, a chain complex K is called an A_∞*-coalgebra* if there are given operations

$$\nabla_n\colon K \to K^{\otimes n+2}, \quad n \geq 0,$$

that raise the dimension by n and satisfy the relations

$$d(\nabla_{n+1}) = \sum_{i=0}^{n} (-1)^\epsilon (1 \otimes \cdots \otimes \nabla_{n-i} \otimes \cdots \otimes 1)\nabla_i,$$

where $\epsilon = nk + ik + ni + i + k + 1$.

An ordinary differential coalgebra can also be regarded as an A_∞-coalgebra with trivial higher operations ∇_n, $n \geq 1$.

To simplify the sign conventions, let us reformulate the above definitions of an A_∞-algebra and an A_∞-coalgebra. Let X be a chain complex. Consider the tensor algebra TX as a coalgebra with the comultiplication $\nabla: TX \to TX \otimes TX$ defined by

$$\nabla(x_1 \otimes \cdots \otimes x_n) = \sum_i (x_1 \otimes \cdots \otimes x_i) \otimes (x_{i+1} \otimes \cdots \otimes x_n).$$

Then giving a differential that makes TX into a differential coalgebra is equivalent to giving operations

$$\pi(n): X^{\otimes n} \to X, \quad n \geq 2,$$

that lower the dimension by one and satisfy the relation

$$d(\pi(n)) + \sum_{i=2}^{n-1} \pi(i)(1 \otimes \cdots \otimes \pi(n-i+1) \otimes \cdots \otimes 1) = 0,$$

or, more concisely, $d(\pi) + \pi \cup \pi = 0$.

THEOREM. *For any chain complex A, giving a family of operations*

$$\pi(n): SA^{\otimes n} \to SA, \quad n \geq 2,$$

that satisfy the relation $d(\pi) + \pi \cup \pi = 0$ is equivalent to giving an A_∞-algebra structure on A.

PROOF. Consider the mutually inverse maps $\xi: A \to SA$, $\eta: SA \to A$. If A is endowed with an A_∞-structure consisting of the operations $\pi_n: A^{\otimes n+2} \to A$, then for $\pi(n+2): (SA)^{\otimes n+2} \to SA$ we take the operations that satisfy the relations $\eta \circ \pi(n+2) = \pi_i \circ \eta^{\otimes n+2}$. We have

$$\begin{aligned}
&\Big\{d(\pi_{n+1}) + \sum_{i=0}^{n} (-1)^{nk+ik+n+k} \pi_i(1 \otimes \cdots \otimes \pi_{n-i} \otimes \cdots \otimes 1)\Big\}\eta^{\otimes n+3} \\
&\qquad = d(\pi_{n+1}\eta^{\otimes n+3}) + \sum_{i=0}^{n} (-1)^{i-k} \pi_i(\eta \otimes \cdots \otimes \eta\pi(n-i+2) \otimes \cdots \otimes \eta) \\
&\qquad = \eta\Big\{d(\pi(n+3)) + \sum_{i=0}^{n} \pi(i+2)(1 \otimes \cdots \otimes \pi(n-i+2) \otimes \cdots \otimes 1)\Big\}.
\end{aligned}$$

Hence the first expression vanishes if and only if the third expression does. □

The differential coalgebra TSA constructed for a given A_∞-algebra A is called the *bar construction on the A_∞-algebra A* and is denoted by BA. It is clear that for ordinary algebras regarded as A_∞-algebras with trivial higher operations π_n for $n \geq 1$, this bar construction coincides with the ordinary one.

Dually, for any complex K, consider the desuspension $S^{-1}K$ and the tensor algebra $TS^{-1}K$. Then endowing K with an A_∞-coalgebra structure is equivalent to giving a differential on $TS^{-1}K$ that endows it with an algebra structure; this algebra is called the *cobar construction on the A_∞-coalgebra K* and is denoted by FK. It is clear that for ordinary coalgebras regarded as A_∞-coalgebras with trivial higher operations ∇_n for $n \geq 1$, this cobar construction coincides with the ordinary one.

Suppose that A', A'' are A_∞-algebras. An *A_∞-algebra map $f: A' \to A''$* is a sequence of maps $f_n: A'^{\otimes n+1} \to A''$ raising dimension by n and such that the corresponding maps $f(n)\colon (SA')^{\otimes n} \to SA''$ satisfy the relations

$$d(f(n)) = \sum_{i=1}^{n-1} f(i)(1 \otimes \cdots \otimes \pi'(n-i+1) \otimes \cdots \otimes 1) - \sum_{n_1+\cdots+n_i=n} \pi''(i)(f(n_1) \otimes \cdots \otimes f(n_i)),$$

or, more concisely, $d(f) = f \cup \pi' - \pi'' \cup f$.

In much the same way as above, it can be shown that giving these operations is equivalent to giving a coalgebra map $B(f)\colon BX' \to BX''$.

Maps $f, g\colon A' \to A''$ of differential A_∞-algebras are said to be *homotopic* (we write $f \simeq g$) if there exists a family $h = \{h_n\}$ of maps $h_n\colon A'^{\otimes n} \to A''$ raising dimension by n and satisfying the relation $d(h) = g - f - \pi'' \cup h - h \cup \pi'$. The family h is called a *homotopy* between f and g. It can be directly verified that if f and g are homotopic, then the corresponding maps of the bar constructions $Bf, Bg\colon BX' \to BX''$ are also homotopic.

A_∞-algebras A', A'' are said to be *homotopy equivalent* (and we write $A' \simeq A''$) if there are maps $f\colon A' \to A''$, $g\colon A'' \to A'$ such that

$$g \circ f \simeq \mathrm{Id}_{A'}, \quad f \circ g \simeq \mathrm{Id}_{A''}.$$

The notions of A_∞-coalgebra map, homotopy, and homotopy equivalence of A_∞-coalgebras are defined in a dual way.

Let K be a coalgebra and A an A_∞-algebra. A *twisting cochain* $\phi\colon K \to A$ is a homomorphism that lowers the dimension by one and satisfies the condition

$$d(\phi) + A_\infty(\phi) = 0,$$

where $A_\infty(\phi) = \sum_{n\geq 0} \pi_n(\phi \otimes \ldots \phi)\nabla(n+2)$.

It is clear that in this case a twisting cochain $\phi\colon K \to A$ induces a coalgebra map $B(\phi)\colon K \to BA$, and for any A_∞-algebra A there is a standard twisting cochain $\beta\colon BA \to A$.

THEOREM. *For any A_∞-algebra A, there is a chain equivalence $FBA \simeq A$.*

PROOF. Consider the inclusion map $i\colon A \to FBA$ and the spectral sequence on FBA induced by the filtration on BA. The first term of this spectral sequence is isomorphic to the homology of FBA, where A is regarded as a trivial algebra. By a theorem proved above, in this case we have a chain equivalence $FBA \simeq A$, and hence the inclusion map $i\colon A \to FBA$ induces an isomorphism in homology. □

Later we will show that this chain equivalence is a chain equivalence of A_∞-algebras.

Similarly, if K is an A_∞-coalgebra and A is an algebra, then a *twisting cochain* $\phi\colon K \to A$ is a homomorphism that lowers the dimension by one and satisfies the relation

$$d(\phi) + A_\infty(\phi) = 0,$$

where $A_\infty(\phi) = \sum_{n\geq 0} \pi(n+2)(\phi \otimes \cdots \otimes \phi)\nabla_n$.

Clearly, in this case a twisting cochain $\phi\colon K \to A$ induces an algebra map $F(\phi)\colon FK \to A$, and for any A_∞-coalgebra K there is a standard twisting cochain $\alpha\colon K \to FK$.

The following theorem is dual to the previous one.

THEOREM. *For any A_∞-coalgebra K of finite type, there is a chain equivalence $BFK \simeq K$.*

Now let us elucidate the question concerning the structure arising on the homology of a differential graded algebra. As we noted above, the Massey products are defined on the homology of a differential algebra. Their serious drawback is the fact that they are only partially defined and multivalued. To eliminate it, we shall use the language of A_∞-algebras.

Indeed, it is worthwhile to consider A_∞-structures not only in the category of chain complexes, but also in the category of graded modules; the latter category can be thought of as a subcategory of the category of chain complexes, where each graded module is regarded as a chain complex with zero differential. We define a *graded A_∞-algebra* as a graded module M endowed with operations $\pi_n\colon M^{\otimes n+2} \to M$ that raise the dimension by n and satisfy the relations

$$\sum_{i=0}^{n}(-1)^{\epsilon}\pi_i(1\otimes\cdots\otimes\pi_{n-i}\otimes\cdots\otimes 1)=0.$$

In particular, for $n=0$, we obtain

$$\pi_0(1\otimes\pi_0-\pi_0\otimes 1)=0,$$

which means the associativity of the operation π_0.

For $n=1$, we obtain

$$\pi_1(\pi_0\otimes 1\otimes 1-1\otimes\pi_0\otimes 1+1\otimes 1\otimes\pi_0)-\pi_0(\pi_1\otimes 1+1\otimes\pi_1)=0,$$

which is the associativity relation between π_0 and π_1, and so on.

As we have seen, to simplify the formulas and to eliminate cumbersome expressions with signs, it is convenient to pass to the suspension SM of M and consider the operations $\pi(n)\colon(SM)^{\otimes n}\to SM$ induced by the operations π_{n-2}. They lower the dimension by one and satisfy the relations

$$\sum_{i=2}^{n-1}\pi(i)(1\otimes\cdots\otimes\pi(n-i+1)\otimes\cdots\otimes 1)=0,$$

or, more concisely, $\pi\cup\pi=0$.

Let us show that such an A_∞-algebra structure arises on the homology of a differential algebra. Namely, the following assertion holds (see [**24**]).

THEOREM. *If A is a differential algebra over a field R, then its homology $A_*=H_*(A)$ possesses an A_∞-algebra structure, and there is a chain equivalence of A_∞-algebras $A\simeq A_*$ inducing a chain equivalence of differential coalgebras $BA\simeq BA_*$.*

We shall list the formulas expressing the higher operations on the homology of a differential algebra. Let $\pi\colon A\otimes A\to A$ be the multiplication in A. Denote by $\widetilde{\pi}\colon SA\otimes SA\to SA$ the map induced by the multiplication π. Let us choose maps $\xi\colon SA_*\to SA$, $\eta\colon SA\to SA_*$ and a homotopy $h\colon SA\to SA$ so that

$$\eta\circ\xi=\mathrm{Id},\quad d(h)=\xi\circ\eta-\mathrm{Id},\quad h\circ\xi=0,\quad \eta\circ h=0,\quad h\circ h=0.$$

Such maps exist since the ground ring R is a field.

Let us define the A_∞-algebra structure $\pi=\{\pi(n)\}$ by setting

$$\pi(n)=\eta\widetilde{\pi}(h\widetilde{\pi}\otimes 1+1\otimes h\widetilde{\pi})\dots(h\widetilde{\pi}\otimes\cdots\otimes 1+\cdots+1\otimes\cdots\otimes h\widetilde{\pi})(\xi\otimes\cdots\otimes\xi).$$

Straightforward computations show that the required relation $\pi \cup \pi = 0$ holds.

Let us also define maps $\xi(n)\colon (SA_*)^{\otimes n} \to SA$, $\eta(n)\colon (SA)^{\otimes n} \to SA_*$ and a homotopy $h(n)\colon (SA)^{\otimes n} \to SA$ by setting

$$\begin{aligned}
\xi(n) &= h\widetilde{\pi}(h\widetilde{\pi} \otimes 1 + 1 \otimes h\widetilde{\pi}) \dots (h\widetilde{\pi} \otimes \cdots \otimes 1 + \cdots + 1 \otimes \cdots \otimes h\widetilde{\pi})(\xi \otimes \cdots \otimes \xi),\\
\eta(n) &= \eta\widetilde{\pi}(h\widetilde{\pi} \otimes 1 + 1 \otimes h\widetilde{\pi}) \dots (h \otimes \xi\eta \cdots \otimes \xi\eta + \cdots + 1 \otimes \cdots \otimes 1 \otimes h),\\
h(n) &= h\widetilde{\pi}(h\widetilde{\pi} \otimes 1 + 1 \otimes h\widetilde{\pi}) \dots (h \otimes \xi\eta \cdots \otimes \xi\eta + \cdots + 1 \otimes \cdots \otimes 1 \otimes h).
\end{aligned}$$

Straightforward computations show that the required relations

$$d(\xi) = \xi \cup \pi - \widetilde{\pi} \cup \xi, \quad d(\eta) = \eta \cup \widetilde{\pi} - \pi \cup \eta, \quad \eta \circ \xi = \mathrm{Id}, \quad d(h) = \xi \circ \eta - \mathrm{Id} - \pi \cup h - h \cup \pi$$

hold.

These families of maps ξ, η, h induce maps of the bar constructions

$$B(\xi)\colon BA_* \to BA, \quad B(\eta)\colon BA \to BA_*, \quad B(h)\colon BA \to BA,$$

which yield a homotopy equivalence between BA and BA_*.

Of course, the above A_∞-algebra structure on A_* is not uniquely defined. It depends on the choice of the maps ξ, η and the homotopy h. Let us define an equivalence relation for A_∞-algebra structures on the graded module A_* by setting $\pi' \sim \pi''$ if there is a family $\psi = \{\psi(n)\}$ of maps $\psi(n)\colon (SA_*)^{\otimes n} \to SA_*$ of dimension 0 such that $\psi(1)$ is an isomorphism and $\pi'' \cup \psi = \psi \cup \pi'$.

The last relation means that ψ induces a map of the bar constructions

$$B(\psi)\colon B'A_* \to B''A_*,$$

where B' and B'' are the bar constructions corresponding to the A_∞-structures π' and π''.

Let us show that different choices of the maps ξ, η and the homotopy h lead to equivalent A_∞-structures on A_*. Indeed, let ξ', η', h' and ξ'', η'', h'' be maps that give rise to A_∞-structures π' and π''. Then we have the maps of the bar constructions $B(\xi')\colon B'A_* \to BA$ and $B(\eta'')\colon BA \to B''A_*$. Their composition is the required map $B(\psi)\colon B'A_* \to B''A_*$.

Passing to the dual situation, we obtain the following theorem.

THEOREM. *If K is a differential coalgebra over a field R, then its homology $K_* = H_*(K)$ possesses a graded A_∞-coalgebra structure, and there is a chain equivalence of A_∞-coalgebras $K \simeq K_*$ inducing a chain equivalence of differential algebras $\hat{F}K \simeq \hat{F}K_*$.*

4.2. A_∞-modules and A_∞-comodules

Let A be an A_∞-algebra. A chain complex M is called an *A_∞-module over A* if there is given a family of maps $\mu_n\colon A^{\otimes n+1} \otimes M \to M$ that raise the dimension by $n \geq 0$ and satisfy the relation

$$d(\mu_{n+1}) = \sum_{i=0}^{n} (-1)^{\epsilon} \mu_i(1 \otimes \cdots \otimes \pi_{n-i} \otimes \cdots \otimes 1 + 1 \otimes \cdots \otimes \mu_{n-i}),$$

where $\epsilon = nk + ik + n + k$.

For example, if $n = 0$, then this formula yields the homotopy associativity relation

$$d(\mu_1) = \mu_0(-\pi_0 \otimes 1 + 1 \otimes \mu_0)$$

for the operation $\mu_0: A \otimes M \to M$.

If $n = 1$, then we have the relation

$$d(\mu_2) = -\mu_0(\pi_1 \otimes 1 + 1 \otimes \mu_1) + \mu_1(\pi_0 \otimes 1 \otimes 1 - 1 \otimes \pi_0 \otimes 1 + 1 \otimes 1 \otimes \mu_0) = 0.$$

And so on.

Dually, let K be an A_∞-coalgebra; a chain complex L is called an *A_∞-comodule over K* if it is endowed with operations $\tau_n : L \to K^{\otimes n+1} \otimes L$ that raise the dimension by $n \geq 0$ and satisfy the relations

$$d(\tau_{n+1}) = \sum_{i=0}^{n} (-1)^\epsilon (1 \otimes \cdots \otimes \nabla_{n-i} \otimes \cdots \otimes 1 + 1 \otimes \cdots \otimes \tau_{n-i})\nabla_i,$$

where $\epsilon = nk + ik + ni + i + k + 1$.

To eliminate difficulties with signs, we reformulate these definitions. Namely, let A be an A_∞-algebra, and let M be a chain complex. Consider the complex $B(A) \otimes M$, where $B(A)$ is the bar construction on the A_∞-algebra A. Then giving a differential on $B(A) \otimes M$ that makes this complex into a comodule over the coalgebra $B(A)$ is equivalent to endowing M with the structure of an A_∞-module over A. The complex $B(A) \otimes M$ with this differential is called the *bar construction on M* and is denoted by $B(A, M)$.

It is readily seen that endowing M with the structure of an A_∞-module over an A_∞-algebra A is equivalent to endowing M with a Lie BA-module structure.

Dually, for any A_∞-coalgebra K and any chain complex L, consider the complex $F(K) \otimes L$, where $F(K)$ is the cobar construction on the A_∞-coalgebra K. Then giving a differential on $F(K) \otimes L$ that makes this complex into a module over the algebra $F(K)$ is equivalent to endowing L with the structure of an A_∞-comodule over K. The complex $F(K) \otimes L$ with this differential is referred to as the *cobar construction on L* and is denoted by $F(K, L)$.

It is readily seen that endowing L with the structure of an A_∞-comodule over an A_∞-coalgebra K is equivalent to endowing L with a Lie FK-comodule structure.

By using bar and cobar constructions, we can define the notions of A_∞-maps for A_∞-modules and A_∞-comodules.

Suppose that M' is an A_∞-module over an A_∞-algebra A', M'' is an A_∞-module over an A_∞-algebra A'', and an A_∞-map $f = \{f^n\}$, $f^n: A'^{\otimes n+2} \to A''$, is given. Then an *$A_\infty$-map of the A_∞-module M' to the A_∞-module M''* is a family $g = \{g^n\}$ of maps $g^n: A'^{\otimes n} \otimes M' \to M''$ that raise the dimension by n and satisfy the condition that the corresponding map of the bar constructions

$$B(f, g): B(A', M') \to B(A'', M'')$$

is a comodule map, i.e., the following diagram is commutative:

$$\begin{array}{ccc} B(A', M') & \xrightarrow{B(f,g)} & B(A'', M'') \\ \tau' \downarrow & & \downarrow \tau'' \\ B(A') \otimes B(A', M') & \xrightarrow{B(f)\otimes B(f,g)} & B(A'', M'') \end{array}$$

Dually, suppose that L' is an A_∞-comodule over an A_∞-coalgebra K', L'' is an A_∞-comodule over an A_∞-coalgebra K'', and $f = \{f^n\}$, $f^n: K' \to K'^{\otimes n+2}$, is an A_∞-map. Then an *A_∞-map of the A_∞-comodule L' to the A_∞-comodule L''* is a family $g = \{g^n\}$ of maps $g^n \colon L' \to K''^{\otimes n} \otimes L''$ that raise the dimension by n and satisfy the condition that the corresponding map

$$F(f,g) \colon F(K',L') \to F(K'',L'')$$

is a module map, i.e., the following diagram is commutative:

$$\begin{array}{ccc} F(K',L') & \xrightarrow{F(f,g)} & F(K'',L'') \\ \mu' \uparrow & & \uparrow \mu'' \\ F(K') \otimes F(K',L') & \xrightarrow{F(f)\otimes F(f,g)} & F(K'',L'') \end{array}$$

The Brown theorem implies the following assertion.

THEOREM. *Suppose that the ground ring is a field and M is a differential graded module over a differential graded algebra A. Then the homology $M_* = H_*(M)$ possesses the structure of an A_∞-module over A, and there is a chain equivalence of A_∞-modules $M \simeq M_*$.*

As in the case of differential algebras and coalgebras, we obtain the following assertions.

THEOREM. *Suppose that the ground ring is a field and M is a differential graded module over a differential graded algebra A. Then the homology $M_* = H_*(M)$ possesses the structure of a graded A_∞-module over the graded A_∞-algebra $A_* = H_*(A)$, and there is a homotopy equivalence of A_∞-modules $M \simeq M_*$ inducing a homotopy equivalence of comodules $B(A,M) \simeq B(A_*,M_*)$.*

THEOREM. *Suppose that the ground ring is a field and L is a differential graded comodule over a differential graded coalgebra K. Then the homology $L_* = H_*(L)$ possesses the structure of a graded A_∞-comodule over the graded A_∞-coalgebra $K_* = H_*(K)$, and there is a homotopy equivalence of A_∞-comodules $L \simeq L_*$ inducing a homotopy equivalence of modules $\hat{F}(K,L) \simeq \hat{F}(K_*,L_*)$.*

4.3. A_∞-$\vee$-algebras and A_∞-$\wedge$-coalgebras

By analogy with the notions of A_∞-algebra and A_∞-coalgebra, we define the notions of A_∞-$\vee$-algebra and A_∞-$\wedge$-coalgebra.

DEFINITION. A chain complex X is called an A_∞-$\vee$-*algebra* if there are given maps

$$\pi_n \colon X^{\vee n+2} \to X, \quad n \geq 0,$$

that raise the dimension by n and satisfy the relations

$$d(\pi_{n+1}) = \sum_{i=0}^{n} (-1)^\epsilon \pi_i(1 \otimes \cdots \otimes \pi_{n-i} \otimes \cdots \otimes 1),$$

where the summation is also over all possible positions (indexed by k) of the factor π_{n-i} in the tensor product, and $\epsilon = nk + ik + n + k$.

For example, if $n = 0$, then we have the relation

$$d(\pi_1) = \pi_0(1 \otimes \pi_0 - \pi_0 \otimes 1),$$

which means the homotopy associativity of π_0.

As in the case of A_∞-algebras, giving such a family of maps π_i is equivalent to endowing the bar construction B_cY with a differential that makes B_cY into a commutative differential coalgebra.

Note that ordinary differential $\vee$-algebras can be thought of as A_∞-$\vee$-algebras with higher multiplications π_i, $i \geq 1$, equal to zero.

DEFINITION. Let X' and X'' be A_∞-$\vee$-algebras. An *A_∞-map from X' to X''* is a sequence of maps

$$f_n \colon X'^{\vee(n+1)} \to X'', \quad n \geq 0,$$

that raise the dimension by n and satisfy the relations

$$\begin{aligned} d(f_{n+1}) &= \sum_{i=0}^{n} (-1)^{\epsilon} f_i(1 \otimes \cdots \otimes \pi'_{n-i} \otimes \cdots \otimes 1) \\ &+ \sum_{i=0}^{n} (-1)^{\delta} \pi''_i (f_{n_1} \otimes \cdots \otimes f_{n_{i+2}}), \end{aligned}$$

where k indicates the position occupied by the factor π'_{n-i}, and $n_1 + \cdots + n_{i+2} = n - i$.

For example, if $n = 0$, then we have the relation

$$d(f_1) = f_0 \pi'_0 - \pi''_0 (f_0 \otimes f_0),$$

which means that the map f_0 is homotopy multiplicative.

Giving an A_∞-map f from X' to X'' is equivalent to giving a differential coalgebra map $B_c(f) \colon B_cY' \to B_cY''$.

Dually, we give the following definition.

DEFINITION. A chain complex Y is called an *A_∞-$\wedge$-coalgebra* if there are given maps

$$\nabla_n \colon X \to X^{\wedge n+2}, \quad n \geq 0,$$

that raise the dimension by n and satisfy the relations

$$d(\nabla_{n+1}) = \sum_{i=0}^{n} (-1)^{\epsilon} (1 \otimes \cdots \otimes \nabla_{n-i} \otimes \cdots \otimes 1) \nabla_i.$$

As in the case of A_∞-coalgebras, defining such a family of maps ∇_i is equivalent to endowing the cobar construction F_sY with a differential that makes F_sY into a commutative differential algebra.

THEOREM. *Suppose that X is a differential graded $\vee$-algebra over a field R. Then its homology X_* possesses an A_∞-$\vee$-algebra structure, and there is a chain equivalence of coalgebras $B_cX \simeq B_cX_*$. Dually, the homology Y_* of a differential graded $\wedge$-coalgebra Y possesses an A_∞-$\wedge$-coalgebra structure, and there is a chain equivalence of algebras $\hat{F}_sY \simeq \hat{F}_sY_*$.*

The proof goes along the same lines as in the case of ordinary differential algebras and coalgebras. The formulas given there are appropriate for the present case as well.

4.4. Commutative A_∞-algebras and A_∞-coalgebras

Here we define commutative A_∞-algebras and A_∞-coalgebras.

DEFINITION. An A_∞-coalgebra K is said to be *commutative* if the maps

$$\nabla_n : K \to K^{\otimes(n+2)}, \quad n \geq 0,$$

raising dimension by n that constitute the A_∞-coalgebra structure additionally satisfy the commutativity relations, i.e., for any $n \geq 0$ and any i, $1 \leq i \leq n+1$, we have

$$\sum_\sigma (-1)^\epsilon \sigma(i, n+2-i)\nabla_n = 0,$$

where the summation is over all $(i, n+2-i)$-shuffles σ, and ϵ is the sign of the shuffle.

Passing to the desuspensions, we see that endowing K with an A_∞-coalgebra structure is equivalent to giving a map $\delta: S^{-1}K \to T_\vee S^{-1}K$ that lowers the dimension by one and satisfies the condition

$$d(\delta) + \delta \cup \delta = 0$$

or, what is the same, is equivalent to endowing $F_\vee K$ with a differential $\vee$-algebra structure.

Dually, we give the following definition.

DEFINITION. An A_∞-algebra A is said to be *commutative* if the maps

$$\pi_n : A^{\otimes(n+2)} \to A, \quad n \geq 0,$$

raising dimension by n that constitute the A_∞-algebra structure additionally satisfy the commutativity relations, i.e., for any $n \geq 0$ and any i, $1 \leq i \leq n+1$, we have

$$\sum_\sigma (-1)^\epsilon \pi_n \sigma(i, n+2-i) = 0,$$

where the summation is over all $(i, n+2-i)$-shuffles σ, and ϵ is the sign of the shuffle.

Passing to the suspensions, we see that endowing A with an A_∞-algebra structure is equivalent to giving a map $\pi: T_\wedge SA \to SA$ that lowers the dimension by one and satisfies the relation $d(\pi) + \pi \cup \pi = 0$, and it is also equivalent to endowing $F_\wedge A$ with a differential $\wedge$-coalgebra structure.

THEOREM. *Suppose that A is a commutative differential graded algebra over a field R. Then its homology A_* possesses a commutative A_∞-algebra structure, and there is a chain equivalence of $\wedge$-coalgebras $B_\wedge A \simeq B_\wedge A_*$. Dually, the homology K_* of a commutative differential graded coalgebra K possesses a commutative A_∞-coalgebra structure, and there is a chain equivalence of $\vee$-algebras $\hat{F}_\vee K \simeq \hat{F}_\vee K_*$.*

The proof goes along the same lines as in the case of ordinary differential algebras and coalgebras. The formulas given there are appropriate for the present case as well.

THEOREM. *Suppose that the ground ring R is a field of characteristic zero. Then the singular chain complex $C_*(X; R)$ possesses a natural commutative A_∞-coalgebra structure, and any two structures of this kind are chain equivalent.*

PROOF. We shall use the method of acyclic models and the following lemma.

LEMMA. *Suppose that R is a field of characteristic zero. Then, for any chain complex X we have the following isomorphisms*:

$$H_*(T_sX) \cong T_sH_*(X), \quad H_*(T_cX) \cong T_cH_*(X),$$
$$H_*(T_\vee X) \cong T_\vee H_*(X), \quad H_*(T_\wedge X) \cong T_\wedge H_*(X).$$

PROOF OF THE LEMMA. Indeed, the Künneth formula yields isomorphisms

$$H_*(X^{\otimes n}) \cong H_*(X)^{\otimes n}.$$

Since the ground ring is a field of characteristic zero, we can choose these isomorphisms so that they commute with the actions of the symmetric groups Σ_n. Hence they induce the required isomorphisms. □

Now consider the singular chain complex $C_*(X;R)$ of a topological space X with coefficients in a field R of characteristic zero. The usual natural comultiplication

$$\nabla : C_*(X;R) \to C_*(X;R) \otimes C_*(X;R)$$

is associative but not commutative. We define the comultiplication

$$\widetilde{\nabla} : C_*(X;R) \to C_*(X;R) \otimes C_*(X;R)$$

by setting $\widetilde{\nabla} = \frac{1}{2}(\nabla + T\nabla)$, where T is the map interchanging the factors of the tensor product. It is clear that the comultiplication $\widetilde{\nabla}$ is commutative but is not associative. By using the method of acyclic models, this map is extended in a standard way to a map $\delta : S^{-1}C_*(X;R) \to T_\vee S^{-1}C_*(X;R)$ satisfying the condition $d(\delta) + \delta \cup \delta = 0$ and hence defining a commutative A_∞-coalgebra structure on $C_*(X;R)$. If $\delta', \delta'' : S^{-1}C_*(X;R) \to T_\vee S^{-1}C_*(X;R)$ are two structures of this kind, then, by using the method of acyclic models, we construct a map $f : S^{-1}C_*(X;R) \to T_\vee S^{-1}C_*(X;R)$ satisfying the condition $d(f) = f \cup \delta' - \delta'' \cup f$ and hence defining an equivalence between the two commutative coalgebra structures on $C_*(X;R)$. □

The commutative A_∞-coalgebra structure on $C_*(X;R)$ allows us to construct a functor of commutative chains of a topological space. Namely, let $\widetilde{C}_*(X;R) = B_cF_\vee C_*(X;R)$. Then the correspondence $X \longmapsto \widetilde{C}_*(X;R)$ defines the required functor from the category of topological spaces to the category of commutative differential coalgebras over a field R of characteristic zero. The projection

$$B_cF_\vee C_*(X;R) \to C_*(X;R)$$

is a chain map inducing an isomorphism in homology.

Passing to the dual situation, we obtain the following assertion.

THEOREM. *Suppose that the ground ring R is a field of characteristic zero. Then the singular cochain complex $C^*(X;R)$ possesses a natural commutative A_∞-algebra structure, and any two structures of this kind are chain equivalent.*

The commutative A_∞-algebra structure on $C^*(X;R)$ allows us to construct the functor of commutative cochains of a topological space. Namely, let $\widetilde{C}^*(X;R) = F_sB_\wedge C^*(X;R)$. Then the correspondence $X \longmapsto \widetilde{C}^*(X;R)$ defines the required functor from the category of topological spaces to the category of commutative differential algebras over a field R of characteristic zero. The inclusion map $C^*(X;R) \to F_s B_\wedge C^*(X;R)$ is a chain map inducing an isomorphism in homology.

4.5. The functor $\mathcal{D}$ for A_∞-algebras

Here we define the functor $\mathcal{D}$ for A_∞-algebras. To avoid difficulties in determining the signs, by an A_∞-algebra structure on a chain complex A we shall understand a family $\pi = \{\pi(n)\}$ of operations

$$\pi(n)\colon A^{\otimes n} \to A, \quad n \geq 2,$$

that lower the dimension by one and satisfy the relations

$$d(\pi(n)) + \sum_{i=2}^{n-1} \pi(i)(1 \otimes \cdots \otimes \pi(n-i+1) \otimes \cdots \otimes 1) = 0$$

or, in a more concise form,

$$d(\pi(n)) + \sum_{i=2}^{n-1} \pi(i) \cup \pi(n-i+1) = 0$$

or, still more concisely, $d(\pi) + \pi \cup \pi = 0$.

Two A_∞-structures π' and π'' on a chain complex A are said to be *equivalent* (we write $\pi' \sim \pi''$) if there is a family $c = \{c(n)\}$ of maps $c(n)\colon A^{\otimes n} \to A$ that preserve dimension and satisfy the relations $c(1) = \mathrm{Id}$ and

$$d(c) = c \cup \pi' - \pi'' \cup c.$$

Let X be a chain complex. Denote the set of equivalence classes of A_∞-structures on X by $\mathcal{D}_\infty(X)$.

Consider the category $\mathcal{X}$ whose objects are chain complexes X and whose morphisms $X \to Y$ are triples (f, g, h), where $f\colon X \to Y$, $g\colon Y \to X$ are chain maps, and $h\colon X \to X$ is a chain homotopy, $d(h) = g \circ f - \mathrm{Id}$. The composition of morphisms $(f', g, h')\colon X \to X'$ and $(f'', g'', h'')\colon X' \to X''$ is the morphism

$$(f'' \circ f', g' \circ g'', h' + g' \circ h'' \circ f')\colon X \to X''.$$

Two morphisms $(f', g', h'), (f'', g'', h'')\colon X \to Y$ are said to be *homotopic* (and we write $(f', g', h') \simeq (f'', g'', h'')$) if there is a triple (u, v, w) such that

$u\colon X \to Y$ is a homotopy between f' and f'', i.e., $d(u) = f' - f''$;

$v\colon Y \to X$ is a homotopy between g' and g'', i.e., $d(v) = g' - g''$;

$w\colon X \to X$ is a map that raises the dimension by two and satisfies the relation

$$d(w) = h' - h'' - v \circ f' - g'' \circ u.$$

The objects X', X'' are said to be *homotopy equivalent* (and we write $X' \simeq X''$) if there are morphisms

$$(f', g', h')\colon X' \to X'', \quad (f'', g'', h'')\colon X'' \to X'$$

such that the compositions $(f'', g'', h'') \circ (f', g', h')$, $(f', g', h') \circ (f'', g'', h'')$ are homotopic to the identity maps.

THEOREM. *The correspondence $X \longmapsto \mathcal{D}_\infty(X)$ defines a functor $\mathcal{D}_\infty\colon \mathcal{X} \to Ens_*$ from the category $\mathcal{X}$ to the category of pointed sets.*

PROOF. To a morphism $(f, g, h)\colon X \to Y$ in the category $\mathcal{X}$ we assign a map

$$\mathcal{D}_\infty(f, g, h)\colon \mathcal{D}_\infty(X) \to \mathcal{D}_\infty(Y)$$

as follows. Let $\pi = \{\pi(n)\}$, $\pi(n)\colon X^{\otimes n} \to X$, be an A_∞-structure on the chain complex X. We define the family $\phi = \{\phi(n)\}$ of operations $\phi(n)\colon Y^{\otimes n} \to Y$ by setting

$$\phi(n) = f\pi(n)g^{\otimes n} + (f\pi \cup h\pi)(n)g^{\otimes n} + (f\pi \cup h\pi \cup h\pi)(n)g^{\otimes n} + \dots.$$

Straightforward computations show that the relations

$$d(\phi) + \phi \cup \phi = 0$$

hold, i.e., ϕ is an A_∞-structure on Y.

Let us show that this correspondence takes equivalent A_∞-structures to equivalent ones. Suppose that $\pi \sim \pi'$ and $c = \{c(n)\}$ is the corresponding family of maps $c(n)\colon X^{\otimes n} \to X$ that satisfies the relation

$$d(c) = c \cup \pi - \pi' \cup c.$$

We define the family $b = \{b(n)\}$ of maps $b(n)\colon Y^{\otimes n} \to Y$ by setting

$$\begin{aligned} b(n) = {} & fc(n)g^{\otimes n} + (fc \cup h\pi - f\pi' \cup hc)(n)g^{\otimes n} \\ & + (f\pi' \cup h\pi' \cup hc - f\pi' \cup hc \cup h\pi + fc \cup h\pi \cup h\pi)(n)g^{\otimes n} + \dots. \end{aligned}$$

Straightforward computations show that the following relations hold:

$$d(b) = b \cup \phi - \phi' \cup b,$$

i.e., the A_∞-structures ϕ and ϕ' are equivalent. □

It is similarly shown that homotopic morphisms

$$(f', g', h'), (f'', g'', h'')\colon X \to Y$$

induce the same map

$$\mathcal{D}_\infty(f', g', h') = \mathcal{D}_\infty(f'', g'', h'')\colon \mathcal{D}_\infty(X) \to \mathcal{D}_\infty(Y).$$

Hence we obtain the following assertion.

THEOREM. *For any chain equivalent complexes X' and X'', there is a bijection*

$$\mathcal{D}_\infty(X') \longleftrightarrow \mathcal{D}_\infty(X'').$$

In particular, if the ground ring is a field, then for any chain complex X there is a bijection

$$\mathcal{D}_\infty(X) \longleftrightarrow \mathcal{D}_\infty(H_*(X)).$$

Now let us define the functor $\mathcal{D}$ for A_∞-algebras. Let A be a differential bigraded A_∞-algebra, $A = \sum_{n,m} A^n_m$. The differential preserves the upper grading and decreases the lower grading by one. The operations $\pi_n\colon A^{\otimes n+2} \to A$ preserve the upper grading and increase the lower grading by n.

A family $\phi = \{\phi_n\}_{n\geq 1}$ of elements $\phi_n \in A^n_{-1}$ is called a *twisting family* if the following relations hold:

$$d(\phi) + \sum_{n\geq 0} \pi_n(\phi \otimes \cdots \otimes \phi) = 0.$$

Note that giving a twisting family ϕ is equivalent to giving an element ψ in $B(A)$ that has lower dimension zero and satisfies the relations

$$d(\psi) = 0, \quad \nabla(\psi) = \psi \otimes \psi.$$

Two twisting families ϕ' and ϕ'' are said to be *equivalent* (and we write $\phi' \sim \phi''$) if there is a family $c = \{c_n\}_{n\geq 0}$ of elements $c_n \in A^n_0$ such that $c_0 = 1$ and

$$d(c) = \sum_{n\geq 0} \pi_n(c \otimes \phi' \otimes \cdots \otimes \phi' + \cdots + \phi'' \otimes \cdots \otimes \phi'' \otimes c).$$

Note that defining an equivalence c is the same as giving an element b in $B(A)$ satisfying the relations

$$d(b) = \psi' - \psi'', \quad \nabla(b) = b \otimes \psi' + \psi'' \otimes b.$$

Denote the set of equivalence classes of twisting families with respect to the equivalence relation defined above by $\mathcal{D}(A)$.

It is readily seen that the correspondence $A \longmapsto \mathcal{D}(A)$ defines a functor $\mathcal{D}$ from the category of A_∞-algebras to the category of pointed sets. Moreover, homotopic A_∞-maps $f, g\colon A' \to A''$ induce the same map

$$\mathcal{D}(f) = \mathcal{D}(g)\colon \mathcal{D}(A') \to \mathcal{D}(A'').$$

Thus we obtain the following assertion.

THEOREM. *If A', A'' are homotopy equivalent A_∞-algebras, then there is a bijection*

$$\mathcal{D}(A') \longleftrightarrow \mathcal{D}(A'').$$

In particular, if the ground ring is a field, then for any A_∞-algebra A there is a bijection

$$\mathcal{D}(A) \longleftrightarrow \mathcal{D}(A_*),$$

where the homology $A_* = H_*(A)$ is regarded as an A_∞-algebra.

Note that if two A_∞-structures π and π' on A_* are equivalent, then $\pi(2) = \pi'(2)$. Hence $\mathcal{D}_\infty(A_*)$ is decomposed into the union of the sets $\mathcal{D}_\infty(A_*, \pi(2))$ of equivalence classes of A_∞-structures with a prescribed algebra structure $\pi(2)$. Let us consider the question of computing these sets.

Let X be a graded module with a prescribed algebra structure. As above, this means that we are given an operation $\pi(2)\colon X \otimes X \to X$ that lowers the dimension by one and satisfies the relation

$$\pi(2)(\pi(2) \otimes 1 + 1 \otimes \pi(2)) = 0$$

or, more concisely, $\pi(2) \cup \pi(2) = 0$.

Denote by

$$C^*(X; X) = \sum_n C^n(X; X), \quad C^n(X; X) = \mathrm{Hom}(X^{\otimes n}; X),$$

the Hochschild complex with the differential defined by

$$d(f) = f \cup \pi(2) + (-1)^{n+1}\pi(2) \cup f,$$

where $n = \dim(f)$.

We have the multiplication

$$\cup: C^n(X;X) \otimes C^m(X;X) \to C^{n+m-1}(X;X)$$

defined by

$$f \cup g = \sum f(1 \otimes \cdots \otimes g \otimes \cdots \otimes 1),$$

where the summation is over all possible positions of the factor g.

However, this multiplication does not commute with the differential and is not associative. To overcome this difficulty, we shall use the notion of $\vee$-algebra.

Direct verification shows that the restriction of the multiplication

$$\cup: C^*(X;X) \otimes C^*(X;X) \to C^*(X;X)$$

to the subcomplex $C^*(X;X) \vee C^*(X;X)$ endows $C^*(X;X)$ with a $\vee$-algebra structure.

It is readily seen that the family $\{\pi(n)\}_{n\geq 2}$ of maps $\pi(n): X^{\otimes n} \to X$ endows X with an A_∞-algebra structure if and only if the family $\pi = \{\pi(n)\}_{n\geq 3}$ is a twisting family for the $\vee$-algebra $C^*(X;X)$, i.e., the relations $d(\pi) + \pi \cup \pi = 0$ hold.

Thus we can employ the methods of computing the functor $\mathcal{D}$.

Denote by $J^n H^*_{-1}(X,X)$, $n \geq 3$, the set of all elements $h(n)$ of the Hochschild homology $H^n_{-1}(X,X)$ with the property that there are elements $h(i)$ of $H^i_{-1}(X,X)$, $n < i < 2n$, such that for any $m \geq 2n$ we have

$$\sum_{k\geq 2} m_{k-2}(h(i_1) \otimes \cdots \otimes h(i_k)) = 0,$$

where the summation is over all $n \leq i_1, \ldots, i_k \leq 2n-1$.

By repeating the arguments from the previous chapter concerning the computation of the functor $\mathcal{D}$, we obtain the following theorem.

THEOREM. *Suppose that X is a finitely generated graded module with a prescribed algebra structure $\pi(2): X \otimes X \to X$. Then for any $n \geq 1$ there is a short exact sequence*

$$* \to \mathcal{D}^{n+1}_\infty(X,\pi(2)) \xrightarrow{i^{n+1}} \mathcal{D}^n_\infty(X,\pi(2)) \xrightarrow{p^n} J^n H^*_{-1}(X,X) \to *,$$

where $\mathcal{D}^n_\infty(X,\pi(2))$ is the subset of $\mathcal{D}_\infty(X,\pi(2))$ generated by the A_∞-structures $\pi = \{\pi(i)\}$ such that $\pi(i) = 0$ for $3 < i < n$.

In particular, $\mathcal{D}_\infty(X,\pi(2)) = 0$ if and only if $J^n H^*_{-1}(X,X) = 0$ for any $n \geq 3$.

This theorem answers the question of whether it is possible to extend a given algebra structure to a nontrivial A_∞-algebra structure.

CHAPTER 5

Operads and Algebras over Operads

5.1. Operads in the category of chain complexes

DEFINITION. A *symmetric family* $\mathcal{E}$ *in the category of chain complexes* is a family $\mathcal{E} = \{\mathcal{E}(j)\}_{j\geq 1}$ of chain complexes $\mathcal{E}(j)$ acted on by the symmetric groups Σ_j.

For symmetric families $\mathcal{E}'$, $\mathcal{E}''$, we define the symmetric families

$$\mathcal{E}' \otimes \mathcal{E}'', \quad \mathcal{E}' \times \mathcal{E}''$$

by setting

$$(\mathcal{E}' \otimes \mathcal{E}'')(j) = \mathcal{E}'(j) \otimes \mathcal{E}''(j)$$

and letting $(\mathcal{E}' \times \mathcal{E}'')(j)$ be the quotient of the free Σ_j-module generated by

$$\sum_{j_1+\cdots+j_k=j} \mathcal{E}'(k) \otimes \mathcal{E}''(j_1) \otimes \cdots \otimes \mathcal{E}''(j_k)$$

modulo the equivalence generated by the relations

$$x'\sigma \otimes x_1'' \otimes \cdots \otimes x_k'' \sim x' \otimes x''_{\sigma^{-1}(1)} \otimes \cdots \otimes x''_{\sigma^{-1}(k)} \cdot \sigma(j_1, \ldots, j_k),$$
$$x' \otimes x_1''\sigma_1 \otimes \cdots \otimes x_k''\sigma_k \sim x' \otimes x_1'' \otimes \cdots \otimes x_k'' \cdot (\sigma_1 \times \cdots \times \sigma_k),$$

where $\sigma(j_1, \ldots, j_k)$ is the permutation of the set of j elements which is defined by partitioning this set into k blocks of $j_1, \ldots, j_k$ elements, respectively, and acting on the blocks by the permutation σ, and $\sigma_1 \times \cdots \times \sigma_k$ is the image of $(\sigma_1, \ldots, \sigma_k)$ under the natural embedding $\Sigma_{j_1} \times \cdots \times \Sigma_{j_k} \to \Sigma_j$.

It is readily seen that for any symmetric families $\mathcal{E}', \mathcal{E}'', \mathcal{F}', \mathcal{F}''$, there is a natural permutation operator

$$T\colon (\mathcal{E}' \otimes \mathcal{E}'') \times (\mathcal{F}' \otimes \mathcal{E}'') \to (\mathcal{E}' \times \mathcal{F}') \otimes (\mathcal{E}'' \times \mathcal{F}'').$$

DEFINITION. An *operad* is a symmetric family $\mathcal{E}$ given together with a map of symmetric families $\gamma\colon \mathcal{E} \times \mathcal{E} \to \mathcal{E}$ that satisfies the relation $\gamma(\gamma \times 1) = \gamma(1 \times \gamma)$ or, equivalently, makes the following diagram commutative:

$$\begin{array}{ccc} \mathcal{E} \times \mathcal{E} \times \mathcal{E} & \xrightarrow{\gamma \times 1} & \mathcal{E} \times \mathcal{E} \\ {\scriptstyle 1\times\gamma}\downarrow & & \downarrow{\scriptstyle \gamma} \\ \mathcal{E} \times \mathcal{E} & \xrightarrow{\gamma} & \mathcal{E} \end{array}$$

If, in addition, we are given an element $1 \in \mathcal{E}(1)$ such that $\gamma(1 \otimes x) = \gamma(x \otimes 1^{\otimes k}) = x$ for all $x \in \mathcal{E}(k)$, then we say that $\mathcal{E}$ is an operad with identity.

DEFINITION. A *map of operads* $f\colon \mathcal{E}' \to \mathcal{E}''$ is a map of symmetric families such that the following diagram is commutative:

$$\begin{array}{ccc} \mathcal{E}' \times \mathcal{E}' & \xrightarrow{\gamma'} & \mathcal{E}' \\ {\scriptstyle f\times f}\downarrow & & \downarrow{\scriptstyle f} \\ \mathcal{E}'' \times \mathcal{E}'' & \xrightarrow{\gamma''} & \mathcal{E}'' \end{array}$$

If $\mathcal{E}'$ and $\mathcal{E}''$ are operads with identities $1'$ and $1''$, respectively, then it is additionally required that $f(1') = 1''$.

DEFINITION. An operad $\mathcal{E}$ is called a *Hopf operad* if, in addition to the operation $\gamma\colon \mathcal{E} \times \mathcal{E} \to \mathcal{E}$, we are given an operation $\nabla\colon \mathcal{E} \to \mathcal{E} \otimes \mathcal{E}$ which is associative and satisfies the Hopf relation, i.e., such that the following diagram is commutative:

$$\begin{array}{ccccc} \mathcal{E} \times \mathcal{E} & \xrightarrow{\nabla\times\nabla} & (\mathcal{E}\otimes\mathcal{E}) \times (\mathcal{E}\otimes\mathcal{E}) & \xrightarrow{T} & (\mathcal{E}\times\mathcal{E}) \otimes (\mathcal{E}\times\mathcal{E}) \\ {\scriptstyle \gamma}\downarrow & & & & \downarrow{\scriptstyle \gamma\otimes\gamma} \\ \mathcal{E} & \xrightarrow{\nabla} & \mathcal{E}\otimes\mathcal{E} & \xrightarrow{=} & \mathcal{E}\otimes\mathcal{E} \end{array}$$

DEFINITION. Let $\mathcal{E}$ be an operad, and let $\mathcal{F}$ be a symmetric family. We say that $\mathcal{E}$ *acts on* $\mathcal{F}$ *on the left* (*on the right*) if we are given a map

$$\mu'\colon \mathcal{E} \times \mathcal{F} \to \mathcal{F} \quad (\mu''\colon \mathcal{F} \times \mathcal{E} \to \mathcal{F})$$

such that

$$\mu'(\gamma \times 1) = \mu'(1 \times \mu') \quad (\mu''(1 \times \gamma) = \mu''(\mu'' \times 1))$$

or, equivalently, the following diagram is commutative:

$$\begin{array}{ccc} \mathcal{E}\times\mathcal{E}\times\mathcal{F} & \xrightarrow{\gamma\times 1} & \mathcal{E}\times\mathcal{F} \\ {\scriptstyle 1\times\mu'}\downarrow & & \downarrow{\scriptstyle \mu'} \\ \mathcal{E}\times\mathcal{F} & \xrightarrow{\mu'} & \mathcal{F} \end{array} \qquad \left(\begin{array}{ccc} \mathcal{F}\times\mathcal{E}\times\mathcal{E} & \xrightarrow{1\times\gamma} & \mathcal{F}\times\mathcal{E} \\ {\scriptstyle \mu''\times 1}\downarrow & & \downarrow{\scriptstyle \mu''} \\ \mathcal{F}\times\mathcal{E} & \xrightarrow{\mu''} & \mathcal{F} \end{array}\right)$$

Let us consider some examples of operads.

1. The Hopf operad $E_0 = \{E_0(j)\}$, where $E_0(j)$ is the free module with a single zero-dimensional generator $e(j)$ and with the trivial action of the symmetric group Σ_j, i.e., $E_0(j) \cong R$. The operation $\gamma\colon E_0 \times E_0 \to E_0$ is given by

$$\gamma(e(k) \otimes e(j_1) \otimes \cdots \otimes e(j_k)) = e(j_1 + \cdots + j_k).$$

The operation $\nabla\colon E_0 \to E_0 \otimes E_0$ is given by

$$\nabla(e(j)) = e(j) \otimes e(j).$$

It is readily seen that the required relations hold.

2. The Hopf operad $A = \{A(j)\}$, where $A(j)$ is a free Σ_j-module with a single zero-dimensional generator $a(j)$, i.e., $A(j) \cong R(\Sigma_j)$. The operation $\gamma\colon A \times A \to A$ is given by

$$\gamma(a(k) \otimes a(j_1) \otimes \cdots \otimes a(j_k)) = a(j_1 + \cdots + j_k).$$

The operation $\nabla\colon A \to A \otimes A$ is given by

$$\nabla(a(j)) = a(j) \otimes a(j).$$

It is readily seen that the required relations hold.

3. Let $\mathcal{E}$ be a symmetric family. We define the suspension $S\mathcal{E}$ by setting $(S\mathcal{E})(j) = S^{j-1}\mathcal{E}(j)$ (the $(j-1)$-fold suspension of $\mathcal{E}(j)$). It is clear that if $\mathcal{E}$ is an operad, then $S\mathcal{E}$ is also an operad.

4. For any operads $\mathcal{E}'$, $\mathcal{E}''$, their tensor product $\mathcal{E}' \otimes \mathcal{E}''$ is clearly an operad.

5. Let $\mathcal{F} = \{\mathcal{F}(j)\}$ be a symmetric family of chain complexes. By $T\mathcal{F}$ denote the free operad generated by $\mathcal{F}$, namely,

$$T\mathcal{F} = \sum_n \mathcal{F}^{\times n} / \sim,$$

where the equivalence $\sim$ is generated by the relation $1 \times x \sim x \times 1 \sim x$, and the operad structure is defined by the maps $\mathcal{F}^{\times n} \times \mathcal{F}^{\times m} \to \mathcal{F}^{\times n+m}$.

The operad $T\mathcal{F}$ possesses the universal property, i.e., for any operad $\mathcal{E}$ and any map of symmetric families $f: \mathcal{F} \to \mathcal{E}$, there is a unique operad map $\widetilde{f}: T\mathcal{F} \to \mathcal{E}$ such that the following diagram is commutative:

$$\begin{array}{ccc} \mathcal{F} & \xrightarrow{i} & T\mathcal{F} \\ & \searrow^{f} & \downarrow \widetilde{f} \\ & & \mathcal{E} \end{array}$$

6. Let $\mathcal{F}$ be a symmetric family. Denote by $\Sigma\mathcal{F}$ the symmetric family consisting of free Σ_j-complexes generated by the complexes $\mathcal{F}(j)$. This family $\Sigma\mathcal{F}$ will be referred to as the *Σ-free family generated by $\mathcal{F}$*. It is clear that if $\mathcal{E}$ is an operad, then $\Sigma\mathcal{E}$ is also an operad. For example, we have $A = \Sigma E_0$.

7. We define the operad A_∞ as the Σ-free operad which is free as a graded operad and which is generated by the elements $\pi_n \in A_\infty(n+2)$ of dimension n. The differential on these elements is defined by

$$d(\pi_{n+1}) = \sum_{i=0}^{n} (-1)^\epsilon \pi_i \otimes (1 \otimes \cdots \otimes \pi_{n-i} \otimes \cdots \otimes 1),$$

where the summation is also over all possible positions (indexed by k) of the factor π_{n-i}, and $\epsilon = nk + ik + n + k$.

In particular, if $n = 0$, then we obtain

$$d(\pi_1) = \pi_0 \otimes (1 \otimes \pi_0 - \pi_0 \otimes 1),$$

which is the homotopy associativity relation for π_0.

If $n = 1$, then we obtain

$$d(\pi_2) = \pi_1 \otimes (\pi_0 \otimes 1 \otimes 1 - 1 \otimes \pi_0 \otimes 1 + 1 \otimes 1 \otimes \pi_0) - \pi_0 \otimes (\pi_1 \otimes 1 + 1 \otimes \pi_1),$$

which is, in a sense, the homotopy associativity relation between π_0 and π_1. And so on.

THEOREM. *There is a natural projection of operads $A_\infty \to A$, and this projection is a chain equivalence.*

PROOF. We define the projection $p: A_\infty \to A$ by setting on the generators $p(\pi_0) = a(2)$ and $p(\pi_n) = 0$ for $n > 0$. Let us show that p is a chain equivalence. To this end, we define a map $q: A \to A_\infty$ by setting

$$q(a(j)) = \pi_0 \otimes (1 \otimes \pi_0) \ldots (1 \otimes \cdots \otimes \pi_0).$$

It is clear that $p \circ q = \mathrm{Id}$. Consider the map $q \circ p \colon A_\infty \to A_\infty$. Let us define a homotopy $h \colon A_\infty \to A_\infty$ between $q \circ p$ and Id. Note that the elements of A_∞ can be uniquely represented as sums of the elements

$$\phi = \pi_{i_1} \otimes (1 \otimes \cdots \otimes \pi_{i_2} \otimes \cdots \otimes 1) \ldots (1 \otimes \cdots \otimes \pi_{i_k} \otimes \cdots \otimes 1),$$

where π_{i_j} occupies the position indexed n_j, and $1 = n_1 \leq \cdots \leq n_k$. We set $h(\phi) = 0$ if $i_1 > 0$, $h(\pi_0) = 0$, $h(\pi_0 \otimes (1 \otimes \phi) = \pi_0 \otimes (1 \otimes h(\phi))$, and

$$\begin{aligned} &h(\pi_0 \otimes (\pi_{i_2} \otimes 1) \ldots (1 \otimes \cdots \otimes \pi_{i_k} \otimes \cdots \otimes 1) \\ &= \pi_{i_2+1} \ldots (1 \otimes \cdots \otimes \pi_{i_k} \otimes \cdots \otimes 1) + h(\pi_0 \otimes (1 \otimes \pi_{i_2}) \ldots (1 \otimes \cdots \otimes \pi_{i_k} \otimes \cdots \otimes 1). \end{aligned}$$

Straightforward computations show that $d(h) = q \circ p - \mathrm{Id}$ and, besides, $h \circ q = 0$, $p \circ h = 0$, $h \circ h = 0$.

We define the map $\nabla \colon A_\infty \to A_\infty \otimes A_\infty$, which makes A_∞ a Hopf operad, by setting

$$\nabla(\pi_0) = \pi_0 \otimes \pi_0; \qquad \nabla(\pi_n) = (1 \otimes h + h \otimes q \circ p)\nabla(d(\pi_n)).$$

Straightforward computations show that the required relations hold. □

8. We define the operad $S_\infty = \{S_\infty(j)\}$ as the quotient of the operad A_∞ by the equivalence relation generated by the relations

$$\sum_\sigma (-1)^\epsilon \pi_n \sigma(i, n+2-i) = 0,$$

where n and i are fixed, and the summation is over all $(i, n+2-i)$-shuffles σ. The above theorem implies that the projection $S_\infty \to E_0$ is a chain equivalence.

9. Let us define an acyclic operad $E_\infty = \{E_\infty(j)\}$ with free actions of symmetric groups which is free as a graded operad.

We shall construct the operad E_∞ by induction over the dimension. To do this, we shall define for each n a family F^n and a free operad E^n which is generated by F^n and is acyclic through dimension $n-1$.

Let $F^0(2) = R(\Sigma_2)$, $F^0(j) = 0$ for $j > 2$, and $E^0 = TF^0$. We have the projection $p \colon E^0 \to E_0$ and the short exact sequence

$$0 \to \mathrm{Ker}(p) \xrightarrow{i} E^0 \xrightarrow{p} E_0 \to 0.$$

We set $F^1_0 = F^0_0$, $F^1_1 = \Sigma\,\mathrm{Ker}(p)$, and $E^1 = TF^1$. Let the differential in E^1 be defined on the generators $u_1 \in F^1$ by the formula $d(u_1) = i(u_1)$. It is clear that the zero-dimensional homology of E^1 is isomorphic to E_0.

Assume that the family F^n and the operad E^n are already constructed. We define the family F^{n+1} and the operad E^{n+1} by setting $F^{n+1}_i = F^n_i$ for $i \neq n+1$ and $F^{n+1}_{n+1} = \Sigma\,\mathrm{Ker}\{d^n \colon E^n_n \to E^n_{n-1}\}$, $E^{n+1} = TF^{n+1}$; the differential d^{n+1} on the elements of dimension less than $n+1$ is determined by d^n, and on the generators $u_{n+1} \in F^{n+1}_{n+1}$ it is defined by the formula $d^{n+1}(u_{n+1}) = i(u_{n+1})$, where $i \colon \mathrm{Ker}(d^n) \to E^n_n$ is the inclusion map. It is clear that the operad E^{n+1} is acyclic though dimension n and, regarded as a graded operad, it is free and Σ-free. Besides, we have an inclusion of operads $E^n \to E^{n+1}$. Passing to the limit, we obtain the required operad E_∞.

Note that all acyclic operads with free actions of symmetric groups consist of Σ_j-homotopy equivalent chain complexes. These operads are called *E_∞-operads*.

The property of the operad E_∞ being free allows us to construct an operad map $f: E_\infty \to E$ for any acyclic operad E. In particular, we can construct an operad map $\nabla: E_\infty \to E_\infty \otimes E_\infty$, which endows E_∞ with a Hopf operad structure.

10. For an arbitrary chain complex X, we define the operads $\mathcal{E}_X$ and $\mathcal{E}^X$ by setting

$$\mathcal{E}_X(j) = \mathrm{Hom}(X^{\otimes j}; X); \quad \mathcal{E}^X(j) = \mathrm{Hom}(X; X^{\otimes j}).$$

The symmetric groups act by permuting the factors in $X^{\otimes j}$, and the operad structure is given by

$$\gamma_X(f \otimes g_1 \otimes \cdots \otimes g_k) = f \circ (g_1 \otimes \cdots \otimes g_k), \quad f \in \mathcal{E}_X(k), \ g_i \in \mathcal{E}_X(j_i);$$
$$\gamma^X(f \otimes g_1 \otimes \cdots \otimes g_k) = (g_1 \otimes \cdots \otimes g_k) \circ f, \quad f \in \mathcal{E}^X(k), \ g_i \in \mathcal{E}^X(j_i).$$

Similarly, for two chain complexes X and Y, we define the symmetric families $\mathcal{F}_{X,Y}$ and $\mathcal{F}^{X,Y}$,

$$\mathcal{F}_{X,Y}(j) = \mathrm{Hom}(X^{\otimes j}; Y), \quad \mathcal{F}^{X,Y}(j) = \mathrm{Hom}(X; Y^{\otimes j}),$$

and the actions

$$\mu': \mathcal{E}_Y \times \mathcal{F}_{X,Y} \to \mathcal{F}_{X,Y}, \qquad \mu'': \mathcal{F}_{X,Y} \times \mathcal{E}_X \to \mathcal{F}_{X,Y};$$
$$\mu': \mathcal{E}^X \times \mathcal{F}^{X,Y} \to \mathcal{F}^{X,Y}, \qquad \mu'': \mathcal{F}^{X,Y} \times \mathcal{E}^Y \to \mathcal{F}^{X,Y}.$$

11. It is readily seen that if $\mathcal{E} = \{\mathcal{E}(j)\}$ is an operad in the category of topological spaces, then the family $C_*(\mathcal{E}) = \{C_*(\mathcal{E}(j))\}$ of singular chain complexes is an operad in the category of chain complexes. In particular, for the Boardman–Vogt operad of little n-cubes E_n, the family $C_*(E_n)$ is an operad in the category of chain complexes. For the operad E_∞, which is the limit of the little n-cubes operads, the family $C_*(E_\infty)$ is an E_∞-operad in the category of chain complexes.

12. Another example of an E_∞-operad is furnished by simplicial resolutions of symmetric groups. Denote by $E\Sigma_*(j)$ the free simplicial resolution of the symmetric group Σ_j, i.e.,

$$E\Sigma_*(j) : \Sigma_j \leftarrow \Sigma_j \times \Sigma_j \leftarrow \dots .$$

The maps $\gamma: \Sigma_k \times \Sigma_{j_1} \times \cdots \times \Sigma_{j_k} \to \Sigma_{j_1+\cdots+j_k}$ induce an operad structure

$$\gamma_*: E\Sigma_*(k) \times E\Sigma_*(j_1) \times \cdots \times E\Sigma_*(j_k) \to E\Sigma_*(j_1 + \cdots + j_k).$$

Thus $E\Sigma_*$ is an acyclic operad with free action of symmetric groups in the category of simplicial sets.

By taking the chain complexes $C_*(E\Sigma_*)$, we obtain an E_∞-operad in the category of chain complexes. Denote it by $E\Sigma$.

Similarly, the homology $H_*(\mathcal{E}) = \{H_*(\mathcal{E}(j))\}$ of a topological operad $\mathcal{E}$ is an operad in the category of graded modules.

Note that for any chain operad $\mathcal{E}$, the operad $\mathcal{E} \otimes E\Sigma$ has the same homology and is endowed with a free action of symmetric groups. The projection $E\Sigma \to E_0$ induces a projection $\mathcal{E} \otimes E\Sigma \to \mathcal{E}$. Hence $\mathcal{E} \otimes E\Sigma$ can be regarded as a Σ-free resolution of $\mathcal{E}$. If $\mathcal{E}$ is an acyclic operad, then $\mathcal{E} \otimes E\Sigma$ is an E_∞-operad.

13. Examples of operads are provided by realizations of simplicial objects in the category of operads. Namely, let $\mathcal{E}_*$ be a simplicial object in the category of symmetric families of chain complexes. We define its *realization* $|\mathcal{E}_*|$ as the

symmetric family with $|\mathcal{E}_*|(j) = |\mathcal{E}_*(j)|$. For two symmetric families $\mathcal{E}'_*$ and $\mathcal{E}''_*$, there is a map

$$|\mathcal{E}'_*(k)| \otimes |\mathcal{E}''_*(j_1)| \otimes \cdots \otimes |\mathcal{E}''_*(j_k)| \to |\mathcal{E}'_*(k) \otimes \mathcal{E}''_*(j_1) \otimes \cdots \otimes \mathcal{E}''_*(j_k)|$$

which is obtained by iterating the Alexander–Whitney map, and which defines a map of symmetric families

$$|\mathcal{E}'_*| \times |\mathcal{E}''_*| \to |\mathcal{E}'_* \times \mathcal{E}''_*|.$$

If $\mathcal{E}_*$ is a simplicial object in the category of operads, $\gamma_*: \mathcal{E}_* \times \mathcal{E}_* \to \mathcal{E}_*$, then the realization $|\mathcal{E}_*|$ is an operad, in which the operad structure is defined by the composite map

$$|\mathcal{E}_*| \times |\mathcal{E}_*| \to |\mathcal{E}_* \times \mathcal{E}_*| \xrightarrow{|\gamma_*|} |\mathcal{E}_*|.$$

By analogy with the notion of bioperad in the category of topological spaces, we can define the notion of bioperad in the category of chain complexes. Namely, for two symmetric families $\mathcal{E} = \{\mathcal{E}(n)\}$ and $\mathcal{F} = \{\mathcal{F}(n)\}$ of chain complexes, we set

$$\mathcal{E}(k) * \mathcal{F}(n) = \mathcal{E}(k) \otimes \mathcal{F}(n)^{\otimes k}.$$

It is readily seen that there are natural permutation operators

$$T: (\mathcal{E}'(k) \otimes \mathcal{E}''(k)) * (\mathcal{F}'(n) \otimes \mathcal{F}''(n)) \to (\mathcal{E}'(k) * \mathcal{F}'(n)) \otimes (\mathcal{E}''(k) * \mathcal{F}''(n)).$$

DEFINITION. We say that a symmetric family $\mathcal{E}$ is a *bioperad* if there are given maps

$$\pi: \mathcal{E}(n) \otimes \mathcal{E}(m) \to \mathcal{E}(n+m), \quad \gamma: \mathcal{E}(k) * \mathcal{E}(n) \to \mathcal{E}(kn),$$

satisfying the following conditions:

1. The maps π and γ are compatible with the actions of the symmetric groups, i.e.,

$$\pi(x\sigma \otimes y\tau) = \pi(x \otimes y)\pi(\sigma \otimes \tau),$$

where $x \in \mathcal{E}(n)$, $y \in \mathcal{E}(m)$, $\sigma \in \Sigma_n$, $\tau \in \Sigma_m$;

$$\gamma(x\sigma \otimes y_1\tau_1 \otimes \cdots \otimes y_k\tau_k) = \gamma(x \otimes y_{\sigma^{-1}(1)} \otimes \cdots \otimes y_{\sigma^{-1}(k)})\gamma(\sigma \otimes \tau_1 \cdots \otimes \tau_k),$$

where $x \in \mathcal{E}(k)$, $y_i \in \mathcal{E}(n)$, $\sigma \in \Sigma_k$, $\tau_i \in \Sigma_n$.

2. The maps π and γ are associative, i.e., the following diagrams are commutative:

$$\begin{array}{ccc} \mathcal{E}(k) \otimes \mathcal{E}(n) \otimes \mathcal{E}(m) & \xrightarrow{\pi \otimes 1} & \mathcal{E}(k+n) \otimes \mathcal{E}(m) \\ {\scriptstyle 1\otimes\pi}\downarrow & & \downarrow{\scriptstyle \pi} \\ \mathcal{E}(k) \otimes \mathcal{E}(n+m) & \xrightarrow{\pi} & \mathcal{E}(k+n+m) \end{array}$$

$$\begin{array}{ccc} \mathcal{E}(k) * \mathcal{E}(n) * \mathcal{E}(m) & \xrightarrow{\gamma * 1} & \mathcal{E}(kn) * \mathcal{E}(m) \\ {\scriptstyle 1*\gamma}\downarrow & & \downarrow{\scriptstyle \gamma} \\ \mathcal{E}(k) * \mathcal{E}(nm) & \xrightarrow{\gamma} & \mathcal{E}(knm) \end{array}$$

3. We are given an identity element $1 \in \mathcal{E}(1)$ such that

$$\gamma(1 * x) = \gamma(x * 1) = x, \quad x \in \mathcal{E}(n).$$

4. The distributivity relations hold, which means the commutativity of the diagrams

$$\begin{array}{ccc}
(\mathcal{E}(k') \otimes \mathcal{E}(k'')) * \mathcal{E}(n) & \xrightarrow{\pi * 1} & \mathcal{E}(k'+k'') * \mathcal{E}(n) \\
\downarrow & & \downarrow{\scriptstyle \gamma} \\
(\mathcal{E}(k') * \mathcal{E}(n)) \otimes (\mathcal{E}(k'') * \mathcal{E}(n)) & & \mathcal{E}((k'+k'')n) \\
{\scriptstyle \gamma\otimes\gamma}\downarrow & & \downarrow{\scriptstyle =} \\
\mathcal{E}(k'n) \otimes \mathcal{E}(k''n) & \xrightarrow{\pi} & \mathcal{E}(k'n+k''n)
\end{array}$$

The diagrams

$$\begin{array}{ccc}
\mathcal{E}(k) * (\mathcal{E}(n) \otimes \mathcal{E}(m)) & \xrightarrow{1 * \pi} & \mathcal{E}(k) * \mathcal{E}(n+m) \\
{\scriptstyle \nabla * 1}\downarrow & & \downarrow{\scriptstyle \gamma} \\
(\mathcal{E}(k) \otimes \mathcal{E}(k)) * (\mathcal{E}(n) \otimes \mathcal{E}(m)) & & \mathcal{E}(k(n+m)) \\
\downarrow & & \uparrow{\scriptstyle \xi\circ\pi} \\
(\mathcal{E}(k) * \mathcal{E}(n)) \otimes (\mathcal{E}(k) * \mathcal{E}(m)) & \xrightarrow{\gamma\otimes\gamma} & \mathcal{E}(kn) \otimes \mathcal{E}(km)
\end{array}$$

are commutative up to certain isomorphisms

$$\xi\colon \mathcal{E}(k(n+m)) \to \mathcal{E}(k(n+m))$$

such that $\xi(x\sigma) = \xi(x)\xi(\sigma)$, $x \in \mathcal{E}(k(n+m))$, $\sigma \in \Sigma_{k(n+m)}$.

DEFINITION. Let $\mathcal{E}$ be a bioperad. We say that a symmetric family $\mathcal{F}$ is an *$\mathcal{E}$-multiplicative family* if there are given maps

$$\pi\colon \mathcal{F}(n) \otimes \mathcal{F}(m) \to \mathcal{F}(n+m), \quad \mu\colon \mathcal{E}(k) * \mathcal{F}(n) \to \mathcal{F}(kn)$$

that are compatible with the actions of the symmetric groups and satisfy the associativity and distributivity relations similar to those listed above.

A bioperad $\mathcal{E}$ is called an *E_∞-bioperad* if the spaces $\mathcal{E}(n)$ are contractible and the symmetric groups act freely on them.

It is clear that if $\mathcal{E}$ is a bioperad in the category of topological spaces, then the functor of singular chains with coefficients in a ring R takes it to a bioperad $C_*(\mathcal{E}, R)$ in the category of chain complexes. Furthermore, if $\mathcal{F}$ is an $\mathcal{E}$-multiplicative family in the category of topological spaces, then $C_*(\mathcal{F}, R)$ is a $C_*(\mathcal{E}, R)$-multiplicative family in the category of chain complexes.

5.2. Algebras and coalgebras over operads

DEFINITION. Let $\mathcal{E}$ be an operad. A chain complex X is called an *algebra* (a *coalgebra*) *over* $\mathcal{E}$ or just an *$\mathcal{E}$-algebra* (an *$\mathcal{E}$-coalgebra*) if an operad map

$$\xi\colon \mathcal{E} \to \mathcal{E}_X \quad (\xi\colon \mathcal{E} \to \mathcal{E}^X)$$

is given.

Let us reformulate this definition in the language of monads and comonads.

For a chain complex X and a symmetric family $\mathcal{E}$, we define the chain complexes $\underline{\mathcal{E}}(X)$, $\overline{\mathcal{E}}(X)$ by setting

$$\underline{\mathcal{E}}(X) = \sum_k \mathcal{E}(k) \otimes_{\Sigma_k} X^{\otimes k}, \quad \overline{\mathcal{E}}(X) = \prod_k \mathrm{Hom}_{\Sigma_k}(\mathcal{E}(k), X^{\otimes k}).$$

If $\mathcal{E}$ is an operad, then the operad structure on $\mathcal{E}$ determines a map

$$\gamma: \underline{\mathcal{E}}^2(X) = \underline{\mathcal{E}}(\underline{\mathcal{E}}(X)) \to \underline{\mathcal{E}}(X).$$

The correspondence $X \longmapsto \underline{\mathcal{E}}(X)$ is a monad on the category of chain complexes.

It is readily seen that a chain complex X is an algebra over an operad $\mathcal{E}$ if it is an algebra over the corresponding monad $\underline{\mathcal{E}}$.

Dually, the operad structure in $\mathcal{E}$ determines a map

$$\overline{\gamma}: \overline{\mathcal{E}}(X) \to \overline{\mathcal{E}}^2(X) = \overline{\mathcal{E}}(\overline{\mathcal{E}}(X)).$$

The correspondence $X \longmapsto \overline{\mathcal{E}}(X)$ is a comonad on the category of chain complexes.

It is readily seen that a chain complex X is a coalgebra over the operad $\mathcal{E}$ if it is a coalgebra over the corresponding comonad $\overline{\mathcal{E}}$.

Let us consider some examples of algebras and coalgebras over operads.

1. The algebras (coalgebras) over the operad A are just algebras (coalgebras).

2. The algebras (coalgebras) over the operad E_0 are just commutative algebras (coalgebras).

3. If X is an algebra (coalgebra) over an operad $\mathcal{E}$, then the suspension SX is an algebra (coalgebra) over the operad $S\mathcal{E}$.

4. If X', X'' are algebras (coalgebras) over operads $\mathcal{E}'$, $\mathcal{E}''$, respectively, then $X' \otimes X''$ is an algebra (coalgebra) over the operad $\mathcal{E}' \otimes \mathcal{E}''$.

5. If $\mathcal{E}$ acts on a topological space X, then the singular chain complex $C_*(X)$ is an algebra over the operad $C_*(\mathcal{E})$. In particular, the singular chain complex $C_*(\Omega^n X)$ of the loop space $\Omega^n X$ is an algebra over the operad $C_*(E_n)$. The singular chain complex $C_*(X)$ of an infinite loop space X is an algebra over the operad $C_*(E_\infty)$.

6. Endowing a complex X with the structure of an algebra over the operad A_∞ means giving an A_∞-algebra structure on X. Dually, endowing a complex X with the structure of a coalgebra over the operad A_∞ means giving an A_∞-coalgebra structure on X.

The presence of the diagonal map $\nabla: A_\infty \to A_\infty \otimes A_\infty$ allows us to extend the notion of twisting cochain to the case of A_∞-algebras and A_∞-coalgebras.

DEFINITION. Let K be an A_∞-coalgebra and A an A_∞-algebra. A *twisting cochain* $\phi: K \to A$ is a homomorphism that lowers the dimension by one and satisfies the relation

$$d(\phi) + A_\infty(\phi) = 0,$$

where $A_\infty(\phi) = \sum_{n \geq 0} \pi'_n(\phi \otimes \cdots \otimes \phi)\pi''_n$ and $\sum \pi'_n \otimes \pi''_n = \nabla(\pi_n)$.

7. Endowing a complex X with the structure of an algebra over the operad S_∞ means defining a commutative A_∞-algebra structure on X. Dually, endowing a complex X with the structure of a coalgebra over the operad S_∞ means defining a commutative A_∞-coalgebra structure on X.

Let us also introduce the notion of Lie algebra over an operad.

Let L be the suboperad of A generated by the element $b(2) = a(2) - a(2)T$, $T \in \Sigma_2$. Then the algebras (coalgebras) over the operad L are the Lie algebras (coalgebras).

Similarly, let L_n be the suboperad of $S^n A$ generated by the element $b_n(2) = s^n a(2) + (-1)^n s^n a(2)T$. Then the algebras (coalgebras) over the operad L_n are the n-Lie algebras (coalgebras).

Let $P_n = E_0 \times L_n$, and let the operad structure γ be defined by the corresponding structures in E_0, L_n and by the formula

$$\gamma(b_n(2) \otimes 1 \otimes e(2)) = e(2) \otimes b_n(2) \otimes 1 + e(2) \otimes 1 \otimes b_n(2) \cdot (213).$$

The operad P_n is called the *n-Poisson operad.* The algebras (coalgebras) over this operad are called *n-Poisson algebras* (*coalgebras*).

Note that since the homology $H_*(E_n)$ of the little n-cubes operad E_n is isomorphic to the n-Poisson operad P_n, the homology $H_*(\Omega^n X)$ of the loop space $\Omega^n X$ is an algebra over P_n.

DEFINITION. Let $\mathcal{E}$ be an operad. A chain complex Y is called a *Lie algebra over $\mathcal{E}$* if there is given a map $\phi\colon \overline{\mathcal{E}}(Y) \to Y$ that lowers the dimension by one and satisfies the relation $d(\phi) + \phi \cup \phi = 0$, where $\phi \cup \phi = \phi \circ \overline{\mathcal{E}}(\phi) \circ \overline{\gamma}(Y)$.

Passing to the dual situation, we define the notion of Lie coalgebra over an operad.

DEFINITION. Let $\mathcal{E}$ be an operad. A chain complex Y is call a *Lie coalgebra over $\mathcal{E}$* if there is given a map $\psi\colon Y \to \underline{\mathcal{E}}(Y)$ that lowers the dimension by one and satisfies the condition $d(\psi) + \psi \cup \psi = 0$.

Consider the operad $G_0 = \{G_0(j)\}$, where $G_0(2) = E_0(2)$ and $G_0(j) = 0$ for $j > 2$. The operad structure is trivial. It is readily seen that Lie algebras over this operad are $\vee$-algebras, and Lie coalgebras ove it are $\wedge$-coalgebras.

Directly by definition, we see that Lie algebras over the operad E_0 are A_∞-$\vee$-algebras, and Lie coalgebras over E_0 are A_∞-$\wedge$-coalgebras.

Let us consider the question of what natural structure there is on the chain complex of a topological space. As was noted above, the singular chain complex $C_*(X)$ of a topological space X possesses a natural coalgebra structure, which is defined on an n-dimensional singular simplex x by the formula

$$\nabla(x) = \sum_{i=0}^{n} d_{i+1} \dots d_n(x) \otimes d_0^i(x).$$

It induces an algebra structure $\cup\colon C^*(X) \otimes C^*(X) \to C^*(X)$ on the singular cochain complex $C^*(X)$ of a topological space X.

Along with the coalgebra structure

$$\nabla\colon C_*(X) \to C_*(X) \otimes C_*(X)$$

on the singular chain complex, there are comultiplications

$$\nabla_i\colon C_*(X) \to C_*(X) \otimes C_*(X)$$

that raise the dimension by i and satisfy the relations

$$d(\nabla_i) = \nabla_{i-1} + (-1)^i T \nabla_{i-1},$$

where $T\colon C_*(X)\otimes C_*(X)\to C_*(X)\otimes C_*(X)$ is the permutation operator.

Dually, along with the algebra structure

$$\cup\colon C^*(X)\otimes C^*(X)\to C^*(X)$$

on the cochain complex $C^*(X)$, there are multiplications

$$\cup_i\colon C^*(X)\otimes C^*(X)\to C^*(X)$$

such that $d(\cup_i)=\cup_{i-1}+(-1)^i\cup_{i-1}T$.

To describe all operations on the singular chain complex $C_*(X)$ of a topological space X and on the dual cochain complex $C^*(X)$, we consider the corresponding operad.

For $n\ge 0$, denote by Δ^n the normalized chain complex of the standard n-simplex. Then $\Delta^*=\{\Delta^n\}$ is a cosimplicial object in the category of chain complexes. Consider also the cosimplicial object $(\Delta^*)^{\otimes j}=\Delta^*\otimes\cdots\otimes\Delta^*$ and its realization

$$E^\Delta(j)=\operatorname{Hom}(\Delta^*;(\Delta^*)^{\otimes j}),$$

where Hom is taken in the category of cosimplicial objects.

The family $E^\Delta=\{E^\Delta(j)\}$ is an operad; the actions of the symmetric groups and the operad structure on it are defined in the same way as for the operad $\mathcal{E}^X$ described above, with the only difference that X must be replaced by Δ^*.

Note that since the complexes Δ^n are acyclic, the operad E^Δ is also acyclic.

THEOREM. *The chain complex $C_*(X)$ of a topological space X possesses a natural E^Δ-coalgebra structure $\tau\colon C_*(X)\to\overline{E}^\Delta(C_*(X))$ that satisfies the following universal property: if, for a certain operad $\mathcal{E}$, the chain complex $C_*(X)$ possesses a natural $\mathcal{E}$-coalgebra structure $\widetilde{\tau}\colon C_*(X)\to\overline{\mathcal{E}}(C_*(X))$, then there is a unique operad map $\xi\colon\mathcal{E}\to E^\Delta$ such that $\widetilde{\tau}=\overline{\xi}(C_*(X))\circ\tau$.*

PROOF. Our aim is to define the operations

$$\tau\colon E^\Delta(j)\otimes C_*(X)\to C_*(X)^{\otimes j}.$$

Let $x_n\in C_n(X)$, and let $y\in E^\Delta(j)=\operatorname{Hom}(\Delta^*;(\Delta^*)^{\otimes j})$. The element $x_n\in C_n(X)$ determines a chain map $\overline{x}_n\colon\Delta^n\to C_*(X)$ that takes the generator $u_n\in\Delta^n$ to x_n.

Note that there are operations $\tau^n\colon E^\Delta(j)\otimes\Delta^n\to(\Delta^n)^{\otimes j}$. We define the required operations τ by setting

$$\tau(y\otimes x_n)=(\overline{x}_n)^{\otimes j}\circ\tau^n(y\otimes u_n).$$

Then we have the commutative diagram

$$\begin{array}{ccc}
E^\Delta(j)\otimes C_*(X) & \xrightarrow{\ \tau\ } & C_*(X)^{\otimes j}\\
{\scriptstyle 1\otimes\overline{x}_n}\uparrow & & \uparrow{\scriptstyle(\overline{x}_n)^{\otimes j}}\\
E^\Delta(j)\otimes\Delta^n & \xrightarrow{\ \tau^n\ } & (\Delta^n)^{\otimes j}
\end{array}$$

It is readily seen that the required relations hold. □

The E^Δ-coalgebra structure on the chain complex $C_*(X)$ of a space X induces an E^Δ-algebra structure on the cochain complex

$$C^*(X)=\operatorname{Hom}(C_*(X);R).$$

The corresponding operations $\mu\colon E^{\Delta}(j)\otimes C^*(X)^{\otimes j}\to C^*(X)$ are defined by

$$\mu(y\otimes f_1\otimes\cdots\otimes f_j)(x)=(f_1\otimes\cdots\otimes f_j)\circ\tau(y\otimes x),$$

where $y\in E^{\Delta}(j)$, $f_i\colon C_*(X)\to R$, $x\in C_*(X)$. Thus we obtain the following theorem.

THEOREM. *The cochain complex $C^*(X)$ of a topological space X possesses a natural E^{Δ}-algebra structure $\mu\colon \underline{E}^{\Delta}(C^*(X))\to C^*(X)$, which satisfies the following universal property: if, for a certain operad $\mathcal{E}$, the chain complex $C_*(X)$ of a topological space X possesses a natural $\mathcal{E}$-algebra structure $\widetilde{\mu}\colon \underline{\mathcal{E}}(C^*(X))\to C^*(X)$, then there is a unique operad map $\xi\colon\mathcal{E}\to E^{\Delta}$ such that $\widetilde{\mu}=\mu\circ\underline{\xi}(C^*(X))$.*

Let $R(\Sigma_2)$ be the Σ_2-free resolution with generators e_i of dimension i and with differential defined by

$$d(e_i)=e_{i-1}+(-1)^ie_{i-1}T,\quad T\in\Sigma_2.$$

Since $E^{\Delta}(2)$ is acyclic, there is a Σ_2-chain map $R(\Sigma_2)\to E^{\Delta}(2)$, and hence a map

$$R(\Sigma_2)\otimes_{\Sigma_2}C^*(X)^{\otimes 2}\to C^*(X).$$

Its restrictions to the elements e_i are generally denoted by

$$\cup_i\colon C^*(X)\otimes C^*(X)\to C^*(X)$$

and referred to as the $\cup_i$-*products.*

The formulas for the differential on the elements e_i imply the following relations for the operations $\cup_i$:

$$d(x\cup_i y)=(-1)^id(x)\cup_i y+(-1)^{i+n}x\cup_i d(y)+x\cup_{i-1}y+(-1)^{i+nm}y\cup_{i-1}x,$$

where $x\in C^n(X)$, $y\in C^m(X)$.

Let $p^n\colon\Delta^n\to S\Delta^{n-1}$ be the projection obtained by contracting the $(n-1)$-dimensional face formed by the vertices indexed $0,1,\dots,n-1$. These projections induce a projection of operads $E^{\Delta}\to SE^{\Delta}$. For a topological space X, the suspension $SC_*(X)$ is a coalgebra over SE^{Δ}, and the following diagram is commutative:

$$\begin{array}{ccc} SC_*(X) & \longrightarrow & \overline{SE}^{\Delta}(SC_*(X)) \\ \downarrow & & \downarrow \\ C_*(SX) & \longrightarrow & \overline{E}^{\Delta}(C_*(SX)) \end{array}$$

By iterating this construction, we obtain projections $E^{\Delta}\to S^nE^{\Delta}$, which make the following diagrams commutative:

$$\begin{array}{ccc} S^nC_*(X) & \longrightarrow & \overline{S^nE}^{\Delta}(S^nC_*(X)) \\ \downarrow & & \downarrow \\ C_*(S^nX) & \longrightarrow & \overline{E}^{\Delta}(C_*(S^nX)) \end{array}$$

In general, the operad E^{Δ} is not Σ-free, and hence it is not an E_∞-operad. To obtain an E_∞-operad, consider the operad $E^{\Delta}\otimes E\Sigma$, and denote it simply by E.

The projection $E\to E^{\Delta}$ induces an E-coalgebra structure on the chain complex $C_*(X)$ of a topological space X. The map $E^{\Delta}\to SE^{\Delta}$ induces an operad map $E\to$

SE, and for the chain complex $C_*(S^n X)$ of the suspension $S^n X$ of a topological space X, we have commutative diagrams similar to those for the operad E^Δ.

The operads E^Δ and E can be used to improve the Eilenberg–Zilber map. Namely, let X_* be a simplicial module or a simplicial object in the category of chain complexes. Although the Eilenberg–Zilber map

$$\psi\colon |X_* \otimes X_*| \to |X_*| \otimes |X_*|$$

is a chain equivalence, it is not compatible with permutations of the factors, and hence is not a Σ_2-chain equivalence.

Similarly, the iteration of the Eilenberg–Zilber map

$$\psi(j)\colon |X_*^{\otimes j}| \to |X_*|^{\otimes j}$$

does not commute with the action of the symmetric group Σ_j, and hence is not a Σ_j-chain equivalence.

THEOREM. *Let X_* be a simplicial object in the category of chain complexes. There is a natural map*

$$\psi(j)\colon |X_*^{\otimes j}| \otimes E^\Delta(j) \to |X_*|^{\otimes j}$$

commuting with the action of the symmetric group Σ_j.

PROOF. Let $x \in X_n^{\otimes j}$, $u \in \Delta^n$, $f \in E^\Delta(j)$. We set

$$\psi(j)(x \otimes u \otimes f) = T(x \otimes f(u)),$$

where $T\colon X_n^{\otimes j} \otimes (\Delta^n)^{\otimes j} \to (X_n \otimes \Delta^n)^{\otimes j}$ is the permutation operator. Straightforward computations show that the required relations hold. □

THEOREM. *The realization of a simplicial object in the category of commutative coalgebras is an E^Δ-coalgebra.*

PROOF. Let K_* be a simplicial object in the category of commutative coalgebras, and let $\nabla_*\colon K_* \to K_* \otimes K_*$ be the corresponding map. We define the E^Δ-coalgebra structure on the realization $|K_*|$ to be the composite map

$$|K_*| \otimes E^\Delta(j) \xrightarrow{|\nabla_*(j)\otimes 1} |K_*^{\otimes j}| \otimes E^\Delta(j) \xrightarrow{\psi(j)} |K_*|^{\otimes j},$$

where $\nabla_*(j)$ is the iteration of the comultiplication ∇_*.

Straightforward computations show that the required relations hold. □

Note that the singular chain complex of a topological space X can be thought of as a realization of the free simplicial module generated by the singular simplicial set $S_*(X)$. The commutative coalgebra structure is given on the generators x_n of this module by the formula $\nabla(x_n) = x_n \otimes x_n$. Therefore, as a special case of the last theorem, we obtain the theorem on the structure of the singular chain complex of a topological space.

5.3. E_∞-structure on cobar constructions

Let K be a commutative Hopf algebra. Consider the cobar construction

$$FK: \quad K \to K \otimes K \to \cdots \to K^{\otimes n} \to K^{\otimes n+1} \to \dots.$$

Then, in addition to the usual multiplication $FK \otimes FK \to FK$ given by

$$[x_1, \dots, x_n] \cdot [x_{n+1}, \dots, x_{n+m}] = [x_1, \dots, x_{n+m}],$$

the cobar construction FK possesses an E_∞-algebra structure. This follows from the fact that FK is the realization of a cosimplicial object in the category of commutative coalgebras. In particular, on FK there are $\cup_i$-products

$$\cup_i: FK \otimes FK \to FK, \quad i \geq 0.$$

Our aim is to obtain formulas for these products. For simplicity, we consider the case of $\mathbb{Z}/2$-coefficients. Let us introduce the notion of $\cup_\infty$-algebra.

DEFINITION. A $\cup_\infty$-*algebra* is a differential module A endowed with operations $\cup_i: A \otimes A \to A$, $i \geq 0$, that raise the dimension by i and satisfy the following conditions:

1. The operation $\cup_0: A \otimes A \to A$ endows A with the structure of a differential algebra. (In what follows, we shall sometimes write just xy instead of $x \cup_0 y$.)
2. For any $i \geq 1$ and any $x \in A_n$, $y \in A_m$, we have

$$d(x \cup_i y) = d(x) \cup_i y + x \cup_i d(y) + x \cup_{i-1} y + y \cup_{i-1} x.$$

3. For $i \geq 1$ and $x_1, x_2, y_1, y_2 \in A$, we have

$$\begin{aligned}(x_1x_2) \cup_i (y_1y_2) = {} & \sum_k (x_1 \cup_{i-k} T^k y_1)(x_2 \cup_k y_2) \\ & + (x_1 \cup_i (y_1y_2))x_2 + x_1(x_2 \cup_i (y_1y_2)) \\ & + ((x_1x_2) \cup_i y_1)y_2 + y_1((x_1x_2) \cup_i y_2) \\ & + x_1(x_2 \cup_i y_1)y_2 + y_1(x_1 \cup_i y_2)x_2,\end{aligned}$$

where $T: A \otimes A \to A \otimes A$ is the permutation operator, and T^k is its kth iteration.

Straightforward computations show that this definition is consistent.

Let us define the $\cup_i$-products $\cup_i: FK \otimes FK \to FK$ so that the cobar construction FK is a $\cup_\infty$-algebra.

On the generators $[x], [y] \in FK$, we set

$$[x] \cup_i [y] = \begin{cases} [x, y], & i = 0, \\ [xy], & i = 1, \\ 0, & i > 1. \end{cases}$$

Further, for $x_1, x_2 \in FK$, $y \in K$, we set

$$(x_1x_2) \cup_i [y] = (x_1 \cup_i [y])x_2 + x_1(x_2 \cup_i [y]).$$

On the remaining elements, the $\cup_i$-products are determined by the $\cup_\infty$-algebra structure.

Straightforward computations show that the required relations hold.

In particular, we have

$$[x_1, \dots, x_n] \cup_1 [y] = \sum_{i=1}^{n} [x_1, \dots, x_i y, \dots, x_n];$$

$$[x] \cup_1 [y_1, y_2] = \sum [x' y_1, x'' y_2],$$

where $\sum x' \otimes x'' = \nabla(x)$.

A $\cup_\infty$-algebra structure on the cobar construction can also be defined in a somewhat more general situation than the case of commutative Hopf algebras.

DEFINITION. A *Hopf* $\cup_\infty$*-algebra* is a $\cup_\infty$-algebra A endowed with a comultiplication $\nabla: A \to A \otimes A$ satisfying the conditions of associativity, $(\nabla \otimes 1)\nabla = (1 \otimes \nabla)\nabla$, and distributivity,

$$\nabla(x \cup_i y) = \sum_k (x' \cup_{i-k} T^k y') \otimes (x'' \cup_k y''),$$

where $\nabla(x) = \sum x' \otimes x''$, $\nabla(y) = \sum y' \otimes y''$.

THEOREM. *The cobar construction FK on a Hopf $\cup_\infty$-algebra K is a $\cup_\infty$-algebra.*

PROOF. Indeed, let us define the $\cup_i$-products by setting, on the generators,

$$[x] \cup_i [y] = \begin{cases} [x \cup_{i-1} y], & i \geq 1, \\ [x, y], & i = 0. \end{cases}$$

We also set

$$(x_1 x_2) \cup_i [y] = (x_1 \cup_i [y]) x_2 + x_1 (x_2 \cup_i [y]), \quad x_1, x_2 \in FK, \; y \in K.$$

On the remaining elements, the $\cup_i$-products are determined by the $\cup_\infty$-algebra structure.

Straightforward computations show that the required relations hold. □

5.4. Bar constructions on operads

Suppose that an operad $\mathcal{E}$ acts on a symmetric family $\mathcal{F}$ on the right and on a symmetric family $\mathcal{G}$ on the left,

$$\nu: \mathcal{F} \times \mathcal{E} \to \mathcal{F}, \quad \mu: \mathcal{E} \times \mathcal{G} \to \mathcal{G}.$$

Consider the simplicial object

$$B_*(\mathcal{F}, \mathcal{E}, \mathcal{G}) = \{B_n(\mathcal{F}, \mathcal{E}, \mathcal{G})\},$$

where $B_n(\mathcal{F}, \mathcal{E}, \mathcal{G}) = \mathcal{F} \times \mathcal{E}^{\times n} \times \mathcal{G}$, and the face and degeneracy operators are defined by

$$d_0 = \nu \times 1^{\times n}, \qquad d_i = 1^{\times i} \times \gamma \times 1^{\times n-i}, \quad 0 < i < n;$$

$$d_n = 1^{\times n} \times \mu, \qquad s_j = 1^{\times j+1} \times i \times 1^{\times n-j+1}, \quad 0 \leq j \leq n.$$

Its realization

$$B(\mathcal{F}, \mathcal{E}, \mathcal{G}) = |B_*(\mathcal{F}, \mathcal{E}, \mathcal{G})| = \sum_n \Delta^n \otimes B_n(\mathcal{F}, \mathcal{E}, \mathcal{G}) / \sim$$

is called the *two-sided bar construction.*

In the case of a trivial family $\mathcal{F}$, i.e., $\mathcal{F}(1) = R$ and $\mathcal{F}(j) = 0$ for $j \geq 2$, the corresponding bar construction is denoted by $B(\mathcal{E}, \mathcal{G})$ and referred to as the *left-sided bar construction.*

In the case of a trivial family $\mathcal{G}$, the corresponding bar construction is denoted by $B(\mathcal{F}, \mathcal{E})$ and referred to as the *right-sided bar construction.*

In the case of trivial families $\mathcal{F}$ and $\mathcal{G}$, the corresponding bar construction is denoted by $B(\mathcal{E})$.

If $\mathcal{F} = \mathcal{E}$, then the simplicial object $B_*(\mathcal{E}, \mathcal{E}, \mathcal{G})$ is a simplicial resolution of the family $\mathcal{G}$, and hence there is a chain equivalence $B(\mathcal{E}, \mathcal{E}, \mathcal{G}) \simeq \mathcal{G}$.

Similarly, if $\mathcal{G} = \mathcal{E}$, then the simplicial object $B_*(\mathcal{F}, \mathcal{E}, \mathcal{E})$ is a simplicial resolution of $\mathcal{F}$, and hence there is a chain equivalence $B(\mathcal{F}, \mathcal{E}, \mathcal{E}) \simeq \mathcal{F}$.

Let $\mathcal{F} \times_{\mathcal{E}} \mathcal{G}$ denote the cokernel of the pair of maps

$$\nu \times 1,\ 1 \times \mu \colon \mathcal{F} \times \mathcal{E} \times \mathcal{G} \to \mathcal{F} \times \mathcal{G}.$$

Then we have

$$B_*(\mathcal{F}, \mathcal{E}, \mathcal{G}) = \mathcal{F} \times_{\mathcal{E}} B_*(\mathcal{E}, \mathcal{E}, \mathcal{G}) = B_*(\mathcal{F}, \mathcal{E}, \mathcal{E}) \times_{\mathcal{E}} \mathcal{G},$$

and hence

$$B(\mathcal{F}, \mathcal{E}, \mathcal{G}) = \mathcal{F} \times_{\mathcal{E}} B(\mathcal{E}, \mathcal{E}, \mathcal{G}) = B(\mathcal{F}, \mathcal{E}, \mathcal{E}) \times_{\mathcal{E}} \mathcal{G}.$$

One of the most important properties of these bar constructions is the fact that, up to a chain equivalence, they are independent of the choice of a simplicial resolution, i.e., if $R_*(\mathcal{F}, \mathcal{E})$ and $R_*(\mathcal{E}, \mathcal{G})$ are arbitrary simplicial resolutions of the families $\mathcal{F}$ and $\mathcal{G}$, respectively, then we have chain equivalences

$$\begin{gathered} B(\mathcal{F}, \mathcal{E}, \mathcal{G}) \simeq |R_*(\mathcal{F}, \mathcal{E}) \times_{\mathcal{E}} \mathcal{G}| = |R_*(\mathcal{F}, \mathcal{E})| \times_{\mathcal{E}} \mathcal{G}, \\ B(\mathcal{F}, \mathcal{E}, \mathcal{G}) \simeq |\mathcal{F} \times_{\mathcal{E}} R_*(\mathcal{E}, \mathcal{G})| = \mathcal{F} \times_{\mathcal{E}} |R_*(\mathcal{E}, \mathcal{G})|, \\ B(\mathcal{F}, \mathcal{E}, \mathcal{G}) \simeq |R_*(\mathcal{F}, \mathcal{E}) \times_{\mathcal{E}} |R_*(\mathcal{E}, \mathcal{G})||. \end{gathered}$$

To exemplify the applications of these properties, let us consider the following two theorems.

THEOREM. *Suppose that an operad $\mathcal{E}$ acts on a family $\mathcal{F}$ on the right and on a family $\mathcal{G}$ on the left. Then, for an arbitrary family $\mathcal{H}$, there is a chain equivalence*

$$B(\mathcal{H} \otimes \mathcal{F}, \mathcal{H} \otimes \mathcal{E}, \mathcal{H} \otimes \mathcal{G}) \simeq \mathcal{H} \otimes B(\mathcal{F}, \mathcal{E}, \mathcal{G}).$$

PROOF. Indeed,

$$\mathcal{H} \otimes B_*(\mathcal{E}, \mathcal{E}, \mathcal{G}), \quad B(\mathcal{H} \otimes \mathcal{E}, \mathcal{H} \otimes \mathcal{E}, \mathcal{H} \otimes \mathcal{G})$$

are $\mathcal{H} \otimes \mathcal{E}$-free simplicial resolutions of the family $\mathcal{H} \otimes \mathcal{G}$. Therefore,

$$\begin{gathered} B(\mathcal{H} \otimes \mathcal{F}, \mathcal{H} \otimes \mathcal{E}, \mathcal{H} \otimes \mathcal{G}) = \mathcal{H} \otimes \mathcal{F} \times_{\mathcal{H} \otimes \mathcal{E}} B(\mathcal{H} \otimes \mathcal{E}, \mathcal{H} \otimes \mathcal{E}, \mathcal{H} \otimes \mathcal{G}) \\ \simeq \mathcal{H} \otimes \mathcal{F} \times_{\mathcal{H} \otimes \mathcal{E}} \mathcal{H} \otimes B(\mathcal{E}, \mathcal{E}, \mathcal{G}) = \mathcal{H} \otimes \mathcal{F} \times_{\mathcal{E}} B(\mathcal{E}, \mathcal{E}, \mathcal{G}) = \mathcal{H} \otimes B(\mathcal{F}, \mathcal{E}, \mathcal{G}). \quad \square \end{gathered}$$

THEOREM. *Suppose that an operad $\mathcal{H}$ acts on a family $\mathcal{F}$ on the right, an operad $\mathcal{E}$ acts on a family $\mathcal{G}$ on the left, and there is given an operad map $\mathcal{E} \to \mathcal{G}$. Then there is a chain equivalence*

$$B(\mathcal{F}, \mathcal{H}, B(\mathcal{H}, \mathcal{E}, \mathcal{G})) \simeq B(\mathcal{F}, \mathcal{E}, \mathcal{G}).$$

PROOF. We have

$$\begin{aligned} B(\mathcal{F}, \mathcal{H}, B(\mathcal{H}, \mathcal{E}, \mathcal{G})) &= B(\mathcal{F}, \mathcal{H}, \mathcal{H}) \times_{\mathcal{H}} B(\mathcal{H}, \mathcal{E}, \mathcal{G}) \\ &\simeq \mathcal{F} \times_{\mathcal{H}} B(\mathcal{H}, \mathcal{E}, \mathcal{G}) = B(\mathcal{F}, \mathcal{E}, \mathcal{G}). \quad \square \end{aligned}$$

THEOREM. *For the operad A above there is a chain equivalence $B(A) \simeq SA$.*

PROOF. We define the map $\xi \colon SA \to B(A)$ by setting

$$\xi(sa(j)) = [a(2), \dots, a(2) \otimes 1 \otimes \dots \otimes 1 - \dots + (-1)^{j+1} 1 \otimes \dots \otimes 1 \otimes a(2)].$$

Let us show that ξ induces an isomorphism in homology. To this end, we consider the product $A \times SA$ on which the operad A acts on the left, and define the differential d_α in it by setting, on the generators,

$$d_\alpha(sa(j)) = \sum_{j_1 + j_2 = j} (-1)^{j_1} a(2) \otimes sa(j_1) \otimes sa(j_2).$$

Straightforward computations show that $d_\alpha \circ d_\alpha = 0$.

Denote the resulting family by $A \times_\alpha SA$. Let us show that it is contractible. We shall define a contracting homotopy $h \colon A \times_\alpha SA \to A \times_\alpha SA$, i.e., a map of families that raises the degree by one and satisfies the relation $d_\alpha h + h d_\alpha = \mathrm{Id}$. Consider the elements

$$a(k) \otimes (sa(j_1) \otimes \dots \otimes sa(j_n) \otimes \dots \otimes sa(j_m) \otimes \dots \otimes sa(j_k)) \in A \times_\alpha SA.$$

If $j_i = 1$ for $i > m$ and $1 \leq n < i < m \leq k$, $j_m = 2$, $j_n \geq 2$, then we set

$$\begin{aligned} &h(a(k) \otimes (sa(j_1) \otimes \dots \otimes sa(j_n) \otimes \dots \otimes sa(j_m) \otimes \dots \otimes sa(j_k)) \\ &= (-1)^{j_1 + \dots + j_{m-1} + m} a(k-1) \otimes sa(j_1) \otimes \dots \otimes sa(j_{m-1} + j_m) \otimes \dots \otimes sa(j_k) \\ &+ (-1)^{j_1 + \dots + j_{m-2} + m} a(k-1) \otimes sa(j_1) \otimes \dots sa(j_{m-2} + j_m) \otimes sa(j_{m-1}) \dots \otimes sa(j_k) \\ &+ \dots \\ &+ (-1)^{j_1 + \dots + j_n + m} a(k-1) \otimes sa(j_1) \otimes \dots sa(j_n + j_m) \dots \otimes sa(j_{m-1}) \otimes \dots \otimes sa(j_k). \end{aligned}$$

In all other cases we set $h(a(k) \otimes sa(j_1) \otimes \dots \otimes sa(j_k)) = 0$.

Straightforward computations show that the map h so defined satisfies the required relations, and hence it is the sought-for chain homotopy.

Note that the right-sided bar construction $B(A, A)$ is an acyclic family and is isomorphic as a graded family to $A \times BA$. The map $\xi \colon SA \to BA$ induces a map of acyclic families

$$1 \times \xi \colon A \times_\alpha SA \to A \times BA \cong B(A, A).$$

Passing to the map of the right-sided bar constructions, we obtain

$$SA \simeq B(A, A \times_\alpha SA) \simeq B(A, B(A, A)) \simeq BA,$$

which is the required result. $\square$

Consider the operad E_0. We define the product $A \times_\phi SE_0$ to be equal as a graded family to $A \times SE_0$, with the differential d_ϕ given on the generators $e_1(j) \in SE_0(j)$ of dimension $j-1$ by the formula

$$d_\phi(e_1(j)) = \sum_{j_1+j_2=j} (-1)^{\epsilon+j_1+1} a(2) \otimes e_1(j_1) \otimes e_1(j_2)\sigma(j_1, j_2),$$

where the summation is over all (j_1, j_2)-shuffles $\sigma(j_1, j_2)$, and ϵ is the sign of the shuffle. Straightforward computations show that $d_\phi \circ d_\phi = 0$.

Note that the complexes $(A \times_\phi SE_0)(j)$ coincide with the Milgram chain complexes $C(j-1)$ [**33**] and hence are acyclic.

The operad A acts freely on $A \times_\phi SE_0$ on the left, and hence there is a chain equivalence $B(A, A \times_\phi SE_0) \simeq SE_0$.

Consider the E_∞-operad E. We have an operad map $p\colon E \to SE$ and an embedding $i\colon A \to E$ such that the composition $p \circ i$ is zero. Therefore, the map p induces a map $B(p)\colon B(A, E) \to SE$.

THEOREM. *The map $B(p)\colon B(A, E) \to SE$ is a chain equivalence.*

PROOF. To prove that this map is a chain equivalence, it suffices to show that it induces an isomorphism in homology. Since $H_*(E) \cong E_0$, it suffices to prove that the map $p_0\colon B(A, E_0) \to SE_0$ is a chain equivalence. Indeed, the induced map

$$B(p_0)\colon B(A, A, E_0) \cong A \times B(A, E_0) \to A \times_\phi SE_0$$

is a chain equivalence. Passing once more to the bar constructions, we see that the initial map p_0 is also a chain equivalence. □

THEOREM. *There is a chain equivalence $B(SE, E) \simeq SA$.*

PROOF. Indeed, the above theorems imply that

$$B(SE, E) \simeq B(B(A, E), E) \simeq B(A) \simeq SA. \quad \square$$

Consider the family $SA \times A$ and define a differential d_ϕ in it by setting

$$\begin{aligned} &d_\phi(sa(k) \otimes a(j_1) \otimes \cdots \otimes a(j_k)) \\ &\quad = \sum_{i=1}^{k-1} (-1)^{i+1} sa(k-1) \otimes a(j_1) \otimes \cdots \otimes a(j_i + j_{i+1})\sigma^{-1}(j_i, j_{i+1}) \otimes \cdots \otimes a(j_k), \end{aligned}$$

where the summation is over all (j_i, j_{i+1})-shuffles $\sigma(j_i, j_{i+1})$. Straightforward computations show that $d_\phi \circ d_\phi = 0$. Denote the resulting family by $SA \times_\phi A$.

THEOREM. *For any positive integer n, there is a chain equivalence*

$$B(S^n E, E) \simeq S^n A \times_{S^{n-1}\phi} S^{n-1} A \times \cdots \times_{S\phi} SA.$$

PROOF. The proof is by induction over n. For $n = 1$, we have a chain equivalence $B(SE, E) \simeq SA$. Assume that for some n the required chain equivalence holds. We shall prove that for $n+1$ the required chain equivalence also holds. Let us use the chain equivalence $B(S^n A, S^n E) \simeq S^{n+1}E$, which is obtained by taking the n-fold suspension of the chain equivalence $B(A, E) \simeq SE$. We have

$$B(S^{n+1}E, E) \simeq B(B(S^n A, S^n E), E) \cong B(S^n A, B(S^n E, E))$$
$$\simeq B(S^n A, S^n A \times_{S^{n-1}\phi} \times \cdots \times_{S\phi} SA) \cong S^{n+1}A \times_{S^n\phi} \times \cdots \times_{S\phi} SA. \quad \square$$

5.5. Bar constructions on algebras over operads

Let X be an algebra over an operad $\mathcal{E}$, $\mu: \mathcal{E}(X) \to X$. Consider the simplicial object $B_*(\mathcal{F}, \mathcal{E}, X) = \{B_n(\mathcal{F}, \mathcal{E}, X)\}$, where $B_n(\mathcal{F}, \mathcal{E}, X) = \mathcal{F}\mathcal{E}^n(X)$ and the face and degeneracy operators are defined by

$$d_0 = \nu 1^n, \qquad d_i = 1^i \gamma 1^{n-i}, \quad 0 < i < n;$$
$$d_n = 1^n \mu, \qquad s_j = 1^{j+1} i 1^{n-j+1}, \quad 0 \le j \le n.$$

Its realization

$$B(\mathcal{F}, \mathcal{E}, X) = |B_*(\mathcal{F}, \mathcal{E}, X)| = \sum_n B_n(\mathcal{F}, \mathcal{E}, X) \otimes \Delta^n / \sim$$

is called the *bar construction on* X.

In the case of a trivial $\mathcal{F}$, i.e., $\mathcal{F}(1) = R$ and $\mathcal{F}(j) = 0$ for $j \geq 2$, the corresponding bar construction is denoted by $B(\mathcal{E}, X)$.

By $\mathcal{F}_{\mathcal{E}}X$ denote the cokernel of the pair of maps $\nu 1$, $1\mu: \mathcal{F}\mathcal{E}X \to \mathcal{F}X$. Then

$$B_*(\mathcal{F}, \mathcal{E}, X) = \mathcal{F}_{\mathcal{E}} B_*(\mathcal{E}, \mathcal{E}, X) = B_*(\mathcal{F}, \mathcal{E}, \mathcal{E})_{\mathcal{E}} X,$$

and hence

$$B(\mathcal{F}, \mathcal{E}, X) = \mathcal{F}_{\mathcal{E}} B(\mathcal{E}, \mathcal{E}, X) = B(\mathcal{F}, \mathcal{E}, \mathcal{E})_{\mathcal{E}} X.$$

One of the most important properties of these bar constructions is the fact that, up to a chain equivalence, they are independent of the choice of the simplicial resolution, i.e., if $R_*(\mathcal{F}, \mathcal{E})$ and $R_*(\mathcal{E}, X)$ are arbitrary simplicial resolutions of the family $\mathcal{F}$ and of the complex X, respectively, then there are chain equivalences

$$B(\mathcal{F}, \mathcal{E}, X) \simeq |R_*(\mathcal{F}, \mathcal{E})_{\mathcal{E}} X| = |R_*(\mathcal{F}, \mathcal{E})|_{\mathcal{E}} X,$$
$$B(\mathcal{F}, \mathcal{E}, X) \simeq |\mathcal{F}_{\mathcal{E}} R_*(\mathcal{E}, X)| = \mathcal{F}_{\mathcal{E}} |R_*(\mathcal{E}, X)|,$$
$$B(\mathcal{F}, \mathcal{E}, X) \simeq |R_*(\mathcal{F}, \mathcal{E})_{\mathcal{E}} |R_*(\mathcal{E}, X)|.$$

The following theorem is proved in the same way as its analog for the case of operads.

THEOREM. *Suppose that an operad $\mathcal{G}$ acts on a family $\mathcal{F}$ on the right, an operad map $\mathcal{E} \to \mathcal{G}$ is given, and a complex X is an algebra over the operad $\mathcal{E}$. Then there is a chain equivalence*

$$B(\mathcal{F}, \mathcal{G}, B(\mathcal{G}, \mathcal{E}, X)) \simeq B(\mathcal{F}, \mathcal{E}, X).$$

THEOREM. *If X is an algebra over the operad A, i.e., just an algebra, then the bar construction $B(A, X)$ is chain equivalent to the desuspension of the ordinary (Adams) bar construction $B(X)$, i.e.,*

$$B(A, X) \simeq S^{-1}BX.$$

PROOF. Indeed, the chain equivalence $FBX \simeq X$ induces a chain equivalence $B(A, FBX) \simeq B(A, X)$. But FBX, regarded as a graded algebra, is a free algebra generated by $S^{-1}BX$. Therefore, we have $B(A, FBX) \simeq S^{-1}BX$, and hence $B(A, X) \simeq S^{-1}BX$. □

A similar assertion holds for commutative algebras. Namely, we have the following theorem.

THEOREM. *If R is a field of characteristic zero and X is an algebra over the operad E_0, i.e., just a commutative algebra, then there is a chain equivalence*

$$B(E_0, X) \simeq S^{-1}B_\wedge X.$$

PROOF. Indeed, for any commutative algebra X, we have the chain equivalence $F_sB_\wedge X \simeq X$, which induces a chain equivalence

$$B(E_0, F_sB_\wedge X) \simeq B(E_0, X).$$

The algebra $F_sB_\wedge X$, regarded as a graded algebra, is a free commutative algebra generated by $S^{-1}B_\wedge X$. Hence there is a chain equivalence

$$B(E_0, F_sB_\wedge X) \simeq S^{-1}B_\wedge X.$$

Therefore, $B(E_0, X) \simeq S^{-1}B_\wedge X$. □

THEOREM. *If X is an A_∞-algebra, then the bar construction $B(A_\infty, X)$ is chain equivalent to the desuspension of the bar construction $B(X)$, i.e.,*

$$B(A_\infty, X) \simeq S^{-1}BX.$$

PROOF. Unfortunately, the chain equivalence $FBX \simeq X$ considered above is not a chain equivalence of A_∞-algebras. To eliminate this drawback, let us improve the definition of the cobar construction. Namely, let K be a coalgebra. We consider the free A_∞-algebra $A_\infty S^{-1}K$ generated by the complex $S^{-1}K$ and define a differential in it so that the natural map $\phi\colon K \to A_\infty S^{-1}K$ is a twisting cochain. Denote the resulting differential A_∞-algebra by $F_\infty K$. The chain equivalence of operads $A_\infty \to A$ induces a chain equivalence $F_\infty K \to FK$.

If X is an A_∞-algebra, then the twisting cochain $\beta\colon BX \to X$ induces an A_∞-algebra map $F_\infty BX \to X$, which is a chain equivalence. Hence there is a chain equivalence $B(A_\infty, F_\infty BX) \simeq B(A_\infty, X)$. But $F_\infty BX$, regarded as an A_∞-algebra, is a free A_∞-algebra generated by $S^{-1}BX$. Therefore, we have $B(A, F_\infty BX) \simeq S^{-1}BX$, and hence $B(A_\infty, X) \simeq S^{-1}BX$. □

The next theorem is proved in a similar way.

THEOREM. *If R is a field of characteristic zero and X is a commutative A_∞-algebra, i.e., an algebra over the operad S_∞, then there is a chain equivalence*

$$B(S_\infty, X) \simeq S^{-1}B_\wedge X.$$

Let us pass to the dual situation. If X is a coalgebra over an operad $\mathcal{E}$, $\tau\colon X \to \overline{\mathcal{E}}(X)$, then we can consider the cosimplicial object

$$F^*(\mathcal{F}, \mathcal{E}, X) = \{F^n(\mathcal{F}, \mathcal{E}, X)\},$$

where $F^n(\mathcal{F},\mathcal{E},X) = \overline{\mathcal{F}\mathcal{E}}^n(X)$ and the coface and codegeneracy operators are defined by

$$\delta^0 = \overline{\nu}1^n, \qquad \delta^i = 1^i\overline{\gamma}1^{n-i}, \quad 0 < i < n;$$
$$\delta^n = 1^n\tau, \qquad \sigma^j = 1^{j+1}p1^{n-j+1}, \quad 0 \leq j \leq n.$$

Its realization

$$F(\mathcal{F},\mathcal{E},X) = |F^*(\mathcal{F},\mathcal{E},X)| = \mathrm{Hom}(\Delta^*; F^*(\mathcal{F},\mathcal{E},X)),$$

where Hom is taken in the category of cosimplicial objects, is referred to as the *cobar construction.*

In the case of a trivial $\mathcal{F}$, the corresponding cobar construction is denoted by $F(\mathcal{E},X)$.

The following theorems are dual to those stated above.

THEOREM. *Suppose that an operad $\mathcal{G}$ acts on a family $\mathcal{F}$ on the right, an operad map $\mathcal{E} \to \mathcal{G}$ is given, and a complex X is a coalgebra over $\mathcal{E}$. Then there is a chain equivalence*

$$F(\mathcal{F},\mathcal{G},F(\mathcal{G},\mathcal{E},X)) \simeq F(\mathcal{F},\mathcal{E},X).$$

THEOREM. *The cobar construction $F(A,X)$ on a coalgebra X of finite type is chain equivalent to the desuspension of the ordinary (Adams) cobar construction $F(X)$, i.e.,*

$$F(A,X) \simeq SFX.$$

THEOREM. *If R is a field of characteristic zero and X is a commutative coalgebra, then there is a chain equivalence*

$$F(E_0,X) \simeq SF_\vee X.$$

THEOREM. *If X is an A_∞-coalgebra of finite type, then the cobar construction $F(A_\infty,X)$ is chain equivalent to the suspension of the cobar construction $F(X)$, i.e.,*

$$F(A_\infty,X) \simeq SFX.$$

THEOREM. *If R is a field of characteristic zero and X is a commutative A_∞-coalgebra of finite type, then there is a chain equivalence*

$$F(S_\infty,X) \simeq SF_\vee X.$$

A commutative coalgebra X can be regarded as an E-coalgebra. The corresponding structure is induced by the projection $E \to E_0$. Moreover, the chain equivalence of operads $SA \simeq B(SE,E)$ induces a chain equivalence $SFX \simeq F(SE,E,X)$ of SE-coalgebras.

Taking into account the fact that the cobar construction FX of a commutative coalgebra is also a commutative coalgebra, we see that this chain equivalence can be iterated, and hence we obtain the following result.

THEOREM. *If X is a commutative coalgebra, then there are chain equivalences*

$$S^nF^nX \simeq F(S^nE,E,X).$$

In particular, if a chain complex X has the trivial E-coalgebra structure, then there are chain equivalences

$$S^n F^n X \simeq F(S^n E, E, X).$$

5.6. Homology and cohomology of an E_∞-operad.

Let E be an E_∞-operad. By the homology (the cohomology) of the operad E we mean the family

$$E_* = \{E_*(j)\},\ E_*(j) = H_*(E(j)/\Sigma_j) \quad (E^* = \{E^*(j)\},\ E^*(j) = H^*(E(j)/\Sigma_j)).$$

Our aim is to compute this homology (cohomology). Of course, it is isomorphic to the homology (cohomology) of the symmetric groups Σ_j.

Actually, we shall consider a more general situation. Namely, let X be a chain complex; consider the complexes

$$\underline{E}(X) = \sum_j \underline{E}(j, X), \quad \underline{E}(j, X) = E(j) \otimes_{\Sigma_j} X^{\otimes j};$$

$$\overline{E}(X) = \prod_j \overline{E}(j, X), \quad \overline{E}(j, X) = \mathrm{Hom}_{\Sigma_j}(E(j), X^{\otimes j}).$$

We shall compute the homology of these complexes. It is clear that for the case in which the ground ring is a field, there are isomorphisms

$$H_*(\underline{E}(X)) \cong H_*(\underline{E}(H_*(X))); \quad H_*(\overline{E}(X)) \cong H_*(\overline{E}(H_*(X))).$$

Therefore, to compute the homology, we can use graded modules M instead of chain complexes X.

If M is a free module with a single zero-dimensional generator, then homology of these complexes is isomorphic to the homology and the cohomology, respectively, of the symmetric groups.

The homology of the symmetric groups does not depend on the choice of a Σ_j-free acyclic family E; therefore, to compute it, we construct another Σ-free acyclic family E^∞.

For $n \geq 1$, we define the symmetric family

$$E^n = A \times_\phi SA \times \cdots \times_\phi S^n E_0,$$

to be equal as a graded family to $A \times SA \times \cdots \times S^n E_0$, with the differential given on the generators $e_m(j) \in S^m A(j)$ by

$$d_\phi(e_m(j)) = \sum (-1)^{\epsilon + j_1 + m} e_{m-1}(2) \otimes e_m(j_1) \otimes e_m(j_2) \sigma(j_1, j_2),$$

where the summation is over all (j_1, j_2)-shuffles $\sigma(j_1, j_2)$, and ϵ is the sign of the shuffle.

THEOREM. *The family $E^n = A \times_\phi SA \times \cdots \times_\phi S^n E_0$ is acyclic.*

PROOF. Indeed, as was shown in the previous section, the family E^1 is acyclic, i.e., $H_*(E^1) \cong E_0$. Assume that the family E^{n-1} is acyclic; then, for the family E^n, we have

$$H_*(E^n) = H_*(A \times_\phi SA \times \cdots \times_\phi S^n E_0) = H_*(A \times_\phi SE^{n-1}) \cong H_*(A \times_\phi SE^0) \cong E_0. \ \square$$

We define the family $E^\infty = A \times_\phi SA \times \ldots$ as the limit of the families E^n. Then E^∞ is a Σ-free acyclic family.

Let us find the homology of the complex $\underline{E}^1(M)$. We have

$$\underline{E}^1(M) = \underline{A \times_\phi SE_0}(M) = \underline{A}_\phi(\underline{SE}_0(M) = TS^{-1}T_s(SM),$$

where $T_s(SM)$ designates the free commutative algebra generated by SM, and the last equality ignores the differential. The algebra $T_s(SM)$ can be endowed with a coalgebra structure $\nabla: T_s(SM) \to T_s(SM) \otimes T_s(SM)$ that makes it into a Hopf algebra, by setting, on the generators $sx \in SM$,

$$\nabla(sx) = sx \otimes 1 + 1 \otimes sx.$$

Comparing the differentials on the cobar construction $FT_s(SM)$ and on the complex $\underline{E}^1(M)$, we can readily see that there is an isomorphism $\underline{E}^1(M) \cong FT_s(SM)$.

If the coefficient ring is the field $\mathbb{Z}/2$, the homology of this cobar construction is a free commutative algebra on the primitive elements of $T_s(SM)$, i.e.,

$$H_*(FT_s(SM)) \cong T_s S^{-1} PT_s(SM),$$

where $PT_s(SM)$ is the module of primitive elements,

$$PT_s(SM) = \{x \in T_s(SM) : \ \nabla(x) = x \otimes 1 + 1 \otimes x\}.$$

It is readily seen that the module of primitive elements is generated by the elements $(sx)^{2^k}$, $sx \in SM$. The corresponding elements in the complex $\underline{E}^1(M)$ are $e_1(2^k) \otimes x^{\otimes 2^k}$.

THEOREM. *In the case of $\mathbb{Z}/2$-coefficients, for any $n \geq 1$, we have*

$$\underline{E}^n_*(M) \cong T_s(S^{-1}PT_s)^n(S^n M).$$

PROOF. For $n = 1$, the formula is already verified. Assume that it is true for $n - 1$. Let us show that it also holds for n. We represent the complex $\underline{E}^n(M)$ as $\underline{E}^n(M) = \underline{A \times_\phi SE^{n-1}}(M)$. Then

$$\begin{aligned} H_*(\underline{E}^n(M)) &\cong H_*(FH_*(\underline{E}^{n-1}(SM))) \\ &\cong H_*(FT_s(S^{-1}PT_s)^{n-1}(S^n M)) \cong T_s(S^{-1}PT_s)^n(S^n M). \ \square \end{aligned}$$

By renaming elements, we obtain the following theorem.

THEOREM. *For $\mathbb{Z}/2$-coefficients, the homology $\underline{E}^n_*(M)$ is a polynomial algebra on generators of the form*

$$e_{i_1} \ldots e_{i_k} x_m, \quad 1 \leq i_1 \leq \cdots \leq i_k \leq n, \quad x_m \in M,$$

with dimensions equal to $i_1 + 2i_2 + \cdots + 2^{k-1} i_k + 2^k m$.

THEOREM. *For $\mathbb{Z}/2$-coefficients, the homology $\underline{E}_*(M)$ is a polynomial algebra on generators of the form*

$$e_{i_1}\dots e_{i_k}x_m,\quad 1\le i_1\le\dots\le i_k,\quad x_m\in M,$$

with dimensions equal to $i_1+2i_2+\dots+2^{k-1}i_k+2^km$.

The elements $e_{i_1}\dots e_{i_k}x_m$ of $\underline{E}_*(M)$ can be expressed in the form

$$Q^{j_1}\dots Q^{j_k}\otimes x_m;\quad j_1\le 2j_2,\dots,j_{k-1}\le 2j_k, m\le j_k,$$

where

$$\begin{aligned} j_k &= i_k+m,\\ j_{k-1} &= i_{k-1}+i_k+2m,\\ &\dots\\ j_1 &= i_1+i_2+2i_3+\dots+2^{k-2}i_k+2^{k-1}m. \end{aligned}$$

The sequences $Q^{j_1}\dots Q^{j_k}$ are elements of the Dyer–Lashof algebra $\mathcal{R}$; see [**14, 17**].

Now consider the case of $\mathbb{Z}/p$-coefficients, $p>2$. Note that the module PT_sSM of primitive elements of the Hopf algebra T_sSM is generated by the elements $sx\in SM$ and the elements $(sx)^{p^k}$ such that $k>0$ and $\dim(sx)$ is even. In the complex $\underline{E}^1(M)$, the corresponding elements are $e_1(p^k)\otimes x^{\otimes p^k}$. However, the homology $H_*(FT_sSM)$ is generated not only by the primitive elements, but also by the elements $\beta(sx)^{p^k}$, where β is the Bockstein homomorphism. Denote the corresponding elements in the complex $\underline{E}^1(M)$ by $\beta(e_1(p^k)\otimes x^{\otimes p^k})$. Denote by $P_\beta T_s(SM)$ the module generated by the primitive elements and the elements $\beta(sx)^{p^k}$. Then the homology of $FT_s(SM)$ is isomorphic to $T_sS^{-1}P_\beta T_s(SM)$. In analogy with the case of $\mathbb{Z}/2$-coefficients, we prove the following theorem.

THEOREM. *In the case of $\mathbb{Z}/p$-coefficients, for any $n\ge 1$, we have*

$$H_*(\underline{E}^n(M))\cong T_s(S^{-1}P_\beta T_s)^n(S^nM).$$

By renaming the elements, we obtain the following assertion.

THEOREM. *For $\mathbb{Z}/p$-coefficients, the homology $\underline{E}^n_*(M)$ is a free commutative algebra on generators of the form*

$$\beta^{\epsilon_1}e_{i_1}\dots\beta^{\epsilon_k}e_{i_k}x_m,\quad 1\le i_1\le\dots\le i_k\ \le n\ ,\ \epsilon_j=\pm1,\ x_m\in M,$$

with dimensions equal to

$$(i_1+pi_2+\dots+p^{k-1}i_k)(p-1)+p^km-(\epsilon_1-p\epsilon_2-\dots-p^{k-1}\epsilon_k).$$

Here it is required that the parity of the indices i_j, $1\le j\le k$, coincide with the parity of the dimension of the element $\beta^{\epsilon_{j+1}}e_{i_{j+1}}\dots\beta^{\epsilon_k}e_{i_k}x_m$.

THEOREM. *For $\mathbb{Z}/p$-coefficients, the homology $\underline{E}_*(M)$ is a free commutative algebra on generators of the form*

$$\beta^{\epsilon_1}e_{i_1}\dots\beta^{\epsilon_k}e_{i_k}x_m,\quad 1\le i_1\le\dots\le i_k,\ \epsilon_j=\pm1,\ x_m\in M,$$

with dimensions equal to

$$(i_1+pi_2+\dots+p^{k-1}i_k)(p-1)+p^km-(\epsilon_1-p\epsilon_2-\dots-p^{k-1}\epsilon_k).$$

Here it is required that the parity of the indices i_j, $1 \le j \le k$, *coincide with the parity of the dimension of the element* $\beta^{\epsilon_{j+1}} e_{i_{j+1}} \dots \beta^{\epsilon_k} e_{i_k} x_m$.

The elements $\beta^{\epsilon_1} e_{i_1} \dots \beta^{\epsilon_k} e_{i_k} x_m$ of $\underline{E}_*(M)$ can be expressed in the form

$$\beta^{\epsilon_1} Q^{j_1} \dots \beta^{\epsilon_k} Q^{j_k} \otimes x_m,$$

where

$$\begin{aligned} 2j_k &= i_k + \dim(x), \\ 2j_{k-1} &= i_{k-1} + (p-1)i_k + p \cdot \dim(x) - \epsilon_k, \\ &\dots \\ 2j_1 &= i_1 + (p-1)i_2 - \epsilon_2 + \dots + p^{k-1} \cdot \dim(x). \end{aligned}$$

The sequences $\beta^{\epsilon_1} Q^{j_1} \dots \beta^{\epsilon_k} Q^{j_k}$ are elements of the mod-p Dyer–Lashof algebra $\mathcal{R}$; see [**14, 17**].

Now let X be a cochain complex, and let E be an E_∞-operad. Consider the cochain complexes

$$\underline{E}(X) = \sum_j \underline{E}(j, X), \quad \underline{E}(j, X) = E(j) \otimes_{\Sigma_j} X^{\otimes j},$$

where the dimension of an element of the form $y \otimes x_1 \otimes \dots \otimes x_j \in E(j, X)$ is the sum of the dimensions of the $x_1, \dots, x_j$ minus the dimension of y.

If, as above, the ground ring is a field, then we have an isomorphism

$$H_*(\underline{E}(X)) \cong H_*(\underline{E}(H^*(X))),$$

and hence we can take graded modules instead of complexes for computing this homology.

Reformulating the theorems proved above, we obtain the following assertion.

THEOREM. *For the case of* $\mathbb{Z}/2$*-coefficients, the homology* $\underline{E}_*(M)$ *is a polynomial algebra on generators of the form*

$$e_{i_1} \dots e_{i_k} x^m, \quad 1 \le i_1 \le \dots \le i_k, \quad x^m \in M,$$

with dimensions equal to $2^k m - (i_1 + 2i_2 + \dots + 2^{k-1} i_k)$.

The elements $e_{i_1} \dots e_{i_k} x^m$ of $\underline{E}_*(M)$ that satisfy the inequality $i_k \le m$ can be expressed as

$$Sq^{j_1} \dots Sq^{j_k} \otimes x^m,$$

where

$$\begin{aligned} j_k &= m - i_k, \\ j_{k-1} &= 2m - i_{k-1} - i_k, \\ &\dots \\ j_1 &= 2^{k-1} m - (i_1 + i_2 + 2i_3 + \dots + 2^{k-2} i_k). \end{aligned}$$

The sequences $Sq^{j_1} \dots Sq^{j_k}$ are elements of the Steenrod algebra $\mathcal{A}$ [**34, 47**].

THEOREM. *For $\mathbb{Z}/p$-coefficients, the homology $\underline{E}_*(M)$ is a free commutative algebra on generators of the form*

$$\beta^{\epsilon_1}e_{i_1}\dots\beta^{\epsilon_k}e_{i_k}x^m,\quad 1\le i_1\le\dots\le i_k,\ \epsilon_j=\pm1,\ x^m\in M,$$

with dimensions equal to

$$(p^km-(i_1+pi_2+\dots+p^{k-1}i_k)(p-1))+(\epsilon_1-p\epsilon_2-\dots-p^{k-1}\epsilon_k).$$

It is required here that the parity of the indices i_j, $1\le j\le k$, coincide with the parity of the dimension of the element $\beta^{\epsilon_{j+1}}e_{i_{j+1}}\dots\beta^{\epsilon_k}e_{i_k}x^m$.

The elements $\beta^{\epsilon_1}e_{i_1}\dots\beta^{\epsilon_k}e_{i_k}x^m$ of $\underline{E}_*(M)$ that satisfy the inequality $i_k\le m$ can be expressed in the form

$$\beta^{\epsilon_1}Sq^{j_1}\dots\beta^{\epsilon_k}Sq^{j_k}\otimes x^m,$$

where

$$\begin{aligned}2j_k&=m-i_k,\\2j_{k-1}&=p\cdot m-i_{k-1}-(p-1)i_k+\epsilon_k,\\&\dots\\2j_1&=p^{k-1}m-(i_1+(p-1)i_2-\epsilon_2+\dots).\end{aligned}$$

The sequences $\beta^{\epsilon_1}Sq^{j_1}\dots\beta^{\epsilon_k}Sq^{j_k}$ are elements of the mod-p Steenrod algebra $\mathcal{A}$ [**34, 47**].

Thus elements of the Dyer–Lashof algebra and the Steenrod algebra are obtained by regrading the elements of one and the same module $\underline{E}_*(M)$.

Passing to the dual situation, we obtain the following theorems.

THEOREM. *For the case of $\mathbb{Z}/2$-coefficients, the homology $\overline{E}_*(M)$ is isomorphic to the free commutative coalgebra on generators of the form*

$$\overline{e_{i_1}\dots e_{i_k}}\times x_m,\quad 1\le i_1\le\dots\le i_k,\quad x_m\in M,$$

with dimensions equal to $2^km-2^{k-1}i_k-\dots-i_1$.

Regrading the elements of $\overline{E}_*(M)$ leads to a coalgebra $\mathcal{K}$ dual to the Steenrod algebra; this coalgebra is referred to as the *Milnor coalgebra.* It is a polynomial algebra on generators ξ_i of dimension 2^i-1, $i\ge0$; see [**34, 47**].

THEOREM. *For the case of $\mathbb{Z}/p$-coefficients, the homology $\overline{E}^n_*(M)$ is a free commutative coalgebra on generators of the form*

$$\overline{\beta^{\epsilon_1}e_{i_1}\dots\beta^{\epsilon_k}e_{i_k}x_m},\quad 1\le i_1\le\dots\le i_k\le n,\ \epsilon_j=\pm1,\ x\in M,$$

with dimensions equal to

$$p^km-(i_1+pi_2+\dots+p^{k-1}i_k)(p-1)+(\epsilon_1-p\epsilon_2-\dots-p^{k-1}\epsilon_k).$$

It is required here that the parity of the indices i_j, $1\le j\le k$, coincide with the parity of the dimension of the element $\beta^{\epsilon_{j+1}}e_{i_{j+1}}\dots\beta^{\epsilon_k}e_{i_k}x_m$.

Regrading the elements of $\overline{E}_*(M)$ leads to the mod-p Milnor coalgebra $\mathcal{K}$, which is dual to the mod-p Steenrod algebra. It is a free commutative algebra generated by the elements ξ_i, τ_i of dimensions $2(p^i-1)$ and $2p^i-1$, respectively; see [**34, 47**].

Now let us consider the question as to what structure is induced on the family E_* by the operad structure on E. For simplicity, we shall only deal with the case of $\mathbb{Z}/2$-coefficients. Denote by $E_*(k) * E_*(n)$ the tensor product $E_*(k) \otimes E_*(n)$, where the dimension of an element of the form $x * y \in E_*(k) * E_*(n)$ is set equal to $\dim(x) + k \cdot \dim(y)$. Note that E is not only an operad, but a bioperad as well. The bioperad structure induces the multiplications

$$\pi_*\colon E_*(n) \otimes E_*(m) \to E_*(n+m), \quad \gamma_*\colon E_*(k) * E_*(n) \to E_*(kn).$$

Besides, we have the comultiplication

$$\nabla_*\colon E_*(n) \to E_*(n) \otimes E_*(n).$$

The definition of a bioperad directly implies that the operations π_*, γ_*, and ∇_* satisfy the following conditions:

1. The operations (π_*, ∇_*) and (γ_*, ∇_*) define Hopf algebra structures on E_*.

2. The operations γ_* and π_* satisfy the distributivity relation, i.e., the following diagrams are commutative:

$$\begin{array}{ccc}
(E_*(k') \otimes E_*(k'')) * E_*(n) & \xrightarrow{\pi_* * 1} & E_*(k'+k'') * E_*(n) \\
\downarrow & & \downarrow{\scriptstyle \gamma_*} \\
(\mathcal{E}_*(k') * E_*(n)) \otimes (E_*(k'') * E_*(n)) & & E_*((k'+k'')n) \\
{\scriptstyle \gamma_*\otimes\gamma_*}\downarrow & & \downarrow{\scriptstyle =} \\
E_*(k'n) \otimes E_*(k''n) & \xrightarrow{\pi_*} & E_*(k'n + k''n)
\end{array}$$

$$\begin{array}{ccc}
E_*(k) * (E_*(n) \otimes E_*(m)) & \xrightarrow{\nabla_* * 1} & (E_*(k) \otimes E_*(k)) * (E_*(n) \otimes E_*(m)) \\
{\scriptstyle 1*\pi_*}\downarrow & & \downarrow \\
E_*(k) * E_*(n+m) & & (E_*(k) * E_*(n)) \otimes (E_*(k) * E_*(m)) \\
{\scriptstyle \gamma_*}\downarrow & & \downarrow{\scriptstyle \gamma_*\otimes\gamma_*} \\
E_*(k(n+m)) & \xleftarrow{\pi_*} & E_*(kn) \otimes E_*(km)
\end{array}$$

Such a structure will be referred to as a *Hopf bialgebra structure*. If the multiplication π_* and the comultiplication ∇_* are commutative, we shall say that the Hopf bialgebra is commutative. The above arguments imply the following assertion.

THEOREM. *The homology E_* of the operad E is a commutative Hopf bialgebra.*

In a dual way, we can define the notion of Hopf bicoalgebra and prove that the cohomology E^* of the operad E is a Hopf bicoalgebra.

Note that the notion of Hopf bialgebra is similar to but does not coincide with the notion of Hopf ring [**41**]. For a Hopf bialgebra, the commutativity of the multiplication γ_* is not required; besides, the distributivity relations are different.

If $\mathcal{F} = \{\mathcal{F}(n)\}$ is an E-multiplicative family, then this structure induces multiplications

$$\pi_*\colon \mathcal{F}_*(n) \otimes \mathcal{F}_*(m) \to \mathcal{F}_*(n+m), \quad \mu_*\colon E_*(k) * \mathcal{F}_*(n) \to \mathcal{F}_*(kn)$$

on the homology

$$\mathcal{F}_* = \{\mathcal{F}_*(n)\}, \quad \mathcal{F}_*(n) = H_*(\mathcal{F}(n)/\Sigma_n).$$

Besides, there is a comultiplication

$$\nabla_*: \mathcal{F}_*(n) \to \mathcal{F}_*(n) \otimes \mathcal{F}_*(n),$$

and $\mathcal{F}_*$ taken with the operations π_*, μ_*, and ∇_* satisfies the following conditions:

1. The operations μ_*, ∇_* define a Hopf E_*-module structure on $\mathcal{F}_*$.
2. The operations π_*, ∇_* define a Hopf algebra structure on $\mathcal{F}_*$.
3. The operations μ_* and π_* satisfy the distributivity relation.

This structure on $\mathcal{F}_*$ will be referred to as the *structure of a Hopf algebra over the Hopf bialgebra* E_*. The above arguments imply the following theorem.

THEOREM. *If $\mathcal{F}$ is an E-multiplicative family, then its homology $\mathcal{F}_*$ possesses the structure of a Hopf algebra over the Hopf bialgebra E_*.*

In particular, this theorem implies the following assertion.

THEOREM. *Suppose that a space X is an E-multiplicative space; then its homology $X_* = H_*(X)$ possesses the structure of a Hopf algebra over the Hopf bialgebra E_*.*

Note also that there is an action of an E_∞-operad E on the cochain complex $C^*(X)$ of an arbitrary topological space X. Therefore, the following assertion holds.

THEOREM. *The cohomology $H^*(X)$ of a space X is an algebra over the Hopf bialgebra E_*.*

Dually, the E-coalgebra structure on the chain complex $C_*(X)$ of a topological space X induces on its homology $H_*(X)$ the structure of a coalgebra over the Hopf bicoalgebra E^*.

Now let us pass to the computation of the operation $\gamma_*: E_* * E_* \to E_*$. To do this, it is more convenient to use the family

$$E^\infty = A \times_\phi SA \times_\phi \dots.$$

The form of the generators of this family implies that $1 \in E_*(1)$ is also an identity element for the multiplication γ_*, and we have $\gamma_*(e_i * e_j) = e_i e_j$ for $i \leq j$. To obtain formulas expressing γ_* for $i > j$, we can use the distributivity relations. Denote the multiplication π_* simply by a dot. For example, for $j = 0$ and $i > 0$, we have

$$\gamma_*(e_i * e_0) = \gamma_*(e_i * (1 \cdot 1)) = \sum_k \gamma_*(e_k * 1) \cdot \gamma_*(e_{i-k} * 1) = \begin{cases} e_0 e_k, & i = 2k, \\ 0, & i = 2k+1. \end{cases}$$

In the general case, it is shown by induction over the dimension that the following relation holds:

$$\gamma_*(e_i * e_j) = \sum_k \binom{k-j-1}{i-k-1} e_{i+2(j-k)} e_k, \quad i > j.$$

Since this formula is derived by using only the relations of a Hopf bialgebra, the following assertion holds.

THEOREM. *In the case of $\mathbb{Z}/2$-coefficients, the homology E_* of the operad E is a free commutative Hopf bialgebra with identity $1 \in E_*(1)$ and the single 1-dimensional generator $e_1 \in E_*(2)$.*

Passing to the elements of the Dyer–Lashof algebra, we obtain from these formulas the following formulas for the multiplication in the Dyer–Lashof algebra:

$$Q^r Q^s = \sum_i \binom{i-s-1}{r-s-i-1} Q^{r+s-i} Q^i, \quad r > 2s.$$

Passing to the elements of the Steenrod algebra, we see that the same formulas yield the following formulas for the multiplication in the Steenrod algebra:

$$Sq^i Sq^j = \sum_k \binom{j-k-1}{i-2k} Sq^{i+j-k} Sq^k, \quad i < 2j.$$

Similar relations hold for $\mathbb{Z}/p$-coefficients.

Dually, the operation $\gamma^*\colon E^* \to E^* * E^*$ on the elements $\overline{e_i e_j} \in E^*(4)$ is given by

$$\gamma^*(\overline{e_i e_j}) = \sum_k \binom{j-i}{k} \overline{e}_{i+2k} * \overline{e}_{j-k}.$$

Passing to the Milnor coalgebra $\mathcal{K}$, we see that these formulas yield formulas for the comultiplication $\nabla\colon \mathcal{K} \to \mathcal{K} \otimes \mathcal{K}$, and the comultiplication on the generators ξ_i takes the form

$$\nabla(\xi_i) = \sum_j \xi_{i-j}^{2^j} \otimes \xi_j.$$

On the remaining elements, the comultiplication is determined by the Hopf relation. Thus the Milnor coalgebra $\mathcal{K}$ is a Hopf algebra.

Similarly, in the case of $\mathbb{Z}/p$-coefficients, the comultiplication $\nabla\colon \mathcal{K} \to \mathcal{K} \otimes \mathcal{K}$ is given on the generators by

$$\nabla(\xi_i) = \sum_j \xi_{i-j}^{p^j} \otimes \xi_j; \quad \nabla(\tau_i) = \tau_i \otimes 1 + \sum_j \xi_{i-j}^{p^j} \otimes \tau_j.$$

On the remaining elements, the comultiplication is determined by the Hopf relation. Thus the Milnor coalgebra $\mathcal{K}$ is a Hopf algebra in this case as well.

Apart from the Milnor coalgebra $\mathcal{K}$, we shall also consider the Milnor coalgebra K, which differs from the former by the additional relation $\xi_0 = 1$. To avoid confusion, we shall call $\mathcal{K}$ the *unstable Milnor coalgebra*, and K the *stable Milnor coalgebra*.

Now let us elucidate the relationship between the homology and the cohomology of the operad E. We have the operad map $\nabla\colon E \to E \otimes E$, i.e., the family $\nabla(j)\colon E(j) \to E(j) \otimes E(j)$, which defines a Hopf operad structure on E. Denote by $\nabla(j,i)\colon E(j) \to E(j)^{\otimes i}$ the iteration of this map, $\nabla(j,2) = \nabla(j)$, which is a Σ_j-map, i.e.,

$$\nabla(j)(x\sigma) = \nabla(j)(x)\sigma^{\otimes j}, \quad \sigma \in \Sigma_j.$$

It does not commute with the permutations of the factors in $E(j)^{\otimes i}$; however, since the operad E is acyclic and Σ-free, this map can be extended to a map $\nabla(j,i)\colon E(i) \otimes E(j) \to E(j)^{\otimes i}$ compatible with the actions of the symmetric groups Σ_i and Σ_j.

Let us rewrite this map as $\nabla(j,i)\colon E(j) \to \overline{E}(i) \otimes E(j)^{\otimes i}$. If the operad E is chosen to be free, then we can assume that this map is compatible with the operad structure. Passing to the dual map, we obtain

$$\overline{\nabla}(j,i)\colon E(i) \otimes \overline{E}(j)^{\otimes i} \to \overline{E}(j).$$

We now define the map

$$T(j,i)\colon E(j) \otimes \overline{E}(i)^{\otimes j} \to E(i) \otimes \overline{E}(j)^{\otimes i}$$

as the composition

$$\begin{aligned} E(j) \otimes \overline{E}(i)^{\otimes j} &\xrightarrow{\nabla(j)\otimes 1^{\otimes j}} E(j) \otimes E(j) \otimes \overline{E}(i)^{\otimes j} \\ &\to E(j) \otimes \overline{E}(i)^{\otimes j} \otimes E(j) \xrightarrow{\overline{\nabla}(j,i)\otimes 1} \overline{E}(i) \otimes E(j) \\ &\xrightarrow{1\otimes\nabla(j,i)} \overline{E}(i) \otimes \overline{E}(i) \otimes E(j)^{\otimes i} \xrightarrow{\overline{\nabla}(i)\otimes 1^{\otimes i}} \overline{E}(i) \otimes E(j)^{\otimes i}. \end{aligned}$$

This family of maps $T(j,i)$ defines the permutation operator $T\colon \underline{E}\circ\overline{E} \to \overline{E}\circ\underline{E}$. Direct verification shows that the relations

$$T \circ \gamma\overline{E} = \overline{E}\gamma \circ T\underline{E} \circ \underline{E}T, \quad \overline{\gamma}\underline{E} \circ T = \overline{E}T \circ T\overline{E} \circ \underline{E}\overline{\gamma}$$

hold; equivalently, the following diagrams are commutative:

$$\begin{array}{ccc} \underline{E}^2 \circ \overline{E} & \xrightarrow{T\underline{E}\circ\overline{E}T} & \overline{E}\circ\underline{E}^2 \\ {\scriptstyle \underline{\gamma}\overline{E}}\Big\downarrow & & \Big\downarrow{\scriptstyle \overline{E}\underline{\gamma}} \\ \underline{E}\circ\overline{E} & \xrightarrow{T} & \overline{E}\circ\underline{E} \end{array} \qquad \begin{array}{ccc} \underline{E}\circ\overline{E} & \xrightarrow{T} & \overline{E}\circ\underline{E} \\ {\scriptstyle \underline{E}\overline{\gamma}}\Big\downarrow & & \Big\downarrow{\scriptstyle \overline{\gamma}\underline{E}} \\ \underline{E}\circ\overline{E}^2 & \xrightarrow{\overline{E}T\circ T\overline{E}} & \overline{E}^2\circ\underline{E} \end{array}$$

The permutation operator $T\colon \underline{E}\circ\overline{E} \to \overline{E}\circ\underline{E}$ induces a permutation operator $T_*\colon E_* * E^* \to E^* * E_*$, which makes the diagrams

$$\begin{array}{ccc} E_*^2 * E^* & \xrightarrow{T_*E_* * E^*T_*} & E^* * E_*^2 \\ {\scriptstyle \gamma_*E^*}\Big\downarrow & & \Big\downarrow{\scriptstyle E^*\gamma_*} \\ E_* * E^* & \xrightarrow{T_*} & E^* * E_* \end{array} \qquad \begin{array}{ccc} E_* * E^* & \xrightarrow{T_*} & E^* * E_* \\ {\scriptstyle E_*\gamma^*}\Big\downarrow & & \Big\downarrow{\scriptstyle \gamma^*E_*} \\ E_* * E^{*2} & \xrightarrow{E^*T_* * T_*E^*} & E^{*2} * E_* \end{array}$$

commutative, and which is given by the formulas

$$T_*(e_i * \overline{e}_j) = \sum_k \binom{j+k}{i-k} \overline{e}_{2(j+k)-i} * e_k.$$

These formulas imply, in particular, that the dual permutation operator $T^*\colon E_* * E^* \to E^* * E_*$ coincides with the original map T_*.

The permutation map T_* induces an action $\mu_*\colon E_* * E^* \to E^*$ which is given on the generators by

$$\mu_*(e_i * \overline{e}_j) = \sum_k \binom{j}{i} \overline{e}_{2j-i}.$$

Taking its dual, we obtain the coaction $\tau_*\colon E_* \to E^* * E_*$ given on the generators by

$$\tau_*(e_i) = \sum_k \binom{k}{i-k} \overline{e}_{2j-i} * e_k.$$

Denote the operations on the unstable Milnor coalgebra that are induced by the action μ_* on the elements e_i by $e_i\colon \mathcal{K} \to \mathcal{K}$. The relations for the permutation map T_* imply the following theorem.

THEOREM. *The operations $e_i\colon \mathcal{K} \to \mathcal{K}$ satisfy the following relations*:
1. $e_0(x) = x^2$;
2. $e_i(xy) = \sum e_k(x)e_{i-k}(y)$;
3. $\nabla e_i(x) = \sum \xi_0^{-k} e_{i-k}(\xi_0^k x') \otimes e_k(x'')$, *where* $\sum x' \otimes x'' = \nabla(x)$.

By using these relations, to compute the operations e_i, it suffices to compute only $e_1(\xi_0)$, which can easily be done, $e_1(\xi_0) = \xi_1\xi_0$. By using relation 3, we obtain the next theorem.

THEOREM. *The following formula holds*:

$$e_i(\xi_k) = \begin{cases} \xi_{m+k}\xi_k, & i = 2^{m+k} - 2^k, \\ \xi_{m+k}\xi_{k-1}, & i = 2^{m+k} - 2^k - 2^{k-1}, \\ \dots & \dots \\ \xi_{m+k}\xi_0, & i = 2^{m+k} - 2^k - \dots - 1, \\ 0 & \textit{otherwise}. \end{cases}$$

By using relation 2, we can obtain formulas for the operations e_i on the products of the elements ξ_k.

Passing from the elements e_i to the elements of the Dyer–Lashof algebra, we see that the action of the Dyer–Lashof algebra on the unstable Milnor coalgebra is defined on the generators ξ_i by the formula

$$Q^{i+2^k-1}(\xi_k) = \begin{cases} \xi_{m+k}\xi_k, & i = 2^{m+k} - 2^k, \\ \xi_{m+k}\xi_{k-1}, & i = 2^{m+k} - 2^k - 2^{k-1}, \\ \dots & \dots \\ \xi_{m+k}\xi_0, & i = 2^{m+k} - 2^k - \dots - 1, \\ 0 & \text{otherwise}. \end{cases}$$

On the remaining elements, this action is defined by using the Hopf relations

$$Q^i(xy) = \sum Q^k(x)Q^{i-k}(y).$$

The formula for the permutation map

$$T_*\colon \mathcal{R} \otimes \mathcal{K} \to \mathcal{K} \otimes \mathcal{R}$$

takes the form

$$T_*(Q^i \otimes x) = \sum_k Q^k(\xi_0^{i-k}x) \otimes Q^{i-k}.$$

CHAPTER 6

Homology of Iterated Loop Spaces

6.1. Chain complex of an iterated loop space

One of the most difficult problems of algebraic topology is the problem of computing the homology of iterated loop spaces. The first steps in solving it were made by Adams [**1**]. He proved that for a simply connected topological space X, there is a chain equivalence of differential algebras $C_*(\Omega X) \simeq FC_*(X)$. In particular, if the coalgebra structure on $C_*(X)$ is trivial (for example, this is the case if X is the suspension SY of a certain space Y), then the chain complex $C_*(\Omega X)$ is chain equivalent to the tensor algebra $TC_*(Y)$ of the chain complex $C_*(Y)$.

However, in passing to the cobar construction we lose the coalgebra structure. The cobar construction FK on a coalgebra K has no natural coalgebra structure. Hence the cobar construction cannot be iterated. Nevertheless, in [**4**] Baues introduced a coalgebra structure on the cobar construction $FC_*(X)$ on the singular chain complex of a topological space by using the collection of operations

$$\nabla_{n,m}: C_*(X) \to C_*(X)^{\otimes n} \otimes C_*(X)^{\otimes m}$$

of dimension $n+m-1$. Next he defined the double cobar construction

$$F^2C_*(X) = FFC_*(X)$$

and proved that for any 2-connected topological space X there is a chain equivalence

$$C_*(\Omega^2 X) \simeq F^2C_*(X).$$

However, this approach failed to provide a coalgebra structure on the double cobar construction, and hence further iterations were not constructed.

In [**29**] the homology $H_*(\Omega^n S^n X)$ of the n-fold loop space of the n-fold suspension of a space X were computed by using operad methods. This homology was also computed in other papers; see [**14, 17, 33**], etc.

We consider the general case of computing the homology $H_*(\Omega^n X)$ of an iterated loop space $\Omega^n X$.

Suppose that X is an n-connected topological space. As was shown above, the singular chain complex $C_*(\Omega^n X)$ possesses an E-coalgebra structure, and hence the chain complex $S^nC_*(\Omega^n X)$ possesses an S^nE-coalgebra structure

$$\tau: S^nC_*(\Omega^n X) \to \overline{S^nE}(S^nC_*(\Omega^n X)).$$

This structure and the map $j: S^n\Omega^n X \to X$ induce a map

$$S^nC_*(\Omega^n X) \to \overline{S^nE}(S^nC_*(\Omega^n X)) \to \overline{S^nE}(C_*(X)),$$

which is a coaugmentation of the cosimplicial object $F^*(S^nE, E, C_*(X))$, and hence induces an S^nE-coalgebra map

$$S^nC_*(\Omega^nX) \to F(S^nE, E, C_*(X)).$$

THEOREM. *For any n-connected topological space X, the map*

$$S^nC_*(\Omega^nX) \to F(S^nE, E, C_*(X))$$

is a chain equivalence of S^nE-coalgebras.

PROOF. For $n = 1$, the fact that the map

$$SC_*(\Omega X) \to F(SE, E, C_*(X))$$

is a chain equivalence follows from the Adams chain equivalence $SC_*(\Omega X) \simeq SFC_*(X)$ and the chain equivalence

$$SFC_*(X) \simeq F(SE, E, C_*(X)).$$

Assume that for any $n-1$-connected topological space X, we have the chain equivalence

$$S^{n-1}C_*(\Omega^{n-1}X) \to F(S^{n-1}E, E, C_*(X)).$$

Then, for any n-connected topological space X, we have the following sequence of chain equivalences:

$$\begin{aligned} S^nC_*(\Omega^nX) &\simeq SF(S^{n-1}E, E, C_*(\Omega X)) \simeq F(S^nE, SE, SC_*(\Omega X)) \\ &\simeq F(S^nE, SE, F(SE, E, C_*(X))) \simeq F(S^nE, E, C_*(X)). \end{aligned}$$

Their composition yields the required chain equivalence

$$S^nC_*(\Omega^nX) \simeq F(S^nE, E, C_*(X)). \quad \square$$

To compute the homology of iterated loop spaces, consider the spectral sequence of the cobar construction $F(S^nE, E, X)$ with respect to the filtration defined by the grading of the operad E. Namely, for any operad $\mathcal{E}$, the degree of each map $f: \mathcal{E}(j) \to X^{\otimes j}$ is set equal to j. This grading provides a grading on $\overline{S^nE}(X)$, $\overline{S^nE}\overline{E}(X)$, etc. Thus the cobar construction $F(S^nE, E, X)$ acquires a grading, which defines a decreasing filtration on it.

Note that the maps $\overline{\mu}: \overline{S^nE} \to \overline{S^nE} \circ \overline{E}$, $\overline{\gamma}: \overline{E} \to \overline{E} \circ \overline{E}$ are degree-preserving, and the maps $\tau: X \to \overline{E}(X)$ are degree-raising. Therefore, the first term of the corresponding spectral sequence is isomorphic to the cobar construction $F(S^nE, E, X)$, where X is regarded as a trivial E-coalgebra.

Thus, to compute the homology of the cobar construction $F(S^nE, E, X)$ on the trivial E-coalgebra X, it suffices to compute the homology of the iterated Adams cobar construction F^nX.

Consider the case of $\mathbb{Z}/2$-coefficients. Let $X_* = H_*(X)$. The homology $H_*(FX)$ is isomorphic to the tensor algebra $TS^{-1}X_*$ of the desuspension $S^{-1}X_*$ of X_*, i.e.,

$$H_*(FX) \cong TS^{-1}X_*.$$

The double cobar construction F^2X is chain equivalent to the cobar construction $FTS^{-1}X_*$, where $TS^{-1}X_*$ is regarded as a Hopf algebra with comultiplication

$$\nabla: TS^{-1}X_* \to TS^{-1}X_* \otimes TS^{-1}X_*$$

defined on the generators $[x] \in S^{-1}X_*$ by the formula

$$\nabla[x] = [x] \otimes 1 + 1 \otimes [x].$$

By $LS^{-1}X_*$ we denote the free Lie algebra generated by the module $S^{-1}X_*$. Then, by the Poincaré–Birkhoff–Witt theorem, there is a coalgebra isomorphism

$$TS^{-1}X_* \cong T_s LS^{-1}X_*,$$

where $T_s LS^{-1}X_*$ denotes the free commutative algebra generated by the module $LS^{-1}X_*$.

The homology $FT_s LS^{-1}X_*$ is isomorphic to the free commutative algebra generated by the module $S^{-1}PT_s LS^{-}X_*$, i.e.,

$$H_*(F^2X) \cong T_s S^{-1}PT_s LS^{-1}X_*,$$

where $PT_s LS^{-1}X_*$ denotes the module of primitive elements in $TS^{-1}X_*$, i.e.,

$$PT_s LS^{-1}X_* = \{y \in T_s LS^{-1}X_* | \ \nabla(y) = y \otimes 1 + 1 \otimes y\}.$$

Note that the module PT_sM of primitive elements of the Hopf algebra T_sM is generated by the elements of the form x^{2^k}, $x \in M$, $k \geq 0$.

Therefore, this construction can be iterated, and by induction we obtain isomorphisms

$$H_*(F^nX) \cong T_s(S^{-1}PT_s)^{n-1}LS^{-1}X_*$$

and hence isomorphisms

$$S^{-n}H_*(F(S^nE, E, X)) \cong T_s(S^{-1}PT_s)^{n-1}LS^{-1}X_*.$$

Recall that a graded module L (over $\mathbb{Z}/2$) is called an *n-Lie algebra* if an operation $[\ ,\ ]\colon L \otimes L \to L$ of dimension n is given (this operation is called the *n-Lie bracket*), and the following relations hold:

$$\begin{aligned} [x, x] &= 0; \\ [x, y] + [y, x] &= 0; \\ [x, [y, z]] &= [[x, y], z] + [y, [x, z]]. \end{aligned}$$

Let M be a graded module; by $\mathcal{E}_n M$ denote the module generated by the elements $e_{i_1} \dots e_{i_k} x$, where $x \in M$, $0 \leq i_1 \leq \dots \leq i_k \leq n$, whose dimension is set equal to $i_1 + 2i_2 + \dots + 2^{k-1}i_k + 2^k \dim(x)$.

These elements $e_{i_1} \dots e_{i_k} x$ of $\mathcal{E}_n M$ can be expressed by means of the Dyer–Lashof algebra $\mathcal{R}$ in the form

$$Q^{j_1} \dots Q^{j_k} x; \quad j_1 \leq 2j_2, \dots, j_{k-1} \leq 2j_k, \quad \dim(x) \leq j_k \leq \dim(x) + n,$$

where

$$\begin{aligned} j_k &= i_k + \dim(x), \\ j_{k-1} &= i_{k-1} + i_k + 2\dim(x), \\ &\dots \\ j_1 &= i_1 + i_2 + 2i_3 + \dots + 2^{k-2}i_k + 2^{k-1}\dim(x). \end{aligned}$$

It is clear that the sequence $Q^{j_1} \dots Q^{j_k}$ is admissible and represents an element of the Dyer–Lashof algebra $\mathcal{R}$.

Denote by $\mathcal{R}_n M$ the submodule of $\mathcal{R} \otimes M$ generated by the elements

$$Q^{j_1} \dots Q^{j_k} \otimes x,$$

where $Q^{j_1} \dots Q^{j_k}$ are admissible sequences, $\dim(x) \leq j_k \leq \dim(x) + n$. Then there is an isomorphism $\mathcal{E}_n M \cong \mathcal{R}_n M$.

The correspondence $M \longmapsto \mathcal{R}_n M$ defines a monad $\mathcal{R}_n$ on the category of graded modules. Algebras over the monad $\mathcal{R}_n$ will be referred to as *$\mathcal{R}_n$-modules.*

Let L_n be an n-Lie algebra; consider the module $\mathcal{R}_n L_n$. Denote by $T_s \mathcal{R}_n L_n$ the quotient algebra of the free commutative algebra generated by $\mathcal{R}_n L_n$ modulo the relations $x \cdot x = e_0 x$.

The above arguments imply that if X is a chain complex (over $\mathbb{Z}/2$) regarded as a trivial E-coalgebra, then there are isomorphisms

$$S^{-n} H_*(F(S^n E, E, X)) \cong T_s \mathcal{R}_{n-1} L_{n-1} S^{-n} X_*.$$

Therefore, the following assertion holds.

THEOREM. *Suppose that X is an n-connected topological space. Then there is a spectral sequence converging to $H_*(\Omega^n X)$, and for its first term (over $\mathbb{Z}/2$) there is an isomorphism $E^1 \cong T_s \mathcal{R}_{n-1} L_{n-1} S^{-n} H_*(X)$.*

Now consider the case of $\mathbb{Z}/p$-coefficients, $p > 2$. Note that the module $PT_s M$ of primitive elements of the Hopf algebra $T_s M$ is generated by the elements $x \in M$ and the elements x^{p^k} such that $k > 0$ and $\dim(x)$ is even. However, the homology $H_*(T_s M)$ is generated not only by the primitive elements, but also by the elements βx^{p^k}, where β is the Bockstein homomorphism. Denote the module generated by the primitive elements and the elements βx^{p^k} by $P_\beta T_s M$. Then the homology of $T_s M$ is isomorphic to $T_s S^{-1} P_\beta T_s M$. Thus, for the homology $H_*(F^n X)$ of the iterated cobar construction $F^n X$ with $\mathbb{Z}/p$-coefficients, we have isomorphisms

$$H_*(F^n X) \cong T_s (S^{-1} P_\beta T_s)^{n-1} L S^{-1} X_*$$

and hence isomorphisms

$$S^{-n} H_*(F(S^n E, E, X)) \cong T_s (S^{-1} P_\beta T_s)^{n-1} L S^{-1} X_*.$$

Recall that a graded module L (over $\mathbb{Z}/p$) is called an *n-Lie algebra* if an operation $[\ ,\ ]: L \otimes L \to L$ of dimension n, called the *n-Lie bracket*, is given, and the following relations hold:

$$[x, y] + (-1)^\epsilon [y, x] = 0;$$
$$[x, [y, z]] = [[x, y], z] + (-1)^\epsilon [y, [x, z]];$$

here $\epsilon = (\dim(x) + n) \cdot (\dim(y) + n)$.

Let M be a graded module; by $\mathcal{E}_n(M)$ we denote the module generated by the sequences $\beta^{\epsilon_1} e_{i_1} \dots \beta^{\epsilon_k} e_{i_k} x$, where $x \in M$, $\epsilon = 0, 1$, $0 \leq i_1 \leq \dots \leq i_k \leq n$. The elements $e_i y$ must satisfy the condition that i and $\dim(y)$ have the same parity, and the dimension of $e_i y$ is set equal to $p \cdot \dim(y) + (p-1)i$.

Using the mod-p Dyer–Lashof algebra $\mathcal{R}$, these elements $\beta^{\epsilon_1} e_{i_1} \dots \beta^{\epsilon_k} e_{i_k} x$ can be expressed as $\beta^{\epsilon_1} Q^{j_1} \dots \beta^{\epsilon_k} Q^{j_k} x$, where

$$\begin{aligned} 2j_k &= i_k + \dim(x), \\ 2j_{k-1} &= i_{k-1} + (p-1)i_k + p \cdot \dim(x) - \epsilon_k, \\ &\dots \\ 2j_1 &= i_1 + (p-1)i_2 - \epsilon_2 + \dots + p^{k-1} \cdot \dim(x). \end{aligned}$$

It is clear that the sequences $\beta^{\epsilon_1} Q^{j_1} \dots \beta^{\epsilon_k} Q^{j_k}$ are admissible and represent elements of the Dyer–Lashof algebra $\mathcal{R}$.

Denote by $\mathcal{R}_n M$ the submodule of $\mathcal{R} \otimes M$ generated by the elements

$$\beta^{\epsilon_1} Q^{j_1} \dots \beta^{\epsilon_k} Q^{j_k} \otimes x,$$

where $\beta^{\epsilon_1} Q^{j_1} \dots \beta^{\epsilon_k} Q^{j_k}$ are admissible sequences, $\dim(x) \leq 2j_k \leq \dim(x) + n$. Then there is an isomorphism $\mathcal{E}_n M \cong \mathcal{R}_n M$.

The correspondence $M \longmapsto \mathcal{R}_n M$ defines a monad $\mathcal{R}_n$ on the category of graded $\mathbb{Z}/p$-modules. Algebras over $\mathcal{R}_n$ will be called $\mathcal{R}_n$-modules.

Let L_n be an n-Lie algebra. Consider the module $\mathcal{R}_n L_n$. Denote by $T_s \mathcal{R}_n L_n$ the quotient algebra of the free commutative algebra generated by $\mathcal{R}_n L_n$ modulo the relations $e_0 x = x^p$.

It follows from the above discussion that if X is a chain complex (over $\mathbb{Z}/p$) regarded as a trivial E-coalgebra, then we have isomorphisms

$$S^{-n} H_*(F(S^n E, E, X)) \cong T_s \mathcal{R}_{n-1} L_{n-1} S^{-n} X_*.$$

Therefore, we obtain the following result.

THEOREM. *For the first term of the spectral sequence converging to* $H_*(\Omega^n X)$ *(over* $\mathbb{Z}/p$*), there is an isomorphism* $E^1 \cong T_s \mathcal{R}_{n-1} L_{n-1} S^{-n} H_*(X)$.

Note that in the case of coefficients in a field of characteristic zero, the module $PT_s M$ of primitive elements $T_s M$ is isomorphic to M. Therefore, we obtain the following assertion.

THEOREM. *The first term of the spectral sequence converging to* $H_*(\Omega^n X)$ *(over a field of characteristic zero) is isomorphic to the* $(n-1)$*-Poisson algebra generated by* $S^{-n} H_*(X)$*, i.e.,*

$$E^1 \cong T_s L_{n-1} S^{-n} H_*(X) = P_{n-1} S^{-n} H_*(X).$$

6.2. Computation of the second term of the spectral sequence of iterated loop spaces

To find the second term of the spectral sequence for the homology of iterated loop spaces, we shall first introduce the notion of $\mathcal{P}_n$-algebra, which generalizes the notion of n-Poisson algebra.

A graded module M (over $\mathbb{Z}/2$) is called a $\mathcal{P}_n$*-algebra* if the following three conditions hold:

1. M is provided with the structure of a commutative algebra

$$x \otimes y \longmapsto x \cdot y, \quad x, y \in M.$$

2. M is provided with the structure of an n-Lie algebra

$$x \otimes y \longmapsto [x, y], \quad x, y \in M,$$

which, together with the commutative algebra structure, constitutes an n-Poisson algebra structure.

3. M is provided with the structure of an $\mathcal{R}_n$-module

$$\mathcal{R}_n M \to M$$

compatible with the n-Poisson algebra structure.

By $\mathcal{P}_n$ we denote the monad assigning to any graded module M the free $\mathcal{P}_n$-algebra generated by M. Then there is an isomorphism

$$\mathcal{P}_n(M) \cong T_s \mathcal{R}_n L_n M,$$

and hence we obtain the following assertion.

THEOREM. *The first term of the spectral sequence for $H_*(\Omega^n X)$ (over $\mathbb{Z}/2$) is isomorphic to the free $\mathcal{P}_{n-1}$-algebra generated by $S^{-n}H_*(X)$, i.e.,*

$$E^1 \cong \mathcal{P}_{n-1} S^{-n} H_*(X).$$

Let $\mathcal{R}$ be the Dyer–Lashof algebra, and let $\mathcal{K}$ be the Milnor coalgebra. We define a homomorphism $\varphi\colon \mathcal{K} \to \mathcal{R}$ of dimension -1 by setting

$$\varphi(x) = \begin{cases} Q^{i-1}, & x = \xi_1^i(1), \\ 0 & \text{otherwise.} \end{cases}$$

It is readily seen that $\varphi \cup \varphi = 0$, and hence φ is a twisting cochain.

This twisting cochain induces on $\mathcal{P}_{n-1} S^{-n} H_*(X)$ a differential d_φ which is given on the generators $s^{-n} x_{i+n} \in H_{i+n}(X)$ by the formula

$$d_\varphi(s^{-n} x_{i+n}) = \sum_{x' < x''} [s^{-n} x', s^{-n} x''] + \varphi \cap s^{-n} x_{i+n},$$

where $\sum x' \otimes x'' = \nabla_*(x_{i+n})$. Denote the corresponding graded $\mathcal{P}_{n-1}$-algebra by $\mathcal{P}_{n-1\varphi} S^{-n} H_*(X)$. Then the following assertion holds.

THEOREM. *For the second term of the spectral sequence for $H_*(\Omega^n X)$ (over $\mathbb{Z}/2$), there is an isomorphism $E^2 \cong H_*(\mathcal{P}_{n-1\varphi} S^{-n} H_*(X))$.*

Now let us consider the case of $\mathbb{Z}/p$-coefficients. A graded module M (over $\mathbb{Z}/p$) will be called a *$\mathcal{P}_n$-algebra* if the following three conditions hold:

1. M is provided with the structure of a commutative algebra

$$x \otimes y \longmapsto x \cdot y, \quad x, y \in M.$$

2. M is provided with the structure of an n-Lie algebra

$$x \otimes y \longmapsto [x, y], \quad x, y \in M,$$

which, taken together with the commutative algebra structure, defines an n-Poisson algebra structure.

3. M is provided with the structure of an $\mathcal{R}_n$-module

$$\mathcal{R}_n M \to M$$

compatible with the n-Poisson algebra structure.

Denote by $\mathcal{P}_n$ the monad assigning to any graded module M the free $\mathcal{P}_n$-algebra generated by M. Then there is an isomorphism $\mathcal{P}_n(M) \cong T_s\mathcal{R}_n L_n M$, and hence we obtain the following result.

THEOREM. *The first term of the spectral sequence for $H_*(\Omega^n X)$ (over $\mathbb{Z}/p$) is isomorphic to the free $\mathcal{P}_{n-1}$-algebra generated by $S^{-n}H_*(X)$, i.e.,*

$$E^1 \cong \mathcal{P}_{n-1}S^{-n}H_*(X).$$

Let $\mathcal{R}$ be the mod-p Dyer Lashof algebra, and let $\mathcal{K}$ be the mod-p Milnor coalgebra. We define the homomorphism $\varphi\colon \mathcal{K} \to \mathcal{R}$ of dimension -1 by setting

$$\varphi(x) = \begin{cases} \beta P^i, & x = \xi_1^i(1), \\ P^i, & x = \xi_1^{i-1}(1)\tau_1(1), \\ 0 & \text{otherwise.} \end{cases}$$

It is readily seen that $\varphi \cup \varphi = 0$, and hence φ is a twisting cochain.

This cochain induces on $\mathcal{P}_{n-1}S^{-n}H_*(X)$ a differential d_φ, which is given on the generators $s^{-n}x_{i+n} \in H_{i+n}(X)$ by the formula

$$d_\varphi(s^{-n}x_{i+n}) = \sum_{x'<x''} (-1)^\epsilon [s^{-n}x', s^{-n}x''] + \varphi \cap s^{-n}x_{i+n},$$

where $\sum x' \otimes x'' = \nabla_*(x_{i+n})$, $\epsilon = n \cdot \dim(x')$. Denote the corresponding differential $\mathcal{P}_{n-1}$-algebra by $\mathcal{P}_{n-1\varphi}S^{-n}H_*(X)$. Then the following theorem holds.

THEOREM. *For the second term of the spectral sequence for $H_*(\Omega^n X)$ (over $\mathbb{Z}/p$), there is an isomorphism $E^2 \cong H_*(\mathcal{P}_{n-1\varphi}S^{-n}H_*(X))$.*

Note that in the case of coefficients in a field of characteristic zero, the $\overline{E}_*$-coalgebra structure on the homology $H_*(X)$ of a space X consists only of the commutative coalgebra structure

$$\nabla_*\colon H_*(X) \to H_*(X) \otimes H_*(X).$$

Therefore, the following theorem holds.

THEOREM. *For the second term of the spectral sequence for $H_*(\Omega^n X)$ (over a field of characteristic zero), there is an isomorphism*

$$E^2 \cong H_*(P_{n-1\varphi}S^{-n}H_*(X)),$$

where the differential d_φ is defined on the generators $s^{-n}x_{i+n} \in H_{i+n}(X)$ by the formula

$$d_\varphi(s^{-n}x_{i+n}) = \sum_{x'<x''} (-1)^\epsilon [s^{-n}x', s^{-n}x''],$$

where $\sum x' \otimes x'' = \nabla_*(x_{i+n})$, $\epsilon = n \cdot \dim(x')$.

6.3. Homology of iterated loop spaces of real projective spaces

Among the most important spaces in algebraic topology, apart from the spheres, are the real and complex projective spaces $\mathbb{RP}^n$, $\mathbb{RP}^\infty$ and $\mathbb{CP}^n$, $\mathbb{CP}^\infty$. In a sense, these spaces are opposite to the spheres. For example, the homology $H_*(\mathbb{RP}^\infty)$ (over $\mathbb{Z}/2$) is a free $\overline{E}_*$-coalgebra with a single 1-dimensional generator, in contrast to the spheres, whose homology is a trivial $\overline{E}_*$-coalgebra.

On the other hand, the Adams spectral sequence for stable homotopy groups of $\mathbb{RP}^\infty$ closely resembles the corresponding spectral sequence for spheres, and one of the problems is to find their relationship.

We start by computing the homology $H_*(\Omega^m(\mathbb{RP}^\infty/\mathbb{RP}^n))$ of iterated loop spaces of $\mathbb{RP}^\infty/\mathbb{RP}^n$ with $\mathbb{Z}/2$-coefficients, $m \leq n$. Denote the i-dimensional generator of $H_*(\mathbb{RP}^\infty)$ by e_i. The coalgebra structure on these generators is given by

$$\nabla(e_i) = \sum_j e_j \otimes e_{i-j}.$$

The action of the Steenrod algebra is defined by

$$Sq^j(e_i) = \binom{i-j}{j} e_{i-j}.$$

Since the homology $H_*(\mathbb{RP}^\infty)$ is a free $\overline{E}_*$-coalgebra, there are no higher $\overline{E}_*$-operations on it. The homology $H_*(\mathbb{RP}^n)$ is an $\overline{E}_*$-subcoalgebra of $H_*(\mathbb{RP}^\infty)$, and hence the $\overline{E}_*$-coalgebra structure on $H_*(\mathbb{RP}^\infty)$ induces an $\overline{E}_*$-coalgebra structure on $H_*(\mathbb{RP}^\infty)/H_*(\mathbb{RP}^n) \cong H_*(\mathbb{RP}^\infty/\mathbb{RP}^n)$. Therefore, there are no higher $\overline{E}_*$-operations on $H_*(\mathbb{RP}^\infty/\mathbb{RP}^n)$. Thus the spectral sequence for the homology of iterated loop spaces of $\mathbb{RP}^\infty/\mathbb{RP}^n$ degenerates, and we obtain the following assertion.

THEOREM. *The homology $H_*(\Omega^m(\mathbb{RP}^\infty/\mathbb{RP}^n))$, $n \geq m$, over $\mathbb{Z}/2$ is isomorphic to the homology of the differential $\mathcal{P}_{m-1}$-algebra $\mathcal{P}_{m-1\varphi}S^{-m}H_*(\mathbb{RP}^\infty/\mathbb{RP}^n)$, where the differential d_φ is defined on the generators $s^{-m}e_i$ by the formula*

$$d_\varphi(s^{-m}e_i) = \sum_{\substack{j \\ 2j<i}} [s^{-m}e_j, s^{-m}e_{i-j}] + \sum_{\substack{j \\ 2j\leq i}} \binom{i-j}{j} e_{2j-i+m-1}(s^{-m}e_{i-j}).$$

Note that in the case $m = 1$, there is an isomorphism

$$\mathcal{P}_0 S^{-1}H_*(\mathbb{RP}^\infty/\mathbb{RP}^n) \cong T_s L S^{-1}H_*(\mathbb{RP}^\infty/\mathbb{RP}^n) \cong TS^{-1}H_*(\mathbb{RP}^\infty/\mathbb{RP}^n),$$

and the differential in the tensor algebra $TS^{-1}H_*(\mathbb{RP}^\infty/\mathbb{RP}^n)$ has a very simple form on the generators $u_i = s^{-1}e_{i+1}$, i.e.,

$$d_\varphi(u_i) = \sum_j u_j \otimes u_{i-j-1}.$$

Therefore, the homology $H_*(\Omega(\mathbb{RP}^\infty/\mathbb{RP}^n))$ (over $\mathbb{Z}/2$) is isomorphic to the algebra defined by the generators u_i of dimension i, $n \leq i \leq 2n$, and the relations

$$\begin{gathered} u_n \cdot u_n = 0; \\ u_n \cdot u_{n+1} + u_{n+1} \cdot u_n = 0; \\ \cdots \\ u_n \cdot u_{2n} + \cdots + u_{2n} \cdot u_n = 0. \end{gathered}$$

Thus the homology module $H_*(\Omega(\mathbb{RP}^\infty/\mathbb{RP}^n))$ is generated by noncommutative products $u_{n_1} \cdot \dots \cdot u_{n_k}$, $n \leq n_1 \leq 2n$, $n < n_2, \dots, n_k \leq 2n$.

Consider the homology $H_*(\Omega^2(\mathbb{RP}^\infty/\mathbb{RP}^2))$. It is isomorphic to the homology of a differential $\mathcal{P}_1$-algebra $\mathcal{P}_{1\varphi}\{u_i|\ i \geq 1\}$, where $u_i = s^{-2}e_{i+2}$ and the differential is given by

$$d_\varphi(u_{2i+1}) = \sum_{j<i}[u_j, u_{2i-j-1}] + \binom{i+2}{i+1} e_0(u_i);$$

$$d_\varphi(u_{2i+2}) = \sum_{j<i}[u_j, u_{2i-j}] + e_1(u_i).$$

In small dimensions, we have:
$d(u_1) = 0;$
$d(u_2) = 0;$
$d(u_3) = u_1u_1;$
$d(u_4) = e_1(u_1);$
$d(u_5) = [u_1, u_2];$
$d(u_6) = [u_1, u_3] + e_1(u_2);$
$d(u_7) = [u_1, u_4] + [u_2, u_3] + u_3u_3;$
$d(u_8) = [u_1, u_5] + [u_2, u_4] + e_1(u_3).$

These formulas imply that the homology $H_i = H_i(\Omega^2\mathbb{RP}^\infty/\mathbb{RP}^2))$ in small dimensions i has the following generators:
$H_1: \quad u_1;$
$H_2: \quad u_2;$
$H_3: \quad u_1u_2;$
$H_4: \quad u_2^2;$
$H_5: \quad u_1u_2^2,\ e_1(u_2);$
$H_6: \quad u_1e_1(u_2),\ u_2^3,\ [u_2, u_3],\ u_3^2;$
$H_7: \quad u_1u_2^3,\ u_1u_3^2,\ u_1[u_2, u_3],\ u_2e_1(u_2),\ e_1(u_3).$

Consider the homology $H_*(\Omega^3(\mathbb{RP}^\infty/\mathbb{RP}^3))$. It is isomorphic to the homology of the differential $\mathcal{P}_2$-algebra $\mathcal{P}_{2\varphi}\{u_i|\ i \geq 1\}$, where $u_i = s^{-3}e_{i+3}$ and the differential is given by

$$d_\varphi(u_{2i+1}) = \sum_{j<i-1}[u_j, u_{2i-j-2}] + \binom{i+3}{i+1} e_0(u_i) + e_2(u_{i-1});$$

$$d_\varphi(u_{2i+2}) = \sum_{j\leq i-1}[u_j, u_{2i-j-1}] + \binom{i+3}{i+2} e_1(u_i).$$

In small dimensions we have:
$d(u_1) = 0;$
$d(u_2) = 0;$
$d(u_3) = 0;$
$d(u_4) = 0;$
$d(u_5) = e_2(u_1);$
$d(u_6) = [u_1, u_2] + e_1(u_2);$
$d(u_7) = [u_1, u_3] + e_2(u_2) + u_3u_3;$
$d(u_8) = [u_1, u_4] + [u_2, u_3].$

These formulas imply that the homology $H_i = H_i(\Omega^3(\mathbb{RP}^\infty/\mathbb{RP}^3))$ in small dimensions i has the following generators:
$H_1: \quad u_1;$

$H_2:\ u_1^2,\ u_2;$
$H_3:\ u_1^3,\ u_1u_2,\ e_1(u_1),\ u_3;$
$H_4:\ u_1^4,\ u_1^2u_2,\ u_1e_1(u_1),\ u_1u_3,\ u_2^2,\ u_4;$
$H_5:\ u_1^5,\ u_1^3u_2,\ u_1^2u_3,\ u_1^2e_1(u_1),\ u_1u_2^2,\ u_1u_4,\ u_2u_3,\ u_2e_1(u_1),\ e_1(u_2);$
$H_6:\ u_1^6,\ u_1^4u_2,\ u_1^3u_3,\ u_1^3e_1(u_1),\ u_1^2u_2^2,\ u_1^2u_4,\ u_1u_2u_3,\ u_1u_2e_1(u_1),\ u_1e_1(u_2),\ u_2u_4,\ u_2^3,\ u_3e_1(u_1),\ u_3^2,\ e_2(u_2),\ e_1(u_1)^2;$
$H_7:\ u_1^7,\ u_1^5u_2,\ u_1^4u_3,\ u_1^4e_1(u_1),\ u_1^3u_2^2,\ u_1^3u_4,\ u_1^2u_2u_3,\ u_1^2u_2e_1(u_1),\ u_1^2e_1(u_2),\ u_1u_2u_4,\ u_1u_2^3,\ u_1u_3e_1(u_1),\ u_1u_3^2,\ u_1e_2(u_2),\ u_1e_1(u_1)^2,\ u_2^2u_3,\ u_2^2e_1(u_1),\ u_2e_1(u_2),\ u_3u_4,\ [u_2,u_3],\ e_1(u_3),\ e_1e_1(u_1).$

We also consider the homology $H_*(\Omega^4(\mathbb{RP}^\infty/\mathbb{RP}^4))$. It is isomorphic to the homology of the differential $\mathcal{P}_3$-algebra $\mathcal{P}_{3\varphi}\{u_i|\ i\geq 1\}$, where $u_i = s^{-4}e_{i+4}$ and the differential is given by

$$d_\varphi(u_{2i+1}) = \sum_{j<i-1}[u_j,u_{2i-j-3}] + \binom{i+4}{i+1}e_0(u_i) + \binom{i+3}{i+2}e_2(u_{i-1});$$

$$d_\varphi(u_{2i+2}) = \sum_{j<i-1}[u_j,u_{2i-j-2}] + \binom{i+4}{i+2}e_1(u_i) + e_3(u_{i-1}).$$

In small dimensions we have:
$d(u_1)=0;$
$d(u_2)=0;$
$d(u_3)=0;$
$d(u_4)=0;$
$d(u_5)=e_2(u_1);$
$d(u_6)=e_1(u_2)+e_3(u_1);$
$d(u_7)=[u_1,u_2]+u_3u_3;$
$d(u_8)=[u_1,u_3]+e_1(u_3)+e_3(u_2).$

These formulas imply that the homology $H_i = H_i(\Omega^4(\mathbb{RP}^\infty/\mathbb{RP}^4))$ in small dimensions i has the following generators:
$H_1:\ u_1;$
$H_2:\ u_1^2,\ u_2;$
$H_3:\ u_1^3,\ u_1u_2,\ e_1(u_1),\ u_3;$
$H_4:\ u_1^4,\ u_1^2u_2,\ u_1e_1(u_1),\ u_1u_3,\ u_2^2,\ u_4;$
$H_5:\ u_1^5,\ u_1^3u_2,\ u_1^2u_3,\ u_1^2e_1(u_1),\ u_1u_2^2,\ u_1u_4,\ u_2u_3,\ u_2e_1(u_1),\ e_1(u_2);$
$H_6:\ u_1^6,\ u_1^4u_2,\ u_1^3u_3,\ u_1^3e_1(u_1),\ u_1^2u_2^2,\ u_1^2u_4,\ u_1u_2u_3,\ u_1u_2e_1(u_1),\ u_1e_1(u_2),\ u_2^3,\ u_2u_4,\ u_3e_1(u_1),\ u_3^2,\ e_2(u_2),\ e_1(u_1)^2;$
$H_7:\ u_1^7,\ u_1^5u_2,\ u_1^4u_3,\ u_1^4e_1(u_1),\ u_1^3u_2^2,\ u_1^3u_4,\ u_1^2u_2u_3,\ u_1^2u_2e_1(u_1),\ u_1^2e_1(u_2),\ u_1u_2^3,\ u_1u_2u_4,\ u_1u_3e_1(u_1),\ u_1u_3^2,\ u_1e_2(u_2),\ u_1e_1(u_1)^2,\ u_2^2u_3,\ u_2^2e_1(u_1),\ u_2e_1(u_2),\ u_3u_4,\ e_1(u_3),\ e_3(u_2),\ e_1e_1(u_1).$

6.4. Homology of iterated loop spaces of complex projective spaces

Now we consider the problem of computing the homology $H_*(\Omega^m(\mathbb{CP}^\infty/\mathbb{CP}^n))$ of the iterated loop spaces of $\mathbb{CP}^\infty/\mathbb{CP}^n$, $m\leq 2n+1$. We start with $\mathbb{Z}/2$-coefficients.

Denote the $2i$-dimensional generator of $H_*(\mathbb{CP}^\infty)$ by c_i. The coalgebra structure on these generators is defined by

$$\nabla(c_i) = \sum_j c_j\otimes c_{i-j}.$$

The action of the Steenrod algebra is defined by

$$Sq^{2j}(c_i) = \binom{i-j}{j} c_{i-j}.$$

Since $H_*(\mathbb{CP}^\infty)$ is a free $\overline{E}_*$-coalgebra, there are no higher $\overline{E}_*$-operations on it. Since $H_*(\mathbb{CP}^n)$ is an $\overline{E}_*$-subcoalgebra of $H_*(\mathbb{CP}^\infty)$, the $\overline{E}_*$-coalgebra structure on $H_*(\mathbb{CP}^\infty)$ induces an $\overline{E}_*$-coalgebra structure on

$$H_*(\mathbb{CP}^\infty)/H_*(\mathbb{CP}^n) \cong H_*(\mathbb{CP}^\infty/\mathbb{CP}^n).$$

Hence there are no higher $\overline{E}_*$-operations on the homology $H_*(\mathbb{CP}^\infty/\mathbb{CP}^n)$. Thus the spectral sequence for the homology of an iterated loop space of $\mathbb{CP}^\infty/\mathbb{CP}^n$ has no higher differentials, so that we obtain the following result.

THEOREM. *The homology $H_*(\Omega^m(\mathbb{CP}^\infty/\mathbb{CP}^n))$, $m \leq 2n+1$, over $\mathbb{Z}/2$ is isomorphic to the homology of the differential $\mathcal{P}_{m-1}$-algebra*

$$\mathcal{P}_{m-1\varphi} S^{-m} H_*(\mathbb{CP}^\infty/\mathbb{CP}^n),$$

where the differential d_φ is defined on the generators $s^{-m}c_i$ by

$$d_\varphi(s^{-m}c_i) = \sum_{\substack{j \\ 2j<i}} [s^{-m}c_j, s^{-m}c_{i-j}] + \sum_{\substack{j \\ 2j \leq i}} \binom{i-j}{j} e_{2(2j-i)+m-1}(s^{-m}c_{i-j}).$$

Note that in the case $m = 1$, there is an isomorphism

$$\mathcal{P}_0 S^{-1} H_*(\mathbb{CP}^\infty/\mathbb{CP}^n) \cong T_s L S^{-1} H_*(\mathbb{CP}^\infty/\mathbb{CP}^n) \cong T S^{-1} H_*(\mathbb{CP}^\infty/\mathbb{CP}^n),$$

and the differential in the tensor algebra $TS^{-1}H_*(\mathbb{CP}^\infty/\mathbb{CP}^n)$ has a very simple form on the generators $v_i = s^{-1}c_{i+1}$ of dimension $2i+1$, i.e.,

$$d_\varphi(v_i) = \sum_j v_j \otimes v_{i-j-1}.$$

It follows that the homology $H_*(\Omega(\mathbb{CP}^\infty/\mathbb{CP}^n))$ (over $\mathbb{Z}/2$) is isomorphic to the algebra defined by the generators v_i, $n \leq i \leq 2n$, and the relations

$$v_n \cdot v_n = 0;$$
$$v_n \cdot v_{n+1} + v_{n+1} \cdot v_n = 0;$$
$$\cdots$$
$$v_n \cdot v_{2n} + \cdots + v_{2n} \cdot v_n = 0.$$

Thus the homology $H_*(\Omega(\mathbb{CP}^\infty/\mathbb{CP}^n))$ is generated as a graded module by the noncommutative products $v_{n_1} \cdot \ldots \cdot v_{n_k}$, $n \leq n_1 \leq 2n$, $n < n_2, \ldots, n_k \leq 2n$.

Consider the homology $H_*(\Omega^2(\mathbb{CP}^\infty/\mathbb{CP}^n))$. It is isomorphic to the homology of the differential $\mathcal{P}_1$-algebra $\mathcal{P}_{1\varphi}\{v_i | \ i \geq n\}$, where $v_i = s^{-2}c_{i+1}$, $\dim(v_i) = 2i$, and the differential is defined by

$$d_\varphi(v_{2i}) = \sum_{j<i} [v_j, v_{2i-j-1}];$$
$$d_\varphi(v_{2i+1}) = \sum_{j<i} [v_j, v_{2i-j}] + e_1(v_i).$$

These formulas imply that the homology $H_*(\Omega^2(\mathbb{CP}^\infty/\mathbb{CP}^n))$ is a $\mathcal{P}_1$-algebra defined by the generators $v_n, \dots, v_{2n}$ and the relations

$$\begin{gathered} e_1(v_n) = 0; \\ [v_n, v_{n+1}] = 0; \\ [v_n, v_{n+2}] + e_1(v_{n+1}) = 0; \\ \dots \\ [v_n, v_{2n}] + [v_{n+1}, v_{2n-1}] + \dots = 0. \end{gathered}$$

Consider the homology $H_*(\Omega^4(\mathbb{CP}^\infty/\mathbb{CP}^2))$. It is isomorphic to the homology of the differential $\mathcal{P}_3$-algebra $\mathcal{P}_{3\varphi}\{v_i|\ i \geq 1\}$, where $v_i = s^{-4}c_{i+2}$, $\dim(v_i) = 2i$, and the differential is defined by

$$\begin{aligned} d_\varphi(v_{2i+1}) &= \sum_{j<i} [v_j, v_{2i-j-1}] + \binom{i+2}{i+1} e_1(v_i); \\ d_\varphi(v_{2i+2}) &= \sum_{j<i} [v_j, v_{2i-j}] + e_3(v_i). \end{aligned}$$

In small dimensions, we have:
$d(v_1) = 0;$
$d(v_2) = 0;$
$d(v_3) = e_1(v_1);$
$d(v_4) = e_3(v_1);$
$d(v_5) = [v_1, v_2];$
$d(v_6) = [v_1, v_3] + e_3(v_2);$
$d(v_7) = [v_1, v_4] + [v_2, v_3] + e_1(v_3);$

These formulas imply that the homology $H_i = H_i(\Omega^4(\mathbb{CP}^\infty/\mathbb{CP}^2))$ in small dimensions i has the following generators:
$H_1 = 0;$
$H_2: \ v_1;$
$H_3 = 0;$
$H_4: \ v_1^2,\ v_2;$
$H_5 = 0;$
$H_6: \ v_1^3,\ v_1v_2,\ e_2(v_1);$
$H_7 = 0;$
$H_8: \ v_1^4,\ v_1^2v_2,\ v_1e_2(v_1),\ v_2^2;$
$H_9: \ e_1(v_2);$
$H_{10}: \ v_1^5,\ v_1^3v_2,\ v_1^2e_2(v_1),\ v_1v_2^2,\ v_2e_2(v_1),\ e_2(v_2);$
$H_{11}: \ v_1e_1(v_2),\ e_3(v_2);$
$H_{12}: \ v_1^6,\ v_1^4v_2,\ v_1^3e_2(v_1),\ v_1^2v_2^2,\ v_1v_2e_2(v_1),\ v_1e_2(v_2);$
$H_{13}: \ v_1^2e_1(v_2),\ v_1e_3(v_2),\ e_1(v_3),\ [v_2, v_3].$

We now consider $\mathbb{Z}/p$-coefficients, $p > 2$. The coalgebra structure is defined on the generators by

$$\nabla(c_i) = \sum_j c_j \otimes c_{i-j}.$$

The action of the mod-p Steenrod algebra is defined by

$$P^j(c_i) = \binom{i-(p-1)j}{j} c_{i-(p-1)j}.$$

Since $H_*(\mathbb{CP}^\infty)$ is a free $\overline{E}_*$-coalgebra, there are no higher $\overline{E}_*$-operations on it. Since $H_*(\mathbb{CP}^n)$ is an $\overline{E}_*$-subcoalgebra of $H_*(\mathbb{CP}^\infty)$, the $\overline{E}_*$-coalgebra structure on $H_*(\mathbb{CP}^\infty)$ induces an $\overline{E}_*$-coalgebra structure on

$$H_*(\mathbb{CP}^\infty)/H_*(\mathbb{CP}^n) \cong H_*(\mathbb{CP}^\infty/\mathbb{CP}^n).$$

Hence there are no higher $\overline{E}_*$-operations on the homology $H_*(\mathbb{CP}^\infty/\mathbb{CP}^n)$. Thus the spectral sequence for the homology of iterated loop spaces of $\mathbb{CP}^\infty/\mathbb{CP}^n$ has no higher differentials, and we obtain the following result.

THEOREM. *The homology $H_*(\Omega^m(\mathbb{CP}^\infty/\mathbb{CP}^n))$, $m \le 2n+1$, over $\mathbb{Z}/p$ is isomorphic to the homology of the differential $\mathcal{P}_{m-1}$-algebra*

$$\mathcal{P}_{m-1\varphi}S^{-m}H_*(\mathbb{CP}^\infty/\mathbb{CP}^n),$$

where the differential d_φ is defined on the generators $s^{-m}e_i$ by

$$\begin{aligned} d_\varphi(s^{-m}c_i) &= \sum_{\substack{j\\ 2j\le i}} [s^{-m}c_j, s^{-m}c_{i-j}] \\ &+ \sum_{\substack{j\\ pj<i}} \binom{i-(p-1)j}{j} \beta e_{2(pj-i)+m}(s^{-m}c_{i-(p-1)j}). \end{aligned}$$

Note that in the case $m=1$ there is an isomorphism

$$\mathcal{P}_0 S^{-1}H_*(\mathbb{CP}^\infty/\mathbb{CP}^n) \cong T_sLS^{-1}H_*(\mathbb{CP}^\infty/\mathbb{CP}^n) \cong TS^{-1}H_*(\mathbb{CP}^\infty/\mathbb{CP}^n),$$

and the differential in the tensor algebra $TS^{-1}H_*(\mathbb{CP}^\infty/\mathbb{CP}^n)$ has a very simple form, i.e.,

$$d_\varphi(s^{-1}c_i) = \sum_j s^{-1}c_j \otimes s^{-1}c_{i-j}.$$

Hence the homology $H_*(\Omega(\mathbb{CP}^\infty/\mathbb{CP}^n))$ (over $\mathbb{Z}/p$) is isomorphic to the algebra defined by the generators $v_i = s^{-1}c_{i+1}$ of dimension $2i+1$, $n \le i \le 2n$, and the relations

$$\begin{gathered} v_n \cdot v_n = 0; \\ v_n \cdot v_{n+1} + v_{n+1} \cdot v_n = 0; \\ \cdots \\ v_n \cdot v_{2n} + \cdots + v_{2n} \cdot v_n = 0. \end{gathered}$$

Thus the homology $H_*(\Omega(\mathbb{CP}^\infty/\mathbb{CP}^n))$ is generated as a graded module by the noncommutative products $v_{n_1} \cdot \ldots \cdot v_{n_k}$, $n \le n_1 \le 2n$, $n < n_2, \ldots, n_k \le 2n$.

Consider the homology $H_*(\Omega^2(\mathbb{CP}^\infty/\mathbb{CP}^n))$. It is isomorphic to the homology of the differential $\mathcal{P}_1$-algebra $\mathcal{P}_{1\varphi}\{v_i |\ i \ge 1\}$, where $v_i = s^{-2}c_{i+1}$, $i \ge n$, and the differential is defined by

$$d_\varphi(v_i) = \sum_{2j<i} [v_j, v_{i-j-1}].$$

These formulas imply that the homology $H_i = H_i(\Omega^2(\mathbb{CP}^\infty/\mathbb{CP}^n))$ is the $\mathcal{P}_1$-algebra generated by the elements $v_n, \dots, v_{2n}$ and the relations

$$\begin{gathered}[v_n, v_n] = 0;\\ [v_n, v_{n+1}] = 0;\\ [v_n, v_{n+2}] + [v_{n+1}, v_{n+1}] = 0;\\ \dots\\ [v_n, v_{2n}] + [v_{n+1}, v_{2n-1}] + \dots = 0.\end{gathered}$$

Consider the homology $H_*(\Omega^4(\mathbb{CP}^\infty/\mathbb{CP}^2))$. It is isomorphic to the homology of the differential $\mathcal{P}_3$-algebra $\mathcal{P}_{3\varphi}\{v_i|\ i \geq 1\}$, where $v_i = s^{-4}c_{i+2}$, $\dim(v_i) = 2i$, and the differential is defined by

$$\begin{aligned} d_\varphi(v_{pi-1}) &= \sum_{2j<pi-2} [v_j, v_{pi-j-3}] + \binom{i+1}{i}\beta e_2(v_{i-1});\\ d_\varphi(v_k) &= \sum_{2j<k-1} [v_j, v_{k-j-2}],\ k \neq pi-1. \end{aligned}$$

In small dimensions, we have:
$d(v_1) = 0;$
$d(v_2) = 0;$
$d(v_3) = 0;$
$d(v_4) = [v_1, v_1];$
$d(v_5) = [v_1, v_2];$
$d(v_6) = [v_1, v_3] + [v_2, v_2];$
$\dots$
$d(v_{2p-1}) = [v_1, v_{2p-4}] + \dots + [v_{p-2}, v_{p-1}] + 3\beta e_2(v_1);$
$\dots$

These formulas imply that the homology $H_i = H_i(\Omega^4(\mathbb{CP}^\infty/\mathbb{CP}^2))$, for example with $\mathbb{Z}/3$-coefficients, in small dimensions i has the following generators:
$H_1 = 0;$
$H_2 :\ \ v_1;$
$H_3 = 0;$
$H_4 :\ \ v_1^2,\ v_2;$
$H_5 = 0;$
$H_6 :\ \ v_1^3,\ v_1v_2,\ v_3;$
$H_7 = 0;$
$H_8 :\ \ v_1^4,\ v_1^2v_2,\ v_1v_3,\ v_2^2;$
$H_9 :\ \ \beta e_2(v_1);$
$H_{10} :\ \ v_1^5,\ v_1^3v_2,\ v_1^2v_3,\ v_1v_2^2,\ v_2v_3,\ e_2(v_1);$
$H_{11} :\ \ v_1\beta e_2(v_1),\ [v_2, v_2].$

CHAPTER 7

Homotopy Theories and E_∞-Structures

7.1. Localization of categories and homotopy theory

In this section we shall outline the necessary preliminaries from category theory and homotopy theory for a closed model category [**39, 40**].

Let $\mathcal{X}$ and $\mathcal{Y}$ be categories. Recall that a functor $G: \mathcal{Y} \to \mathcal{X}$ is said to be *adjoint to a functor* $F: \mathcal{X} \to \mathcal{Y}$ if there are given natural transformations of functors

$$\alpha: \mathrm{Id} \to G \circ F, \quad \beta: F \circ G \to \mathrm{Id}$$

such that for any object X of the category $\mathcal{X}$ and any object Y of the category $\mathcal{Y}$, the following relations hold:

$$\beta(F(X)) \circ F(\alpha(X)) = \mathrm{Id}_{F(X)}, \quad G(\beta(Y)) \circ \alpha(G(Y)) = \mathrm{Id}_{G(Y)} .$$

If a functor $G: \mathcal{Y} \to \mathcal{X}$ is adjoint to a functor $F: \mathcal{X} \to \mathcal{Y}$, then for any objects $X \in \mathcal{X}$, $Y \in \mathcal{Y}$, the natural transformations

$$\alpha(X): X \to G \circ F(X), \quad \beta(Y): F \circ G(Y) \to Y$$

induce a bijection

$$\mathrm{Hom}_{\mathcal{X}}(X, G(Y)) \longleftrightarrow \mathrm{Hom}_{\mathcal{Y}}(F(X), Y).$$

A functor $F: \mathcal{X} \to \mathcal{Y}$ is said to be *completely injective* if for any objects $X', X'' \in \mathcal{X}$ the map

$$\mathrm{Hom}(X', X'') \to \mathrm{Hom}(F(X'), F(X''))$$

induced by the functor is bijective.

If a functor G is adjoint to a functor F, then the natural transformation $\alpha: \mathrm{Id} \to G \circ F$ is an isomorphism if and only if the functor F is completely injective. Indeed, in this case, for any objects $X', X'' \in \mathcal{X}$, we have the commutative diagram

$$\begin{array}{ccc} \mathrm{Hom}(X', X'') & \xrightarrow{\qquad\qquad} & \mathrm{Hom}(F(X'), F(X'')) \\ & \searrow^{\cong} \quad {}^{\cong}\swarrow & \\ & \mathrm{Hom}((GF(X')); X'') & \end{array}$$

Let $\mathcal{X}$ be a category and $\mathcal{U}$ a class of morphisms in $\mathcal{X}$. The *localization of* $\mathcal{X}$ *with respect to* $\mathcal{U}$ consists of a category $\mathcal{U}^{-1}\mathcal{X}$ together with a functor $P: \mathcal{X} \to \mathcal{U}^{-1}\mathcal{X}$ that takes the morphisms of class $\mathcal{U}$ to isomorphisms of the category $\mathcal{U}^{-1}\mathcal{X}$ and is universal with respect to this property, i.e., for any functor $Q: \mathcal{X} \to \mathcal{Y}$ that takes the morphisms of class $\mathcal{U}$ to isomorphisms of the category $\mathcal{Y}$, there exists a unique functor $J: \mathcal{U}^{-1}\mathcal{X} \to \mathcal{Y}$ such that $Q = J \circ P$. The universal property of the functor P directly implies that P is a bijection on the objects of $\mathcal{X}$, and any morphism of

$\mathcal{U}^{-1}\mathcal{X}$ can be represented as a finite composition of morphisms of the forms $P(f)$ and $P(u)^{-1}$, where f is a morphism of $\mathcal{X}$ and $u \in \mathcal{U}$.

Suppose that $\mathcal{X}$ and $\mathcal{Y}$ are categories, $\mathcal{U}$ and $\mathcal{V}$ are classes of morphisms in $\mathcal{X}$ and in $\mathcal{Y}$, respectively, and a functor $G: \mathcal{Y} \to \mathcal{X}$ is adjoint to a functor $F: \mathcal{X} \to \mathcal{Y}$. Suppose F takes the morphisms of class $\mathcal{U}$ to morphisms of class $\mathcal{V}$ and G takes the morphisms of class $\mathcal{V}$ to morphisms of class $\mathcal{U}$. Then F and G induce functors

$$\widetilde{F}: \mathcal{U}^{-1}\mathcal{X} \to \mathcal{V}^{-1}\mathcal{Y}, \quad \widetilde{G}: \mathcal{V}^{-1}\mathcal{Y} \to \mathcal{U}^{-1}\mathcal{X}.$$

Further, if the morphism $\alpha(X): X \to GF(X)$ belongs to $\mathcal{U}$ for any object $X \in \mathcal{X}$, then the functor $\widetilde{F}: \mathcal{U}^{-1}\mathcal{X} \to \mathcal{V}^{-1}\mathcal{Y}$ is completely injective. Indeed, in this case the functor $\widetilde{G}$ is adjoint to $\widetilde{F}$, and for any object $X \in \mathcal{X}$ the morphism $\widetilde{\alpha}(X): X \to \widetilde{G}\widetilde{F}(X)$ is an isomorphism.

A *closed model category* consists of a category $\mathcal{X}$ together with three distinguished classes of morphisms, called *cofibrations*, *fibrations*, and *weak equivalences*, which satisfy the following conditions:

1. The category $\mathcal{X}$ is closed with respect to taking finite projective and injective limits.

2. Suppose that f and g are morphisms in $\mathcal{X}$, and the composition $g \circ f$ is defined; then if two of the three morphisms f, g, and $g \circ f$ are weak equivalences, so is the third.

3. Suppose that $f_k: X'_k \to X''_k$, $k = 1, 2$, are morphisms in $\mathcal{X}$, and f_1 is a retract of f_2, i.e., there are morphisms $i^k: X^k_1 \to X^k_2$, $j^k: X^k_2 \to X^k_1$, $k = {}'{,}''$, such that $j^k \circ i^k = \mathrm{Id}$ and the following diagrams are commutative:

$$\begin{array}{ccc} X'_1 & \xrightarrow{f_1} & X''_1 \\ {\scriptstyle i'}\downarrow\uparrow{\scriptstyle j'} & & {\scriptstyle i''}\downarrow\uparrow{\scriptstyle j''} \\ X'_2 & \xrightarrow{f_2} & X''_2 \end{array}$$

Then f_1 is a cofibration (a fibration, a weak equivalence) whenever f_2 is.

A morphism which is both a fibration (a cofibration) and a weak equivalence is referred to as a *trivial fibration* (*cofibration*).

4. Suppose that a diagram

$$\begin{array}{ccc} A & \longrightarrow & Y \\ {\scriptstyle i}\downarrow & & \downarrow{\scriptstyle p} \\ B & \longrightarrow & X \end{array}$$

is commutative, and one of the following two conditions holds:

a) p is a fibration and i is a trivial cofibration;

b) p is a trivial fibration and i is a cofibration.

Then there exists a diagonal morphism $g: B \to Y$ that can be included into the commutative diagram.

5. Each morphism f can be decomposed in the following two ways:

a) $f = p \circ i$, where i is a cofibration and p is a trivial fibration;

b) $f = p \circ i$, where i is a trivial cofibration and p is a fibration.

Let $i: A \to B$ and $p: Y \to X$ be arbitrary morphisms. We say that the morphism $i: A \to B$ satisfies the *left lifting property with respect to the morphism* $p: Y \to X$

and the morphism $p: Y \to X$ satisfies the *right lifting property with respect to the morphism* $i: A \to B$ if for any commutative diagram

$$\begin{array}{ccc} A & \longrightarrow & Y \\ {\scriptstyle i}\downarrow & & \downarrow{\scriptstyle p} \\ B & \longrightarrow & X \end{array}$$

there exists a diagonal morphism $g: B \to Y$.

It turns out that the cofibrations (the trivial cofibrations) are just the morphisms that satisfy the left lifting property with respect to all trivial fibrations (all fibrations). Similarly, the fibrations (the trivial fibrations) are just the morphisms that satisfy the right lifting property with respect to all trivial cofibrations (all cofibrations).

An object X of category $\mathcal{X}$ is said to be *cofibrant* if the morphism $e \to X$, where e is the initial object, is a cofibration. An object Y is said to be *fibrant* if the morphism $Y \to f$, where f is the final object, is a fibration.

Denote by $X \vee X$ the direct sum of two copies of an object X. A *cylinder over* X consists of an object X_I together with maps

$$d_i: X \to X_I, \ i = 1, 2, \quad \sigma: X_I \to X$$

such that $d_1 + d_2: X \vee X \to X_I$ is a cofibration, σ is a weak equivalence, and $\sigma \circ d_i = \mathrm{Id}_X$, $i = 1, 2$.

Let $f, g: X \to Y$ be morphisms; a *left homotopy from f to g* is a morphism $h: X_I \to Y$ such that $h \circ d_0 = f$, $h \circ d_1 = g$. In this case the morphism f is said to be *left homotopic to g*. If X is a cofibrant object, then the left homotopy relation is an equivalence relation on $\mathrm{Hom}(X; Y)$.

Dually, for any object $Y \in \mathcal{X}$, we define the path object Y^I and the notion of right homotopy $h: X \to Y^I$. If X is a cofibrant object and Y is a fibrant object, then the left and right homotopy relations on $\mathrm{Hom}(X, Y)$ coincide; in this case the set of equivalence classes in $\mathrm{Hom}(X, Y)$ is denoted by $[X, Y]$.

Denote by $\mathcal{X}_{cf}$ the category whose objects are the objects of $\mathcal{X}$ that are both cofibrant and fibrant, and whose morphisms are given by $\mathrm{Hom}_{\pi\mathcal{X}_{cf}}(X, Y) = [X, Y]$. Denote by $Ho\mathcal{X}$ the localization of the category $\mathcal{X}$ with respect to the class of weak equivalences. Then the functor $P: \mathcal{X} \to Ho\mathcal{X}$ induces a functor $\pi\mathcal{X}_{cf} \to Ho\mathcal{X}$, and the following assertions hold:

1. The functor $\pi\mathcal{X}_{cf} \to Ho\mathcal{X}$ is an equivalence of categories.
2. If X is a cofibrant object and Y is a fibrant object, then

$$[X, Y] \cong \mathrm{Hom}_{Ho\mathcal{X}}(P(X), P(Y)).$$

3. $P(f)$ is an isomorphism if and only if f is a weak equivalence.

An example of a closed model category is furnished by the category of simplicial sets, where cofibrations are injective maps, fibrations are Kan fibrations, and weak equivalences are weak homotopy equivalences.

7.2. Curtis spectral sequence

Here we consider the Curtis spectral sequence converging to the homotopy groups of a simplicial set or a simplicial group [**16**].

For the case of simplicial groups, we have a useful alternative definition of the homotopy groups. Namely, suppose that G_* is a simplicial group. Let

$$K_n(G_*) = G_n \cap \operatorname{Ker} d_0 \cap \dots \cap \operatorname{Ker} d_{n-1}.$$

We have the inclusion $d_{n+1}(K_{n+1}(G_*)) \subset K_n(G_*)$, and the image of the map $d_{n+1}: K_{n+1}(G_*) \to K_n(G_*)$ is contained in the kernel of $d_n: K_n(G_*) \to K_{n-1}(G_*)$. Let us show that this image is a normal subgroup of G_n. Indeed, suppose that $x \in K_{n+1}(G_*)$ and $z \in G_n$. Let $w = s_n(z)xs_n(z^{-1})$. It is readily seen that $d_i(w) = e$ for $0 \le i \le n$, and $d_{n+1}w = zd_{n+1}(x)z^{-1}$; therefore, we have $zd_{n+1}(x)z^{-1} \in d_{n+1}(K_{n+1}(G_*))$. This completes the proof.

Thus $K(G_*) = \{K_n(G_*)\}$ is a differential graded group (generally, nonabelian).

Let us show that the homology of this differential group coincides with the homotopy groups, i.e., $H_n(K(G_*)) = \pi_n(G_*)$. Indeed, the set of the simplices $x \in G_n$ such that $d_i(x) = e$, $0 \le i \le n$, coincides with the set of cycles in $K_n(G_*)$. Suppose that such simplices x and y are homotopic, and z is the corresponding homotopy; then, for the element $w = s_n(x^{-1})z \in G_{n+1}$, we have $d_iw = e$ for $0 \le i \le n$, and $d_{n+1}w = x^{-1}y$. This means that x and y represent the same homology class. Therefore, we have an epimorphism $i_n: \pi_n(G_*) \to H_n(K(G_*))$. Assume that $i_n[x] = e$. Then there exists an element $z \in K_{n+1}(G_*)$ such that $d_{n+1}z = x$. But this means that x is homotopic to zero in $\pi_n(G_*)$. Therefore, i_n is an isomorphism.

In the case of an abelian simplicial group G_*, we can take the normalized chain complex $N(G_*)$ instead of the differential graded group $K(G_*)$. There is a natural map $K(G_*) \to N(G_*)$, which is easily shown to be an isomorphism. Therefore, for any simplicial set X_* and any ring R, the homotopy groups $\pi_*(RX_*)$ of the simplicial module RX_* are isomorphic to the homology groups $H_*(X_*; R)$.

The inclusion map $X_* \to RX_*$ induces a homomorphism

$$\pi_*(X_*) \to \pi_*(RX_*) \cong H_*(X_*; R),$$

which is called the *Hurewicz homomorphism*.

This fact can be used to construct Eilenberg–MacLane spaces. Indeed, to construct a space of type $K(R, n)$, we first consider the simplicial set S^n_* with a single zero-dimensional generator $*$ and a single n-dimensional generator u_n. Next, we consider the simplicial module RS^n_* and its realization $|RS^n_*|$. The homotopy groups of this space are isomorphic to the homotopy groups of the simplicial module RS^n_*, which, in turn, are isomorphic to the homology groups of the simplicial set S^n_* with coefficients in R. Since these homology groups are zero in all dimensions except for dimension n, in which the homology group is equal to R, we see that $|RS^n_*|$ is an Eilenberg–MacLane space of type $K(R, n)$.

Let us recall some information concerning commutators in groups. Suppose that G is a group. For any elements $x, y \in G$, let $x^y = y^{-1}xy$, and let $[x, y] = x^{-1}y^{-1}xy$. The last expression is called the *commutator of the elements* $x, y \in G$. Straightforward computations show that for any elements $x, y, z \in G$ the following relation holds:

$$[x^y, [y, z]][y^z, [z, x]][z^x, [x, y]] = 1.$$

Let A and B be subgroups of G. Denote by $[A, B]$ the subgroup of G generated by the elements of the form $[a, b]$, where $a \in A$, $b \in B$. If A, B are normal subgroups of G, so is $[A, B]$. The above relation implies that for any normal subgroups A, B, C of G, we have the inclusion

$$[A, [B, C]] \subset [B, [C, A]][C, [A, B]].$$

The *lower central series of* G is the filtration

$$G = \Gamma^1 \supset \Gamma^2 \supset \cdots \supset \Gamma^m \supset \Gamma^{m+1} \supset \dots,$$

where Γ^{m+1} is the subgroup of Γ^m generated by the commutators of the elements of G and of Γ^m, i.e., $\Gamma^{m+1} = [G, \Gamma^m]$.

Let us show that we have the inclusion $[\Gamma^n, \Gamma^m] \subset \Gamma^{n+m}$ for any n and m.

Indeed, for $n = 1$, the inclusion directly follows from the definition. Assume that the inclusion holds for $n - 1$; let us show that it is true for n. We have

$$\begin{aligned}[\Gamma^n, \Gamma^m] = [[G, \Gamma^{n-1}], \Gamma^m] &\subset [G, [\Gamma^{n-1}, \Gamma^m]][\Gamma^{n-1}, [G, \Gamma^m]] \\ &\subset [G, \Gamma^{n+m-1}][\Gamma^{n-1}, \Gamma^{m+1}] \subset \Gamma^{n+m}\Gamma^{n+m} \subset \Gamma^{n+m}.\end{aligned}$$

For each m, we set $\operatorname{gr}^m G = \Gamma^m/\Gamma^{m+1}$ and $\operatorname{gr} G = \sum_m \operatorname{gr}^m G$. The group $\operatorname{gr} G$ is referred to as the *graded group associated to* G. It is clear that it is an abelian group, and the correspondence $(x, y) \longmapsto [x, y]$ induces a homomorphism $[\ ,\ ]\colon \operatorname{gr} G \otimes \operatorname{gr} G \to \operatorname{gr} G$, which defines a Lie algebra structure on $\operatorname{gr} G$.

Let $G = FX$ be the free group generated by a set X. Denote by $\mathbb{Z}X$ the free abelian group generated by X, and by $L\mathbb{Z}X$ the free Lie algebra generated by $\mathbb{Z}X$. This Lie algebra is the Lie subalgebra of $TAX = \sum_n \mathbb{Z}X^{\otimes n}$ generated by $\mathbb{Z}X$. The canonical map $X \to \operatorname{gr}^1 FX$ induces a Lie algebra isomorphism $L\mathbb{Z}X \cong \operatorname{gr} FX$. Moreover, if $L^m\mathbb{Z}X$ denotes a subgroup of $L\mathbb{Z}X$ generated by the elements consisting of m-fold tensor products, then we have isomorphisms $L^m\mathbb{Z}X \cong \operatorname{gr}^m FX$.

Now suppose that X_* is a simply connected simplicial set, and GX_* is the simplicial loop group. In this group, we have the filtration

$$GX_* = \Gamma^1 \supset \Gamma^2 \supset \cdots \supset \Gamma^m \supset \Gamma^{m+1} \supset \dots,$$

where Γ^{m+1} is the simplicial subgroup of Γ^m generated by the commutators of the elements of GX_* and of Γ^m, i.e., $\Gamma^{m+1} = [GX_*, \Gamma^m]$.

This filtration gives rise to an exact couple

$$\begin{gathered} E_1^{m,n} = \pi_{n-m}(\Gamma^m/\Gamma^{m+1}), \quad D_1^{m,n} = \pi_{n-m}(\Gamma^m), \\ i\colon D_1^{m,n} \to D_1^{m-1,n-1}, \quad j\colon D_1^{m,n} \to E_1^{m,n}, \quad k\colon E_1^{m,n} \to D_1^{m+1,n}. \end{gathered}$$

The corresponding spectral sequence is referred to as the *Curtis spectral sequence.*

The higher differentials d^r are determined by the composite maps

$$j \circ (i^{-1})^{r-1} \circ k\colon E_1^{m,n} \to E_1^{m+r,n+r-1},$$

which are partial and multivalued maps for $r \geq 2$.

For $r > m$, we have the inclusion $E_{r+1}^{m,n} \subset E_r^{m,n}$, and hence the terms

$$E_\infty^{m,n} = \bigcap_{r>m} E_r^{m,n}$$

are defined.

In the case of a simply connected simplicial set X_*, the Curtis spectral sequence converges to the homotopy groups $\pi_*(GX_*)$.

Let us elucidate the structure of the E_1-term of this spectral sequence. For any simplicial abelian group A_*, let $S^{-1}A_*$ denote the simplicial abelian group such that $(S^{-1}A)_n = A_{n+1}/\operatorname{Im} s_0$ and the face and degeneracy operators are given by

$$(S^{-1}d)_0 = d_1 - d_0; \quad (S^{-1}d)_i = d_{i+1},\ i > 0; \quad (S^{-1}s)_i = s_{i+1}, \qquad i \geq 0.$$

There is a chain equivalence $S^{-1}N(A_*) \simeq N(S^{-1}A_*)$. It follows directly from the definition that $S^{-1}\mathbb{Z}X_*$ is the quotient of the group GX_* by its commutant. Hence the projection $GX_* \to S^{-1}\mathbb{Z}X_*$ induces the Hurewicz homomorphism $\pi_*(GX_*) \to S^{-1}H_*(X_*)$. Thus we have an isomorphism

$$E_1^{1,n} = \pi_{n-1}(\Gamma^1/\Gamma^2) \cong H_{n-1}(X_*).$$

Denote the free simplicial Lie algebra generated by a simplicial abelian group A_* by LA_*. By definition, LA_* is the simplicial Lie subalgebra of $TA_* = \sum_n A_*^{\otimes n}$ generated by A_*. Let L^mA_* be the subgroup of elements of filtration m in LA_*, i.e., the subgroup consisting of m-fold tensor products. Then there are isomorphisms $\Gamma^m/\Gamma^{m+1} \cong L^mS^{-1}\mathbb{Z}X_*$ and hence isomorphisms

$$E_1^{m,n} \cong \pi_{n-m}(L^mS^{-1}\mathbb{Z}X_*),$$

which, taken together, yield an isomorphism $E_1 \cong \pi_*(LS^{-1}\mathbb{Z}X_*)$.

Thus the E_1-term of the Curtis spectral sequence for a simplicial set X_* is isomorphic to the homotopy groups of the free simplicial Lie algebra generated by the simplicial abelian group $S^{-1}\mathbb{Z}X_*$.

In the computations it is more convenient to use the mod-p version of this spectral sequence, which is referred to as the *Curtis–Rector spectral sequence.* To define it, consider the *lower p-central series of G*, i.e.,

$$G = \Gamma^1 \supset \Gamma^2 \supset \cdots \supset \Gamma^m \supset \Gamma^{m+1} \supset \dots,$$

where Γ^{m+1} is the subgroup of Γ^m generated by the elements $[x_1, \dots, x_k]^{p^i}$, where $kp^i \geq m$ and $x_1, \dots, x_k \in G$. The symbol $[\,,\dots,\,]$ denotes the iterated commutator.

For example, if $p = 2$, then we have

$$\Gamma^2 = [G, G] + G^2,$$
$$\Gamma^3 = [[G, G], G] + G^4,$$

and so on.

The definition directly implies that there is an isomorphism

$$G/\Gamma^2 \cong G/[G, G] \otimes \mathbb{Z}/p.$$

For any $\mathbb{Z}/p$-module M, we consider the tensor algebra TM and the free restricted Lie algebra LM generated by M, i.e., the smallest submodule of TM which contains M and is closed with respect to the operations $[x, y] = x \otimes y - y \otimes x$ and $x^{[p]} = x^{\otimes p}$, $x, y \in TM$.

Consider the free group $G = FX$ generated by a set X. The canonical map $X \to \operatorname{gr}^1 FX$ induces a Lie algebra isomorphism $L\mathbb{Z}/pX \cong \operatorname{gr} FX$. Furthermore, denote the submodule of $L\mathbb{Z}/pX$ generated by the elements consisting of m-fold tensor products by $L^m\mathbb{Z}/pX$; then there are isomorphisms $L^m\mathbb{Z}/pX \cong \operatorname{gr}^m FX$.

For any simplicial set X_* and the corresponding filtration of the simplicial group GX_*, the E_1-term of the Curtis spectral sequence is isomorphic to the homotopy groups of the free simplicial restricted Lie algebra generated by the simplicial abelian group $S^{-1}\mathbb{Z}/pX_*$.

In the case of a simply connected simplicial set X_*, this spectral sequence converges to the p-component of the homotopy groups $\pi_*(GX_*)$, i.e., the subgroup generated by the elements of infinite order and of order equal to a power of p.

7.3. Bousfield–Kan spectral sequence

Another spectral sequence converging to the homotopy groups of a simplicial set is the Bousfield–Kan spectral sequence [**9**].

Suppose that X_* is a pointed simplicial set, R is a commutative ring, and RX_* is the free simplicial R-module generated by the simplicial set X_*.

We have the inclusion map $i_*\colon X_* \to RX_*$ and the simplicial module map $\gamma_*\colon R^2X_* \to RX_*$, $\gamma_*(1\cdot y) = y$, $y \in RX_*$. Besides, the functor $R\colon Ens_s \to Ens_s$ is a monad on the category of simplicial sets.

The *cosimplicial R-resolution of the simplicial set* X_* is the augmented cosimplicial object

$$R^*X_* : X_* \xrightarrow{\delta^0} RX_* \xrightarrow{\delta^0,\delta^1} R^2X_* \to \dots,$$

where $(R^*X_*)^n = R^{n+1}X_*$ and the coface and codegeneracy operators are defined by

$$\delta^i = R^i i_* R^{n-i}\colon R^nX_* \to R^{n+1}X_*,$$
$$\sigma^i = R^i\gamma_* R^{n-i}\colon R^{n+2}X_* \to R^{n+1}X_*.$$

For an arbitrary augmented cosimplicial object X^* in the category of simplicial sets, we shall construct an augmented cosimplicial object DX^*, which is called the *derived object*, by setting $DX^n = X^n \times_{\phi^n} GX^{n+1}$, where the twisting function $\phi^n\colon X^n \to GX^{n+1}$ is the composite map

$$X^n \xrightarrow{\delta^0} X^{n+1} \xrightarrow{\alpha} GX^{n+1}.$$

Thus the twisted Cartesian product $X^n \times_{\phi^n} GX^{n+1}$ is the pullback of the principal twisted Cartesian product $X^{n+1} \times_{\phi^n} GX^{n+1}$ by the map $\delta^0\colon X^n \to X^{n+1}$.

The cofaces $\delta^i\colon DX^{n-1} \to DX^n$ and the codegeneracies $\sigma^i\colon DX^{n+1} \to DX^n$ are induced by the coface and codegeneracy operators in X^*.

The projections $p^n\colon X^n \times_{\phi^n} GX^{n+1}$ induce a map of cosimplicial objects $p^*\colon DX^* \to X^*$.

By iterating the derived objects, we obtain the sequence of cosimplicial objects

$$X^* \xleftarrow{p^*} D^1X^* = DX^* \xleftarrow{p^*} D^2X^* \leftarrow \dots \leftarrow D^nX^* \xleftarrow{p^*} D^{n+1}X^* \leftarrow \dots.$$

In particular, we have the sequence of simplicial sets

$$X^{-1} \xleftarrow{p^{-1}} D^1X^{-1} \xleftarrow{p^{-1}} D^2X^{-1} \leftarrow \dots \leftarrow D^nX^{-1} \xleftarrow{p^{-1}} D^{n+1}X^{-1} \leftarrow \dots,$$

which can be rewritten as a tower of fibrations, i.e.,

$$\begin{array}{ccccc}
 & & \cdots & & \\
 & & \downarrow & & \\
GD^mX^0 & \xrightarrow{i} & D^{m+1}X^{-1} & \xrightarrow{\delta^0} & D^{m+1}X^0 \\
 & & \downarrow p & & \\
GD^{m-1}X^0 & \xrightarrow{i} & D^mX^{-1} & \xrightarrow{\delta^0} & D^mX^0 \\
 & & \downarrow & & \\
 & & \cdots & & \\
 & & \downarrow & & \\
GX^0 & \xrightarrow{i} & D^1X^{-1} & \xrightarrow{\delta^0} & D^1X^0 \\
 & & \downarrow p & & \\
 & & X^{-1} & \xrightarrow{\delta^0} & X^0
\end{array}$$

This tower of fibrations determines a homotopy exact couple

$$E_1^{m,n} = \pi_{n-m}(D^mX^0), \quad D_1^{m,n} = \pi_{n-m}(D^mX^{-1});$$
$$i^{m,n}\colon D_1^{m,n} \to D_1^{m-1,n-1}, \quad j^{m,n}\colon D_1^{m,n} \to E_1^{m,n}, \quad k^{m,n}\colon E_1^{m,n} \to D_1^{m+1,n}.$$

The differentials $d^r\colon E_1^{m,n} \to E_1^{m+r,n+r-1}$ are the partial and multivalued maps

$$\delta^0 \circ (p^{-1})^{r-1} \circ i\colon \pi_{n-m}(D^mX^{-1}) \to \pi_{n-m+1}(D^{m+r}X^{-1}).$$

Clearly, for $r > m$, we have the inclusion $E_{r+1}^{m,n} \subset E_r^{m,n}$. Therefore, the terms

$$E_\infty^{m,n} = \bigcap_{r>m} E_r^{m,n}$$

are defined.

Consider the Adams filtration

$$F^l\pi_k(D^mX^{-1}) = \operatorname{Im}\{\pi_k(D^{m+l}X^{-1}) \to \pi_k(D^mX^{-1})\},$$
$$F^\infty\pi_k(D^mX^{-1}) = \bigcap_l F^l\pi_k(D^mX^{-1}).$$

There is a short exact sequence

$$0 \to (F^m/F^{m+1})\pi_{n-m}(X^{-1}) \xrightarrow{\delta} E_\infty^{m,n} \xrightarrow{i} F^\infty\pi_{n-m-1}(D^{m+1}X^{-1}) \cap \operatorname{Ker} p \to 0.$$

This is proved by a standard diagram chase in the tower of fibrations.

This short exact sequence implies that if the relations $F^\infty\pi_k(D^mX^{-1}) = 0$ hold, then the spectral sequence converges to the homotopy groups $\pi_*(X^{-1})$.

If the cosimplicial object in the above construction is the cosimplicial resolution R^*X_* of a simplicial set X_*, then the corresponding spectral sequence is called the *Bousfield–Kan spectral sequence*.

Note that the Adams spectral sequence for the stable homotopy groups of a simplicial set X_* differs from the Bousfield-Kan spectral sequence only in that the following tower is taken for the tower of fibrations:

$$\begin{array}{ccccc}
 & & \cdots & & \\
 & & \downarrow & & \\
GRD^{(m)}X_* & \longrightarrow & D^{(m+1)}X_* & \longrightarrow & RD^{(m+1)}X_* \\
 & & \downarrow & & \\
GRD^{(m-1)}X_* & \longrightarrow & D^{(m)}X_* & \longrightarrow & RD^{(m)}X_* \\
 & & \downarrow & & \\
 & & \cdots & & \\
 & & \downarrow & & \\
GRX_* & \longrightarrow & D^{(1)}X_* & \longrightarrow & RD^{(1)}X_* \\
 & & \downarrow & & \\
 & & X_* & \longrightarrow & RX_*
\end{array}$$

where $D^{(m+1)}X_* = D^{(m)}X_* \times_{\phi^{(m)}} GRD^{(m)}X_*$.

If $R = \mathbb{Z}$ and the simplicial set X_* is simply connected, then the Bousfield–Kan spectral sequence which is constructed from the cosimplicial resolution $\mathbb{Z}^*X_*$ satisfies the above condition and hence converges to the homotopy groups of X_*.

If $R = \mathbb{Z}/p$ and the simplicial set X_* is simply connected, then the corresponding Bousfield–Kan spectral sequence satisfies the relations

$$F^\infty \pi_k(D^m X^{-1}) = \bigcap_l p^l \pi_k(D^m X^{-1}),$$

and hence it converges to the p-component of the homotopy groups of X_*.

These assertions are proved by comparing the Bousfield–Kan and Curtis–Rector spectral sequences and by establishing an isomorphism between them.

Let us elucidate the structure of the E_2-term of the Bousfield–Kan spectral sequence. Consider the cosimplicial resolution of a simplicial set X_*,

$$R^*X_*\colon X_* \xrightarrow{\delta^0} RX_* \xrightarrow{\delta^0,\delta^1} R^2X_* \to \dots,$$

and the cosimplicial object in the category of graded abelian groups which is obtained by applying the functor of homotopy groups to this resolution,

$$\pi_*(R^*X_*)\colon \pi_*(RX_*) \xrightarrow{\delta^0,\delta^1} \pi_*(R^2X_*) \to \dots.$$

For each cosimplicial object $A^* = \{A^n\}$ in the category of abelian groups, the *cohomotopy groups* $\pi^*(A^*)$ are defined as the homology of A^* with respect to the differential

$$\delta = \sum_{i=0}^{n} (-1)^i \delta^i\colon A^{n-1} \to A^n.$$

In the case of the Bousfield–Kan spectral sequence, there are natural maps

$$\pi_n(R^{m+1}X_*) \cong \pi_{n-m}(G^m R^{m+1}X_*) \to \pi_{n-m}((D^m R^* X_*)^{-1}) = E_1^{m,n}$$

inducing an isomorphism $E_2^{m,n} \cong \pi^m \pi_n(R^*X_*)$. Hence, to describe the E_2-term of the Bousfield–Kan spectral sequence, it is necessary to compute the homotopy groups $\pi_n(R^{m+1}X_*)$ of the simplicial modules $R^{m+1}X_*$, and then to compute the cohomotopy groups of the resulting cosimplicial object in the category of abelian groups.

The computation of the homotopy groups $\pi_n(R^{m+1}X_*)$ or, what is the same, of the homology groups $H_*(R^m X_*; R)$ for $R = \mathbb{Z}/p$ leads to the Milnor coalgebra. For simplicity, we shall consider the case $R = \mathbb{Z}/2$.

Recall that the unstable Milnor coalgebra $\mathcal{K}$ is a polynomial algebra on generators ξ_i of dimension $2^i - 1$, $i \geq 0$.

The comultiplication $\nabla\colon \mathcal{K} \to \mathcal{K} \otimes \mathcal{K}$ is defined on these generators by

$$\nabla(\xi_i) = \sum_{k=0}^{i} \xi_{i-k}^{2^k} \otimes \xi_k.$$

On the remaining elements, the comultiplication ∇ is determined by the condition that $\mathcal{K}$ is a Hopf algebra.

We define the *excessiveness* $e(x)$ of an element $x \in \mathcal{K}$ by setting $e(\xi_i) = 1$ and letting the excessiveness of any product be the sum of the excessivenesses of its factors.

For any graded module L, by $\mathcal{K} \times L$ we denote the submodule of $\mathcal{K} \otimes L$ generated by the elements $x \otimes y_n$, $x \in \mathcal{K}$, $y_n \in L_n$, such that $e(x) = n$.

A module L is called an *unstable comodule over the Milnor coalgebra* or just an *unstable $\mathcal{K}$-comodule* if a coaction $\tau\colon L \to \mathcal{K} \times L$ is given such that the following diagram is commutative:

$$\begin{array}{ccc} L & \xrightarrow{\tau} & \mathcal{K} \times L \\ {\scriptstyle\tau}\downarrow & & \downarrow{\scriptstyle 1\times\tau} \\ \mathcal{K} \times L & \xrightarrow{\nabla\times 1} & \mathcal{K} \times \mathcal{K} \times L \end{array}$$

Note that in the definition of an unstable comodule over the Milnor coalgebra we can use the Milnor coalgebra K in which $\xi_0 = 1$. Namely, for any graded module L, let $K \times L$ denote the submodule of $K \otimes L$ generated by the elements $x \otimes y_n$ such that $e(x) \leq n$.

It is readily seen that there is an isomorphism $\mathcal{K} \times L \cong K \times L$. Hence L is an unstable comodule over the Milnor coalgebra if a coaction $\tau\colon L \to K \times L$ is given.

Dually, a graded module M is called an *unstable module over the Steenrod algebra* $\mathcal{A}$ if M is endowed with an action of $\mathcal{A}$ satisfying the condition $Sq^i(x_n) = 0$ for $i > n$.

For any graded module L, by $C \times K \times L$ we denote the quotient of the free commutative coalgebra $C(K \times L)$ generated by $K \times L$ modulo the equivalence relation generated by the relations $y_n \otimes y_n \sim \xi_1^n \otimes y_n$, where $y_n \in L_n$.

The correspondence $L \longmapsto C \times K \times L$ is a comonad on the category of graded modules.

A module L is called an *unstable coalgebra over the Milnor coalgebra* or just an *unstable K-coalgebra* if it is a coalgebra over the comonad $C \times K$.

Dually, an unstable module M over the Steenrod algebra is called an *unstable algebra over the Steenrod algebra* if M is an algebra whose multiplication commutes with the action of the Steenrod algebra and satisfies the relation $x_n \cdot x_n = Sq^n(x_n)$.

For any unstable module M over the Steenrod algebra, let $U(M)$ denote the free unstable algebra generated by M. It is the quotient of the free commutative algebra $S(M)$ generated by M modulo the equivalence relation generated by the relations $x_n \cdot x_n \sim Sq^n(x_n)$.

A simplicial set or a topological space X is said to be *good in the Massey–Peterson sense* [**32**] if the cohomology of $H^*(X)$ is isomorphic to $U(M)$ for a certain M. In particular, the spheres S^n are good spaces in the Massey–Peterson sense.

Now let us return to the Bousfield–Kan spectral sequence. There are natural isomorphisms

$$\pi_*((\mathbb{Z}/2)^2X_*) = H_*(\mathbb{Z}/2X_*; \mathbb{Z}/2) \cong C \times K \times \pi_*(\mathbb{Z}/2X_*) = C \times K \times H_*(X_*; \mathbb{Z}/2),$$

and hence there are isomorphisms $\pi_*((\mathbb{Z}/2)^{m+1}X_*) \cong (C \times K)^m \times H_*(X_*; \mathbb{Z}/2)$.

The map $\delta^0 \colon \mathbb{Z}/2X_* \to (\mathbb{Z}/2)^2X_*$ induces the structure of an unstable coalgebra over the Milnor coalgebra on the homology $H_*(X_*; \mathbb{Z}/2)$.

Consider the cosimplicial object

$$F^*(C \times K, H_*(X_*; \mathbb{Z}/2) \colon H_*(X_*; \mathbb{Z}/2) \to C \times K \times H_*(X_*; \mathbb{Z}/2) \to \dots$$

and the corresponding cochain complex $F(C \times K, H_*(X_*; \mathbb{Z}/2))$. The cohomology of this cochain complex is isomorphic to the E_2-term of the Bousfield–Kan spectral sequence for the case $R = \mathbb{Z}/2$.

If X is a good space in the Massey–Peterson sense, then the inclusion map $K \times H_*(X; \mathbb{Z}/2) \to C \times K \times H_*(X; \mathbb{Z}/2)$ induces a chain equivalence

$$F(K, H_*(X; \mathbb{Z}/2)) \simeq F(C \times K, H_*(X; \mathbb{Z}/2)),$$

and thus the E_2-term of the Bousfield–Kan spectral sequence for unstable homotopy groups of a good space is isomorphic to the homology of the cobar construction $F(K, H_*(X; \mathbb{Z}/2))$.

In the stable dimensions it coincides with the E_2-term of the Adams spectral sequence for stable homotopy groups of topological spaces, which is isomorphic to the homology of the cobar construction

$$F(K, H_*(X; \mathbb{Z}/2)) \colon \quad H_*(X; \mathbb{Z}/2) \to K \otimes H_*(X; \mathbb{Z}/2) \to \dots.$$

In particular, the E_2-term of the Adams spectral sequence for the stable homotopy groups of the spheres is isomorphic to the homology of the cobar construction

$$FK \colon \quad K \to K \otimes K \to \dots \to K^{\otimes n} \to \dots$$

Since the cobar construction FK is a differential algebra, it follows that the E_2-term of the Adams spectral sequence is a graded A_∞-algebra or, in another language, a graded algebra with Massey products.

By definition, the term $E_2^{1,*}$ of the Adams spectral sequence is equal to the kernel of the comultiplication ∇ on the coalgebra K. It is readily seen that this kernel is generated by the elements of the form $\xi_1^{2^i}$. The corresponding elements $[\xi_1^{2^i}]$ are denoted by h_i. They have dimension $2^i - 1$ and generate the term $E_2^{1,*}$.

It is not hard to show that the term $E_2^{2,*}$ is generated by the products $h_i \cdot h_j$ subject to the relations

$$h_i \cdot h_j = h_j \cdot h_i, \quad h_i \cdot h_{i+1} = 0.$$

The relations of the first type follow from the relations

$$\nabla(\xi_1^{2^{i+j}}) = \xi_1^{2^i} \otimes \xi_1^{2^j} + \xi_1^{2^j} \otimes \xi_1^{2^i},$$

and the relations of the second type from the relations

$$\nabla(\xi_2^{2^i}) = \xi_1^{2^{i+1}} \otimes \xi_1^{2^i}.$$

In the term $E_2^{3,*}$, the products $h_i \cdot h_j \cdot h_k$ satisfy, in addition to the above relations, the relations

$$h_i \cdot h_{i+2}^2 = 0, \quad h_i^3 = h_{i-1}^2 \cdot h_{i+1}.$$

However, except for these products, there are also other elements, the first of which, i.e., the Massey product $\pi_1(h_1, h_0, h_2^2)$, appears in the term $E_2^{3,11}$.

As is shown in [**22**], the iterated products and Massey products of the elements h_i completely generate the E_2-term of the Adams spectral sequence. In what follows, we reformulate this result with the help of the A_∞-algebra structure and find all the relations between these generators. Thus we obtain a description of the E_2-term of the Adams spectral sequence.

7.4. Homotopy groups of symmetric products

Suppose that Y is a simplicial module; denote the submodule of $Y^{\otimes j}$ consisting of all elements invariant under the permutations of the factors by $C(j, Y)$, and let $CY = \sum_j C(j, Y)$. Our aim is to compute the homotopy groups of these simplicial modules.

Let $N: Mod_s \to Chain$ be the normalized chain functor from the category of simplicial modules to the category of chain complexes; this functor assigns to each simplicial module Y the normalized chain complex NY, where

$$(NY)_n = Y_n / \sum_{i=0}^{n-1} \operatorname{Im}\{s_i: Y_{n-1} \to Y_n\}$$

and the differential $d: NY \to NY$ is induced by the map

$$\sum_{i=0}^{n} (-1)^i d_i: Y_n \to Y_{n-1}.$$

Further, let $N^{-1}: Chain \to Mod_s$ be the functor (adjoint to N) that assigns to each chain complex X the simplicial module

$$N^{-1}X = \operatorname{Hom}(\Delta^*; X) = \{\operatorname{Hom}(\Delta^n, X)\},$$

where Δ^n is the normalized chain complex of the standard n-dimensional simplex, and Hom is taken in the category of chain complexes. The face and degeneracy operators

$$d_i: \operatorname{Hom}(\Delta^n, X) \to \operatorname{Hom}(\Delta^{n-1}, X); \quad s_i: \operatorname{Hom}(\Delta^n, X) \to \operatorname{Hom}(\Delta^{n+1}, X)$$

are induced by the corresponding coface and codegeneracy operators in the cosimplicial object $\Delta^* = \{\Delta^n\}$.

For each chain complex X, consider the Σ_j-chain complex

$$E(j, X) = N(N^{-1}(X)^{\otimes j}).$$

The correspondence $X \longmapsto E(j,X)$ defines a functor from the category of chain complexes to the category of Σ_j-chain complexes. The maps

$$f = f^1 \otimes \cdots \otimes f^j \colon \Delta^n \otimes \cdots \otimes \Delta^n \to X \otimes \cdots \otimes X,$$

where $f^1, \ldots, f^j \colon \Delta^n \to X$ are chain maps, are generators of dimension n in $E(j,X)$. The differential in $E(j,X)$ is induced by the coface operator $\delta^i \colon \Delta^{n-1} \to \Delta^n$,

$$d(f^1 \otimes \cdots \otimes f^j) = \sum_{i=0}^{n} (-1)^i (f^1\delta^i \otimes \cdots \otimes f^j\delta^i).$$

The action of the symmetric group Σ_j on $E(j,X)$ is defined by the formula

$$(\sigma f)(x) = \sigma f(\sigma^{-1}x).$$

THEOREM. *Suppose that $f, g \colon X' \to X''$ are homotopic maps of chain complexes. Then the induced maps $E(j,f), E(j,g) \colon E(j,X') \to E(j,X'')$ are Σ_j-homotopic.*

PROOF. Indeed, under the assumptions of the theorem, the induced maps

$$N^{-1}(f), N^{-1}(g) \colon N^{-1}(X') \to N^{-1}(X'')$$

are simplicially homotopic. Hence their tensor products

$$N^{-1}(f)^{\otimes j}, N^{-1}(g)^{\otimes j} \colon N^{-1}(X')^{\otimes j} \to N^{-1}(X'')^{\otimes j}$$

are Σ_j-simplicially homotopic. Applying the functor N to these products yields Σ_j-chain homotopic maps

$$E(j,f), E(j,g) \colon E(j,X') \to E(j,X''). \quad \square$$

COROLLARY 1. *Suppose that X' and X'' are homotopy equivalent chain complexes. Then the corresponding chain complexes $E(j,X')$ and $E(j,X'')$ are Σ_j-homotopy equivalent.*

COROLLARY 2. *Suppose that R is a field. Then there is a Σ_j-homotopy equivalence $E(j,X) \simeq E(j,H_*(X))$, where $H_*(X)$ is the homology of X regarded as a chain complex with zero differential.*

THEOREM. *For any simplicial module Y, there is a Σ_j-homotopy equivalence $N(Y^{\otimes j}) \simeq E(j,NY)$.*

PROOF. The simplicial homotopy equivalence $Y \simeq N^{-1}NY$ induces a Σ_j-simplicial homotopy equivalence $Y^{\otimes j} \simeq (N^{-1}NY)^{\otimes j}$. By applying the functor N to this equivalence, we obtain the required Σ_j-equivalence $N(Y^{\otimes j}) \simeq E(j,NY)$. $\square$

For any chain complex X, let

$$\overline{E}(j,X) = \operatorname{inv}_{\Sigma_j} E(j,X), \quad \overline{E}(X) = \sum_j \overline{E}(j,X).$$

The above theorem implies the following corollaries.

COROLLARY 1. *For any simplicial module Y, there are chain equivalences*

$$NC(j,Y) \simeq \overline{E}(j,NY), \quad NC(Y) \simeq \overline{E}(NY).$$

COROLLARY 2. *Suppose that R is a field. For any simplicial module Y, there are chain equivalences*

$$NC(j,Y) \simeq \overline{E}(j, \pi_*(Y)), \quad NC(Y) \simeq \overline{E}(\pi_*(Y)).$$

Thus the problem of computing the homotopy groups of the simplicial modules $C(j,Y)$ and $C(Y)$ is reduced to computing the homology of the complexes

$$\overline{E}(j, \pi_*(Y)), \quad \overline{E}(\pi_*(Y)).$$

THEOREM. *Suppose that a chain complex X is the direct sum of complexes X' and X'', i.e., $X \cong X' \oplus X''$. Then there are chain equivalences*

$$\overline{E}(j,X) \simeq \sum_{i=0}^{j} \overline{E}(i,X') \otimes \overline{E}(j-i,X''), \quad \overline{E}(X) \simeq \overline{E}(X') \otimes \overline{E}(X'').$$

PROOF. The isomorphism $N^{-1}(X) \cong N^{-1}(X') \oplus N^{-1}(X'')$ induces an isomorphism

$$(N^{-1}(X))^{\otimes j} \cong \sum_{i=0}^{j} \sigma(i,j-i)(N^{-1}(X'))^{\otimes i} \otimes (N^{-1}(X''))^{\otimes j-i},$$

where $\sigma(i,j-i)$ are all possible $(i,j-i)$-shuffles of $1,2,\ldots,j$. Now, by applying to this isomorphism the functor N, and by using the Eilenberg–Zilber chain equivalence, we obtain the required chain equivalences. $\square$

If R is a field, then, by decomposing $H_*(X)$ into a sum of graded modules with a single generator, we see that to compute the homology of the complexes $\overline{E}(j,X)$ and $\overline{E}(X)$ it suffices to compute the homology of the complexes $\overline{E}(j,M)$ and $\overline{E}(M)$ for graded modules M with a single generator. Namely, if the graded module $H_*(X)$ is generated by the elements $\{x_\alpha\}$, which are arbitrarily ordered, then the homology of the complexes $\overline{E}(j,X)$ and $\overline{E}(X)$ is computed by the formulas

$$H_*(\overline{E}(j,X)) \cong \sum H_*(\overline{E}(j_1, M(x_{\alpha_1})) \otimes \cdots \otimes H_*(\overline{E}(j_k, M(x_{\alpha_k})),$$
$$H_*(\overline{E}(X)) \cong \sum H_*(\overline{E}(M(x_{\alpha_1})) \otimes \cdots \otimes H_*(\overline{E}(M(x_{\alpha_k})),$$

where $M(x_\alpha)$ is the graded module with a single generator x_α, the summation is over all collections $j_1,\ldots,j_k$ such that $j_1+\cdots+j_k=j$, and $x_{\alpha_1} < \cdots < x_{\alpha_k}$.

Let M be a graded module; we again consider the chain complex $E(j,M)$ whose n-dimensional generators are the maps

$$f = f^1 \otimes \cdots \otimes f^j : \Delta^n \otimes \cdots \otimes \Delta^n \to M \otimes \cdots \otimes M,$$

where $f^1,\ldots,f^j: \Delta^n \to M$ are chain maps.

Note that giving a chain map $f: \Delta^n \to M$ is equivalent to giving a dimension-preserving map $\widetilde{f}: \widetilde{\Delta}^n \to M$, where the graded module $\widetilde{\Delta}^n$ is the submodule of Δ^n generated by the simplices $(n_0,\ldots,n_m)$, $0 \le n_0 < \cdots < n_m \le n$, such that $n_0 = 0$. The module $\widetilde{\Delta}^n$ can also be represented as the quotient of the module Δ^n by the equivalence relation generated by the relations

$$\sum_{i=0}^{m} (-1)^i (n_0,\ldots,\hat{n}_i,\ldots,n_m) = 0, \quad \text{where } n_0 = 0.$$

The family $\widetilde{\Delta}^* = \{\widetilde{\Delta}^n\}$ forms a cosimplicial object in the category of graded modules; the coface and codegeneracy operators in it are induced by the corresponding operators in Δ^n, i.e.,

$$\begin{aligned}
\delta^i(0, n_1, \dots, n_m) &= (0, \delta^i(n_1), \dots, \delta^i(n_m)), \quad 1 \le i \le n;\\
\delta^0(0, n_1, \dots, n_m) &= (0, n_1 + 1, \dots, n_m + 1)\\
&\quad - (0, 1, n_2 + 1, \dots, n_m + 1)\\
&\quad + \dots + (-1)^m (0, 1, n_1 + 1, \dots, n_{m-1} + 1);\\
\sigma^i(0, n_1, \dots, n_m) &= (0, \sigma^i(n_1), \dots, \sigma^i(n_m)).
\end{aligned}$$

The elements $(0, n_1, \dots, n_m)$, $0 < n_1 < \dots < n_m \le n$, can be represented by sequences $(a_1, \dots, a_n)$ consisting of zeros and ones. The indices of the positions occupied by ones in these sequences are the corresponding integers $n_1, \dots, n_m$. In this notation, the coface operators are given by

$$\begin{aligned}
\delta^i(a_1, \dots, a_n) &= (a_1, , \dots, a_{i-1}, 0, a_i, \dots, a_n), \quad 1 \le i \le n;\\
\delta^0(a_1, \dots, a_n) &= (0, a_1, \dots, a_n) - \sum (-1)^\epsilon (1, a_1, \dots, \widetilde{a}_i, \dots, a_n),
\end{aligned}$$

where the summation is over all indices i such that $a_i = 1$, $\widetilde{a}_i = 0$, and ϵ is the number of ones preceding a_i in the sequence $(a_1, \dots, a_n)$.

For a graded module M with a single m-dimensional generator, giving a map $f\colon \Delta^n \to M$ is equivalent to giving a sequence $(a_1, \dots, a_n)$ of zeros and ones that contains exactly m ones and such that f takes it to the generator of the module M. Giving a map

$$f^1 \otimes \dots \otimes f^j\colon \Delta^n \otimes \dots \otimes \Delta^n \to M \otimes \dots \otimes M$$

is equivalent in this case to giving a matrix

$$\begin{pmatrix}
a_{11} & a_{12} & \dots & a_{1n}\\
a_{21} & a_{22} & \dots & a_{2n}\\
\dots & \dots & \dots & \dots\\
a_{j1} & a_{j2} & \dots & a_{jn}
\end{pmatrix}$$

of zeros and ones so that each row of the matrix contain exactly m ones.

Consider the cosimplicial module $E^*_m(j)$ generated by such matrices; the symmetric group acts on it by permuting the rows, and the coface and codegeneracy operators are induced by the corresponding operators in Δ^*. We define the Σ_j-chain complexes

$$\mathcal{E}_m(j) = \sum_n \mathcal{E}^n_m(j); \quad \mathcal{E}^n_m(j) = \mathcal{E}^n_m(j) / \sum_{i>0} \operatorname{Im} \delta^i,$$

where the differential is induced by the operator δ^0.

It is clear that a matrix $A(j, n)$ belongs to the image of δ^i, $i > 0$, if and only if all the elements of its ith column are zero. Hence $\mathcal{E}^n_m$ is generated by the matrices $A(j, n)$ such that each column of $A(j, n)$ contains one element 1. Let

$$\underline{\mathcal{E}}_m(j) = \operatorname{coinv}_{\Sigma_j} \mathcal{E}_m(j); \quad \underline{\mathcal{E}}_m = \sum_j \underline{\mathcal{E}}_m(j).$$

If R is a field, then the homology of these complexes is dual to the homology of the complexes $\overline{E}(j; M)$ and $\overline{E}(M)$, where M is a graded module with a single

m-dimensional generator. Thus we have

$$H_*(\overline{E}(j;M)) \cong H_*(\underline{\mathcal{E}}_m(j)), \quad H_*(\overline{E}(M)) \cong H_*(\underline{\mathcal{E}}_m).$$

Therefore, to compute the homotopy groups of the simplicial modules $C(j,Y)$ and $C(Y)$, it suffices to compute the homology of the complexes

$$\underline{\mathcal{E}}(j) = \sum_m \underline{\mathcal{E}}_m(j), \quad \underline{\mathcal{E}} = \sum_m \underline{\mathcal{E}}_m.$$

Consider the complexes $\underline{\mathcal{E}}(j)$, $R = \mathbb{Z}/2$. Denote by $\mathcal{C}$ the family of matrices $C(j,n)$, $j \geq 1$, $n \geq 1$, which is inductively defined as follows:

1) Let $1(i,k)$ be the matrix that consists of i rows and k columns with all elements equal to 1 and the number of rows in which is a power of two; then $1(i,k) \in \mathcal{C}$.

2) If a matrix $C(i,k)$ belongs to $\mathcal{C}$, then the matrix

$$\begin{pmatrix} C(i,k) & 0(i,k) & \dots & 0(i,k) & 1(i,l) \\ 0(i,k) & C(i,k) & \dots & 0(i,k) & 1(i,l) \\ \dots & \dots & \dots & \dots & \dots \\ 0(i,k) & 0(i,k) & \dots & C(i,k) & 1(i,l) \end{pmatrix}$$

belongs to $\mathcal{C}$, provided that the number of rows in it is a power of two.

THEOREM. *The homology $H_*(\underline{\mathcal{E}})$ of the complex $\underline{\mathcal{E}} = \sum_j \underline{\mathcal{E}}(j)$, $R = \mathbb{Z}/2$, is the polynomial algebra on generators $C(j,n)$ of dimension n. The homology $H_*(\underline{\mathcal{E}}(j))$ is generated by the products $C(j_1,n_1)\dots C(j_k,n_k)$ such that $j_1 + \dots + j_k = j$.*

PROOF. Consider the spectral sequence of the complex $\underline{\mathcal{E}}(j)$ with respect to the filtration induced by ordering the matrices $A(j,n)$ in the lexicographic order on the numbers of elements 1 in the columns, starting with the last column. Namely, if the last column of a matrix $A'(j,n)$ contains more elements 1 than the last column of a matrix $A''(j,n)$, then $A'(j,n) < A''(j,n)$. If the numbers of elements 1 in the last columns of these two matrices are equal, then we compare the next to last columns, and so on. It is clear that the differential in $\underline{\mathcal{E}}(j)$ lowers the filtration, and hence we obtain a spectral sequence of the complex $\underline{\mathcal{E}}(j)$.

The first nonzero differential in this spectral sequence acts on the first column of the matrices $A(j,n)$. It is readily seen that a matrix $A(j,n)$ is a cycle not homological to zero with respect to this differential if and only if the number of ones in its first column is a power of two, and the rows in which they stand are equal, i.e., the matrix $A(j,n)$ has the form

$$\begin{pmatrix} 1(i) & a_{12}(i) & \dots & a_{1n}(i) \\ 0 & a_{i+12} & \dots & a_{i+1n} \\ \dots & \dots & \dots & \dots \\ 0 & a_{j2} & \dots & a_{jn} \end{pmatrix}$$

where $a(i)$ denotes the column of elements a.

Cycles are homologous if they are represented by matrices that are obtained from each other by permuting the first and the second columns.

Assume that certain cycles with respect to the differentials acting on the 1st, ..., lth column in this spectral sequence are represented by matrices

$$A(j,n)=\begin{pmatrix} C_1(i_1,l_1) & 0(i_1,l_2) & \dots & 0(i_1,l_k) & A_1(i_1,n-l) \\ 0(i_2,l_1) & C_2(i_2,l_2) & \dots & 0(i_2,l_k) & A_2(i_2,n-l) \\ \dots & \dots & \dots & \dots & \dots \\ 0(i_k,l_1) & 0(i_k,l_2) & \dots & C_k(i_k,l_k) & A_k(i_k,n-l) \\ 0(j-i,l_1) & 0(j-i,l_2) & \dots & 0(j-i,l_k) & A(j-i_k,n-l) \end{pmatrix}$$

where $i = i_1 + \dots + i_k$, $l = l_1 + \dots + l_k$, and each of the matrices $A_1, \dots, A_k$ has equal rows.

The above matrices $A(j,n)$ represent homologous cycles if they can be obtained from one another by permuting the 1st, ..., lth columns.

Consider the differential acting on the $(l+1)$th column; we show that the cycles with respect to it are represented by matrices of the same form but for the $(l+1)$th columns.

It is readily seen that if the first column of the matrix $A(j-i, n-l)$ contains at least one element 1, then the number of such elements should be a power of two, and there are no elements 1 in the first columns of the matrices $A_1, \dots, A_k$. Besides, the rows of $A(j-i, n-l)$ that begin with 1 must coincide. Thus in this case the matrix $A(j,n)$ has the required form. Assume that all elements of the first column of $A(j-i, n-l)$ are zeros; consider those matrices among $A_1, \dots, A_k$ whose first columns consist of elements 1. By permuting the rows and the first l columns of the matrix $A(j,n)$, we can ensure that the matrices $A_1, \dots, A_{k'}$ start with zeros and $A_{k'+1}, \dots, A_k$ start with ones. It is readily seen that for such a matrix to represent a cycle we should have $C_{k'+1} = \dots = C_k$ and $A_{k'+1} = \dots = A_k$, and $k - k'$ should be a power of two. Thus in this case the matrix $A(j,n)$ also has the required form.

By induction, it follows that a matrix $A(j,n)$ represents a cycle with respect to all differentials if it has the form

$$A(j,n)=\begin{pmatrix} C_1(j_1,n_1) & 0(j_1,n_2) & \dots & 0(j_1,n_k) \\ 0(j_2,n_1) & C_2(j_2,n_2) & \dots & 0(j_2,n_k) \\ \dots & \dots & \dots & \dots \\ 0(j_k,n_1) & 0(j_k,n_2) & \dots & C_k(j_k,n_k) \end{pmatrix}$$

where $j = j_1 + \dots + j_k$, $n = n_1 + \dots + n_k$.

Two matrices of this form represent homologous cycles if they can be obtained from each other by permuting the rows and the columns. □

By renaming the elements and taking into account the fact that the polynomial algebra is a free commutative coalgebra, we obtain the following result.

THEOREM. *For the case of $\mathbb{Z}/2$-coefficients, the homotopy groups $\pi_*(CY)$ form a free commutative coalgebra on the generators*

$$e_{i_1} \dots e_{i_k} x_m, \quad 1 \le i_1 \le \dots \le i_k \le m, \quad x_m \in \pi_m(Y),$$

of dimension $2^k \dim(x) - (i_1 + 2i_2 + \dots + 2^{k-1} i_k)$.

In the case of $\mathbb{Z}/p$-coefficients, the following theorem is proved in a similar way.

THEOREM. *For the case of $\mathbb{Z}/p$-coefficients, the homotopy groups $\pi_*(CY)$ form a free commutative coalgebra on the generators*

$$\beta^{\epsilon_1} e_{i_1} \dots \beta^{\epsilon_k} e_{i_k} x_m, \quad 1 \le i_1 \le \dots \le i_k \le m, \quad x_m \in \pi_m(Y),$$

of dimension $p^k m - (i_1 + pi_2 + \cdots + p^{k-1} i_k)(p-1) + \epsilon_1 + p\epsilon_2 + \cdots + p^{k-1}\epsilon_k$. *It is required here that the parity of the indices* i_j, $1 \le j \le k$, *coincide with the parity of the dimension of* $\beta^{\epsilon_{j+1}} e_{i_{j+1}} \dots \beta^{\epsilon_k} e_{i_k} x_m$.

7.5. Homotopy theory of simplicial coalgebras

Suppose that Y is a simplicial module with filtration

$$Y = G^0 Y \supset G^1 Y \supset \cdots \supset G^n Y \supset G^{n+1} Y \supset \dots.$$

The *completion of* Y *with respect to this filtration* is the simplicial module $\hat{Y} = \varprojlim Y/G^n Y$.

A simplicial module Y is said to be *complete* if the projection $Y \to \hat{Y}$ is an isomorphism.

Let Y' and Y'' be complete simplicial modules; their tensor product $Y' \hat{\otimes} Y''$ is defined as the completion of the ordinary tensor product $Y' \otimes Y''$ with respect to the filtration

$$G^n(Y' \otimes Y'') = \operatorname{Im}\Big\{ \sum_{i+j=n} G^i Y' \otimes G^j Y'' \to Y' \otimes Y'' \Big\}.$$

A *complete simplicial coalgebra* is a complete simplicial module Y together with a map of complete simplicial modules $\nabla\colon Y \to Y \hat{\otimes} Y$ such that $(\nabla \hat{\otimes} 1)\nabla = (1 \hat{\otimes} \nabla)\nabla$.

A complete simplicial coalgebra Y is said to be *commutative* if $T\nabla = \nabla$, where $T\colon Y\hat{\otimes} Y \to Y \hat{\otimes} Y$ is the permutation operator.

Denote the category of complete commutative simplicial coalgebras by $\mathcal{K}_s$.

Suppose that Y is a complete simplicial module; in the tensor algebra of Y consider the submodule CY consisting of all elements that are invariant under the permutations of the factors, and let $\hat{C}Y$ be the completion of CY with respect to the ordinary filtration of the tensor algebra. Then $\hat{C}Y$ is a complete commutative simplicial coalgebra with the comultiplication induced by the ordinary comultiplication in the tensor algebra,

$$\nabla(y_1 \otimes \cdots \otimes y_n) = \sum_{i=0}^{n} (y_1 \otimes \cdots \otimes y_i) \otimes (y_{i+1} \otimes \cdots \otimes y_n).$$

It is clear that $\hat{C}Y$ is a cofree object in the category $\mathcal{K}_s$, i.e., for any complete simplicial coalgebra Z and any map of complete simplicial modules $f\colon Z \to Y$, there exists a unique map of complete simplicial coalgebras $\hat{f}\colon Z \to \hat{C}Y$ such that the following diagram is commutative:

$$\begin{array}{ccc} Z & \xrightarrow{\ \hat{f}\ } & \hat{C}Y \\ & {\scriptstyle f}\searrow \quad \swarrow{\scriptstyle p} & \\ & Y & \end{array}$$

The functor $\hat{C}$ assigning to each complete simplicial module Y the corresponding complete commutative simplicial coalgebra $\hat{C}Y$ is a comonad on the category of complete simplicial modules. Any complete commutative simplicial coalgebra Y is a coalgebra over the comonad $\hat{C}$. Thus, for any complete commutative simplicial coalgebra Y, we have a cosimplicial object

$$F^*(\hat{C}, \hat{C}, Y) \colon \hat{C}Y \to \hat{C}^2Y \to \cdots \to \hat{C}^nY \to \hat{C}^{n+1}Y \to \ldots,$$

which is a cosimplicial resolution of Y. Denote its realization by $F(\hat{C}, \hat{C}, Y)$.

For any complete commutative simplicial coalgebra Y, denote the simplicial set of its primitive elements by PY, i.e.,

$$PY = \{y \in Y \mid \nabla(y) = y \otimes y\}.$$

The correspondence $Y \longmapsto PY$ defines a functor $P \colon \mathcal{K}_s \to Ens_s$ from the category of complete commutative simplicial coalgebras to the category of simplicial sets. It is clear that $P\hat{C}Y = Y$. Thus, applying the functor P to the cosimplicial object $F^*(\hat{C}, \hat{C}, Y)$, we get a cosimplicial object $F^*(\hat{C}, Y)$, whose realization is denoted by $F(\hat{C}, Y)$ and is referred to as the *left-sided cobar construction on* Y.

The free simplicial module RX generated by a simplicial set X is a complete commutative simplicial coalgebra with the comultiplication defined on the generators $x \in X$ by the formula $\nabla(x) = x \otimes x$. Thus the correspondence $X \longmapsto RX$ defines a functor $R \colon Ens_s \to \mathcal{K}_s$ from the category of simplicial sets to the category of complete simplicial coalgebras. The inclusion map $i \colon X \to RX$ induces a natural transformation of functors $\alpha \colon \mathrm{Id} \to P \circ R$.

For any complete commutative simplicial coalgebra Y, the map of simplicial modules $RPY \to Y$ induces a map of complete simplicial coalgebras $RPY \to F(\hat{C}, \hat{C}, Y)$. Therefore, we have a natural transformation of functors $\beta \colon R \circ P \to I$, where $I \colon \mathcal{K}_s \to \mathcal{K}_s$ is the functor defined by $I(Y) = F(\hat{C}, \hat{C}, Y)$.

It can be shown that the category $\mathcal{K}_s$ of complete commutative simplicial coalgebras, with the class of cofibrations formed by the embeddings, the class of weak equivalences formed by the maps that induce an isomorphism in homotopy groups, and the class of fibrations formed by the maps satisfying the right lifting property, is a closed model category. Each object $Y \in \mathcal{K}_s$ is cofibrant, and the object $F(\hat{C}, \hat{C}, Y)$ is fibrant for any $Y \in \mathcal{K}_s$.

Let us elucidate the relationship between the Curtis spectral sequence and the Bousfield–Kan spectral sequences constructed for the cosimplicial objects R^*X and $F^*(\hat{C}, RX)$. The natural projection of simplicial modules $R^2X \to RX$ induces a map of simplicial coalgebras $R^2X \to \hat{C}RX$. Similarly, we can construct maps of simplicial coalgebras $R^nX \to \hat{C}^{n-1}RX$, which induce a map of cosimplicial objects

$$\begin{array}{cccccccccc}
R^*X\colon & RX & \longrightarrow & R^2X & \longrightarrow \ldots & R^nX & \longrightarrow & R^{n+1}X & \longrightarrow & \ldots \\
\downarrow & \downarrow & & \downarrow & & \downarrow & & \downarrow & & \\
F^*(\hat{C}, RX)\colon & RX & \longrightarrow & CRX & \longrightarrow \ldots & \hat{C}^{n-1}RX & \longrightarrow & \hat{C}^nRX & \longrightarrow & \ldots
\end{array}$$

This map induces a map of the spectral sequences of the realizations of these cosimplicial objects, i.e.,

$$\pi_*(|R^*X|) \to \pi_*(F(\hat{C}, RX)).$$

The above formulas for the homotopy groups $\pi_*(\hat{C}RX)$ directly imply that in the case $R = \mathbb{Z}/p$, $p \geq 2$, this map of cosimplicial objects induces an isomorphism of the E^1-terms of the spectral sequences, and hence, in case the Bousfield–Kan spectral sequence converges, an isomorphism of the homotopy groups themselves.

The convergence takes place, for example, for simply connected simplicial sets. Therefore, we obtain the following result.

THEOREM. *The homotopy groups* $\pi_*(F(\hat{C}, \mathbb{Z}X))$ *of a simply connected simplicial set* X *are isomorphic to the homotopy groups* $\pi_*(X)$. *In the case of* $\mathbb{Z}/p$-*coefficients, the homotopy groups* $\pi_*(F(\hat{C}, \mathbb{Z}/pX))$ *are isomorphic to the* p-*component of the homotopy groups* $\pi_*(X)$.

This theorem allows us to extend the definition of the homotopy groups to commutative simplicial coalgebras. Namely, suppose that Y is a complete commutative simplicial coalgebra; then we define its homotopy groups $\pi_*^c(Y)$ to be the homotopy groups of the simplicial set $F(\hat{C}, Y)$ which is the realization of the cosimplicial object $F^*(\hat{C}, Y)$. If $R = \mathbb{Z}$ and $Y = RX$, then we obtain the ordinary homotopy groups of the simplicial set X.

THEOREM. *A map* $f\colon Y' \to Y''$ *of simply connected complete simplicial coalgebras induces an isomorphism on the homotopy groups* $\pi_*^c(Y') \to \pi_*^c(Y'')$ *if and only if* $\pi_*(Y') \to \pi_*(Y'')$ *is an isomorphism.*

PROOF. Indeed, by expressing the homotopy groups of the simplicial module $\hat{C}Y$ in terms of the homotopy groups of Y, we see that the map $f\colon Y' \to Y''$ induces an isomorphism $\pi_*(Y') \cong \pi_*(Y'')$ if and only if it induces an isomorphism $\pi_*(\hat{C}Y') \cong \pi_*(\hat{C}Y'')$. By using the expressions for the E^1-terms of the Bousfield–Kan spectral sequences of the cosimplicial objects $F^*(\hat{C}, Y')$ and $F(\hat{C}, Y'')$, we prove the required assertion in a standard way, by induction over the dimension. □

Let $Ho\mathcal{K}_s$ be the localization of the category of simply connected complete simplicial coalgebras with respect to the class of morphisms inducing an isomorphism in homotopy groups. Denote by $Ho_R Ens_s$ the localization of the category of simply connected simplicial sets with respect to the class of morphisms inducing an isomorphism in homology with coefficients in R. Then the functors

$$R\colon Ens_s \to \mathcal{K}_s, \quad Q = F(\hat{C}, -)\colon \mathcal{K}_s \to Ens_s$$

induce adjoint functors

$$HoR\colon Ho_R Ens_s \to Ho\mathcal{K}_s, \quad HoQ\colon Ho\mathcal{K}_s \to HoEns_s.$$

The above theorems imply the following assertion.

THEOREM. *If* $R = \mathbb{Z}$ *or* $R = \mathbb{Z}/p$, *then the functor*

$$HoR\colon Ho_R Ens_s \to HoK_s$$

is completely injective.

7.6. Homotopy theory of E_∞-coalgebras

Consider the category $\mathcal{K}_E$ of coalgebras over an E_∞-operad E. The correspondence under which the chain complex NY is assigned to each complete commutative simplicial coalgebra Y defines a functor $N\colon \mathcal{K}_s \to \mathcal{K}_E$ from the category of complete commutative simplicial coalgebras to the category of E-coalgebras. We want to define the inverse functor $N^*\colon \mathcal{K}_E \to \mathcal{K}_s$. Let the functor N^{-1} from the category of chain complexes to the category of simplicial modules be adjoint to N. The map

$N\hat{C}Y \to \overline{E}(NY)$ constructed above induces a map $\hat{C}Y \to N^{-1}\overline{E}(NY)$ which is the composite map

$$\hat{C}Y \to N^{-1}N\hat{C}Y \to N^{-1}\overline{E}(NY).$$

Thus, for any chain complex X, there is a map

$$\psi\colon \hat{C}N^{-1}X \to N^{-1}\overline{E}(X)$$

which is the composite map

$$\hat{C}N^{-1}X \to N^{-1}\overline{E}NN^{-1}X \to N^{-1}\overline{E}(X),$$

and which commutes with the comonad structures, i.e., the following diagram is commutative:

$$\begin{array}{ccc} \hat{C}N^{-1}X & \xrightarrow{\psi} & N^{-1}\overline{E}X \\ \downarrow & & \downarrow \\ \hat{C}^2N^{-1}X & \xrightarrow{\psi} & N^{-1}\overline{E}^2(X) \end{array}$$

Therefore, for any E-coalgebra K, we have a cosimplicial object

$$F^*(\hat{C}N^{-1},\overline{E},K) : \hat{C}N^{-1}K \xrightarrow{\delta^0,\delta^1} \hat{C}N^{-1}\overline{E}(K) \to \cdots \to \hat{C}N^{-1}\overline{E}^n(K) \to \dots$$

in the category $\mathcal{K}_s$. Denote the realization of this cosimplicial object by $N^*(K)$. Then $N^*(K) \in \mathcal{K}_s$, and the correspondence $K \longmapsto N^*(K)$ defines the required functor N^*.

THEOREM. *There is a natural transformation of functors $\alpha\colon \mathrm{Id} \to N^* \circ N$, and the map $\alpha(Y)\colon Y \to N^*(NY)$ induces an isomorphism in homotopy groups for any $Y \in \mathcal{K}_s$.*

PROOF. For any $Y \in \mathcal{K}_s$, the conjugation map $Y \to N^{-1}NY$ in the category of simplicial modules induces a map of simplicial coalgebras $\alpha\colon Y \to \hat{C}N^{-1}NY$ such that the following diagram is commutative:

$$\begin{array}{ccc} Y & \xrightarrow{\alpha} & \hat{C}N^{-1}NY \\ {\scriptstyle\alpha}\downarrow & & \downarrow{\scriptstyle\delta^0} \\ \hat{C}N^{-1}NY & \xrightarrow{\delta^1} & N^{-1}\overline{E}(NY) \end{array}$$

This means that the map $\alpha\colon Y \to \hat{C}N^{-1}NY$ is an augmentation of the cosimplicial object $F^*(\hat{C}N^{-1},\overline{E}(NY)$ and hence induces the required map $\alpha(Y)\colon Y \to N^*NY$.

The map $\hat{C}Y \to N^{-1}\overline{E}(NY)$ induces a map of cosimplicial objects

$$F^*(\hat{C},\hat{C},Y) \to F^*(\hat{C}N^{-1},\overline{E},NY)$$

and a map of their realizations

$$|F^*(\hat{C},\hat{C},Y)| \to N^*NY.$$

Besides, the following diagram is commutative:

$$\begin{array}{ccc} Y & \xrightarrow{=} & Y \\ \downarrow & & \downarrow \\ |F^*(\hat{C},\hat{C},Y)| & \longrightarrow & N^*NY \end{array}$$

The cosimplicial object $F^*(\hat{C},\hat{C},Y)$ is a cosimplicial resolution of Y, and hence the map $Y \to |F^*(\hat{C},\hat{C},Y)|$ induces an isomorphism in homotopy groups.

By comparing the homotopy groups $\hat{C}Y$ and $N^{-1}\overline{E}(NY)$, we see that the map $|F^*(\hat{C},\hat{C},Y)| \to N^*NY$ also induces an isomorphism in homotopy groups. Therefore, the composite map $Y \to |F^*(\hat{C},\hat{C},Y)| \to N^*NY$ induces an isomorphism in homotopy groups. □

THEOREM. *For any E-coalgebra K, there is a chain equivalence $K \simeq NN^*K$.*

PROOF. Consider the cosimplicial resolution of the E-coalgebra K,

$$F^*(\overline{E},\overline{E},K) : \overline{E}(K) \to \overline{E}^2(K) \to \cdots \to \overline{E}^n(K) \to \dots .$$

The map $N\hat{C}N^{-1}K \to \overline{E}(K)$ induces a map of cosimplicial objects

$$F^*(N\hat{C}N^{-1},\overline{E},K) \to F^*(\overline{E},\overline{E},K),$$

and hence we have a map $NN^*K \to |F^*(\overline{E},\overline{E},K)|$ inducing an isomorphism in homology. Since the map $K \to |F^*(\overline{E},\overline{E},K)|$ also induces an isomorphism in homology, we obtain the required chain equivalence. □

THEOREM. *A map of E-coalgebras $f\colon K' \to K''$ induces an isomorphism in homology if and only if the map $N^*(f)\colon N^*(K') \to N^*(K'')$ induces an isomorphism in homotopy.*

PROOF. Note that the map $f\colon K' \to K''$ induces an isomorphism in homology if and only if the map $\overline{E}(f)\colon \overline{E}(K') \to \overline{E}(K'')$ induces an isomorphism in homology. By considering the spectral sequences for the homotopy groups of the cosimplicial objects $F^*(\hat{C}N^{-1},\overline{E},K)$ and $F^*(\hat{C}N^{-1},\overline{E},K'')$, we obtain the required result. □

By $Ho\mathcal{K}_E$ denote the localization of the category of simply connected E-coalgebras with respect to the class of morphisms inducing an isomorphism in homology. By the above, we obtain the following theorem.

THEOREM. *There is an equivalence of categories $Ho\mathcal{K}_s \cong Ho\mathcal{K}_E$.*

COROLLARY. *The weak homotopy type of a simply connected topological space (simplicial set) X is completely determined by the E-coalgebra structure on the chain complex $C_*(X,R)$.*

Consider the category $\widetilde{\mathcal{K}}_E$ whose objects are E-coalgebras and whose morphisms $\widetilde{f}\colon K' \to K''$ are E-coalgebra maps $f\colon K' \to F(\overline{E},\overline{E},K'')$. The definition of the cobar construction as the realization of the corresponding cosimplicial object implies that giving such a map f is equivalent to giving a family of E-coalgebra maps $f^n\colon K'\otimes\Delta^n \to \overline{E}^{n+1}(K'')$ such that the following diagrams are commutative:

$$(*)\qquad \begin{array}{ccc} K'\otimes\Delta^n & \xrightarrow{f^n} & \overline{E}^{n+1}(K'') \\ {\scriptstyle 1\otimes\delta^i}\downarrow\uparrow{\scriptstyle 1\otimes\sigma^i} & & {\scriptstyle \delta^i}\downarrow\uparrow{\scriptstyle \sigma^i} \\ K'\otimes\Delta^{n+1} & \xrightarrow{f^{n+1}} & \overline{E}^{n+2}(K'') \end{array}$$

Since giving an E-coalgebra map $f^n\colon K'\otimes\Delta^n \to \overline{E}^{n+1}(K'')$ is equivalent to giving a chain map $\overline{f}^n\colon K'\otimes\Delta^n \to \overline{E}^n(K'')$, we conclude that giving a morphism $\widetilde{f}\colon K' \to$

K'' is equivalent to giving a family of maps $\overline{f}^n\colon K' \otimes \Delta^n \to \overline{E}^n(K'')$ such that the corresponding maps f^n are chain maps and the diagrams $(*)$ are commutative.

Two morphisms $\widetilde{f}, \widetilde{g}\colon K' \to K''$ of category $\widetilde{\mathcal{K}}_E$ are said to be *homotopic* (and we write $\widetilde{f} \simeq \widetilde{g}$) if there exists a morphism $\widetilde{h}\colon K' \otimes \Delta^1 \to K''$ (called a *homotopy*) such that

$$\widetilde{h}|_{K'\otimes 0} = \widetilde{f}, \quad \widetilde{h}|_{K'\otimes 1} = \widetilde{g}.$$

As above, giving such a morphism $\widetilde{h}$ is equivalent to giving a family of maps $\overline{h}^n\colon K' \otimes \Delta^1 \otimes \Delta^n \to \overline{E}^n(K'')$ satisfying relations similar to those listed above.

Denote by $\pi\mathcal{K}_E$ the category whose objects are the objects of the category $\widetilde{\mathcal{K}}_E$ and whose morphisms are the homotopy classes of morphisms in $\widetilde{\mathcal{K}}_E$.

THEOREM. *There is an equivalence of categories $Ho\mathcal{K}_E \cong \pi\mathcal{K}_E$.*

PROOF. We have a natural functor $\pi\mathcal{K}_E \to Ho\mathcal{K}_E$; to prove that it is an equivalence of categories, it suffices to show that an E-coalgebra map $f\colon K' \to K''$ induces an isomorphism in homology if and only if it is a homotopy equivalence in the category $\widetilde{\mathcal{K}}_E$. So, suppose that a map $f\colon K' \to K''$ induces an isomorphism in homology. Then there is a chain map $g\colon K'' \to K'$ which is chain homotopy inverse to f. However, it is not an E-coalgebra map. The diagram

$$\begin{array}{ccc} K'' & \xrightarrow{\;g\;} & K' \\ {\scriptstyle\tau''}\downarrow & & \downarrow{\scriptstyle\tau'} \\ \overline{E}(K'') & \xrightarrow{\;\overline{g}\;} & \overline{E}(K') \end{array}$$

is not commutative, but it is homotopy commutative, i.e., there is a chain homotopy $\overline{g}^1\colon K'' \otimes \Delta^1 \to \overline{E}(K')$. Continuing this process, we obtain a system of maps $\overline{g}^n\colon K'' \to \overline{E}^n(K')$ raising dimension by n; this system defines a morphism $\widetilde{g}\colon K'' \to K'$ in $\widetilde{\mathcal{K}}_E$ which is homotopy inverse to $\widetilde{f}\colon K' \to K''$. □

CHAPTER 8

Operad Methods in Cobordism Theory

In this chapter we study algebraic structures on generalized homology and cohomology of a topological space, in particular, in cobordism theories. Clearly, such structures are induced by structures on the corresponding spectra. For example, if a spectrum $\mathbb{H}$ is multiplicative, then the corresponding cohomology theory possesses a multiplication

$$\mathbb{H}^*(X) \otimes \mathbb{H}^*(X) \to \mathbb{H}^*(X)$$

that makes the generalized cohomology of a topological space X into an algebra. If the spectrum $\mathbb{H}$ is homotopy commutative, then this algebra is commutative.

Since the operad language is a convenient tool for describing various structures, it seems natural to use operad methods in the study of generalized homology and cohomology theories. Attempts to extend the operad methods to generalized homology and cohomology theories have been made by many authors; see [**13, 31**], etc. The main difficulty here is that there is no natural action of an E_∞-operad on the Thom spectra MO, MU, and MSp.

To overcome this difficulty, we replace the notion of an operad by the notion of a bioperad and define E_∞-bioperads, which act on the Thom spectra and generate the required E_∞-multiplicative structure on cobordisms.

We define analogs of the Dyer–Lashof algebra $\mathbb{R}$ and the Steenrod algebra $\mathbb{A}$ for the generalized homology and cohomology corresponding to an E_∞-multiplicative spectrum $\mathbb{H}$. We show that there is an action

$$\mathbb{A} \otimes \mathbb{H}^*(X) \to H^*(X)$$

of the Steenrod algebra on the generalized cohomology $\mathbb{H}^*(X)$ of a topological space X. On the other hand, if X is an E_∞-space, then there is an action of the Dyer–Lashof algebra

$$\mathbb{R} \otimes \mathbb{H}_*(X) \to \mathbb{H}_*(X)$$

on the generalized homology $\mathbb{H}_*(X)$. We compute these actions for cobordisms of topological spaces, and we shall relate the Steenrod operations to the Landweber–Novikov operations [**26, 35**].

8.1. Bioperads and classical groups

Recall that a symmetric family $\mathcal{E} = \{\mathcal{E}(j)\}$ of topological spaces $\mathcal{E}(j)$ is called a *bioperad* if we are given maps

$$\pi: \mathcal{E}(n) \times \mathcal{E}(m) \to \mathcal{E}(n+m), \quad \gamma: \mathcal{E}(k) * \mathcal{E}(n) \to \mathcal{E}(kn)$$

satisfying the conditions of associativity, distributivity, and compatibility with the actions of the symmetric groups.

A symmetric family $\mathcal{F} = \{\mathcal{F}(n)\}$ is called a *$\mathcal{E}$-multiplicative family* (where $\mathcal{E}$ is a bioperad) if there are given maps

$$\pi\colon \mathcal{F}(n) \times \mathcal{F}(m) \to \mathcal{F}(n+m), \quad \mu\colon \mathcal{E}(k) * \mathcal{F}(n) \to \mathcal{F}(kn)$$

compatible with the actions of the symmetric groups and satisfying the associativity and distributivity relations.

The considered bioperads and the families on which they act are related to classical groups.

Let $\mathbb{F}$ be the field of real numbers $\mathbb{R}$, or the field of complex numbers $\mathbb{C}$, or the skew field of quaternions $\mathbb{H}$. Consider the family $GL = \{GL(\mathbb{F}, n)\}_{n\geq 1}$ of general linear groups $GL(\mathbb{F}, n)$ of the vector spaces $\mathbb{F}^n$. There is a free action of the symmetric group Σ_n on $GL(\mathbb{F}, n)$ induced by permuting the components of elements of $\mathbb{F}^n$, i.e.,

$$(\sigma f)(x) = f(\sigma^{-1}x), \quad f \in GL(\mathbb{F}, n), \quad x \in \mathbb{F}^n,\ \sigma \in \Sigma_n.$$

If the elements of $GL(\mathbb{F}, n)$ are represented by matrices, then the symmetric group Σ_n acts by permuting the columns of these matrices.

We have natural maps $\pi\colon GL(\mathbb{F}, n) \times GL(\mathbb{F}, m) \to GL(\mathbb{F}, n+m)$ that assign to matrices (A, B) the matrix $A \oplus B = \begin{pmatrix} A & O \\ O & B \end{pmatrix}$, where O denotes the zero matrix.

Let us define maps

$$\gamma\colon GL(\mathbb{F}, k) * GL(\mathbb{F}, n) \to GL(\mathbb{F}, kn).$$

To do this, consider matrices $A \in GL(\mathbb{F}, k)$ and $B_i \in GL(\mathbb{F}, n)$, $1 \leq i \leq k$,

$$A = \begin{pmatrix} a_{11} & \dots & a_{1k} \\ \dots & \dots & \dots \\ a_{k1} & \dots & a_{kk} \end{pmatrix}.$$

Denote by $A(n)$ the matrix in $GL(\mathbb{F}, kn)$ made up of the $n \times n$-matrices

$$\begin{pmatrix} a_{ij} & 0 & \dots & 0 \\ 0 & a_{ij} & \dots & 0 \\ \dots & \dots & \dots & \dots \\ 0 & 0 & \dots & a_{ij} \end{pmatrix},$$

$1 \leq i, j \leq k$. By using tensor multiplication of matrices, we can express $A(n)$ as $A(n) = A \otimes I_n$, where I_n is the identity matrix in $GL(\mathbb{F}, n)$.

We set $\gamma(A \times B_1 \times \dots \times B_k) = A(n) \cdot (B_1 \oplus \dots \oplus B_k)$.

Direct verification shows that the following assertion holds.

THEOREM. *The family $GL = \{GL(\mathbb{F}, n)\}$ with the operations*

$$\pi\colon GL(\mathbb{F}, n) \times GL(\mathbb{F}, m) \to GL(\mathbb{F}, n+m), \quad \gamma\colon GL(\mathbb{F}, k) * GL(\mathbb{F}, n) \to GL(\mathbb{F}, kn)$$

defined above is a bioperad.

The corresponding map $\xi\colon GL(\mathbb{F}, k(n+m)) \to GL(\mathbb{F}, k(n+m))$ is defined as follows. Consider the permutation $\xi \in \Sigma_{k(n+m)}$ defined in Section 1.6. It induces an isomorphism $\xi\colon \mathbb{F}^{k(n+m)} \to \mathbb{F}^{k(n+m)}$. We define the isomorphism

$$\xi\colon GL(\mathbb{F}, k(n+m)) \to GL(\mathbb{F}, k(n+m))$$

by setting $\xi(f) = \xi^{-1} \circ f \circ \xi$.

In particular, we have the following bioperads:
1. the bioperad of orthogonal groups $O = \{O(n)\}$;
2. the bioperad of unitary groups $U = \{U(n)\}$;
3. the bioperad of symplectic groups $Sp = \{Sp(n)\}$.

For the groups $G(n) = GL(\mathbb{F}, n)$, consider the spaces $EG(n)$ which are the realizations of the simplicial objects

$$B_*(*, G(n), G(n)) : G(n) \leftarrow G(n) \times G(n) \leftarrow \ldots \leftarrow G(n)^{\times m+1} \leftarrow \ldots .$$

Then the bioperad structure on GL induces a bioperad structure on the family $EG = \{EG(n)\}$. Taking into account the fact that the spaces $EG(n)$ are contractible spaces with a free action of the symmetric groups, we obtain the following result.

THEOREM. *The bioperad EG is an E_∞-bioperad. In particular, the following are E_∞-bioperads: $EO = \{EO(n)\}$, $EU = \{EU(n)\}$, and $ESp = \{ESp(n)\}$.*

For the groups $G(n) = GL(\mathbb{F}, n)$, we also consider the classifying spaces $BG(n)$, which are the realizations of the simplicial objects

$$B_*(*, G(n), *) : * \leftarrow G(n) \leftarrow G(n) \times G(n) \leftarrow \ldots \leftarrow G(n)^{\times m+1} \leftarrow \ldots .$$

The bioperad structure on GL induces an action of the bioperad EG on $BG = \{BG(n)\}$. Therefore, the following assertion holds.

THEOREM. *The family BG is an EG-multiplicative family. In particular, BO is an EO-multiplicative family; BU is an EU-multiplicative family; and BSp is an ESp-multiplicative family.*

Our aim is to elucidate the algebraic structure on the homology of the classifying spaces of the classical groups.

For example, let us consider the EO-multiplicative family $BO = \{BO(n)\}$. We want to compute the operations $\mu_*: EO_*(k) * BO_*(n) \to BO_*(kn)$. It is known that $BO_*(n)$ is a free commutative coalgebra generated by elements x_i of dimension i, $1 \leq i \leq n$. The multiplications $\pi_*: BO_*(n) \otimes BO_*(m) \to BO_*(n+m)$ make the family $BO_* = \{BO_*(n)\}$ into a free commutative Hopf algebra with a single 1-dimensional generator $x_1 \in BO_*(1)$. The multiplication π_* is given on the elements x_i by

$$\pi_*(x_i \otimes x_j) = x_{i+j} + \text{ nonprimitive elements.}$$

The definition of $\mu_*: EO_*(2) * BO_*(1) \to BO_*(2)$ implies that $\mu_*(e_1 * 1) = x_1$. On the remaining elements, μ_* is determined by the Hopf algebra structure on BO_*.

Dually, consider the cohomology $EO^*(k) = H^*(EO(k)/_{\Sigma_k})$ of the E_∞-bioperad EO and the cohomology $BO^*(n) = H^*(BO(n))$. It is known that $BO^*(n)$ is a free commutative algebra generated by elements $\overline{x}_i$ of dimension i, $1 \leq i \leq n$.

We have the operations $\pi^*: BO^*(n) \to BO^*(i) \otimes BO^*(n-i)$, which make $BO^* = \{BO^*(n)\}$ into a free commutative Hopf algebra with a single 1-dimensional generator $\overline{x}_1 \in BO^*(1)$.

Besides, we have the coactions $\mu^*: BO^*(kn) \to EO^*(k) * BO^*(n)$. The definition of $\mu^*: BO^*(2) \to EO^*(2) * BO^*(1)$ implies that $\mu^*(\overline{x}_1) = \overline{e}_1 * 1$. On the remaining elements, the coaction μ^* is determined by the Hopf algebra structure on BO^*.

By way of example, for $\mu^*\colon BO^*(2) \to EO^*(2) * BO^*(1)$ we have the formula

$$\mu^*(\overline{x}_1^n \cdot \overline{x}_2^m) = \overline{e}_n * \overline{x}_1^m.$$

Let us compute the action $\mu^*\colon BO^*(2n) \to EO^*(2) * BO^*(n)$. Since μ^* is an algebra map, it suffices to compute it on the generators $\overline{x}_m \in BO^*(2n)$. To do this, we use the commutative diagram

$$\begin{array}{ccc} BO^*(2n) & \xrightarrow{\mu^*} & EO^*(2) * BO^*(n) \\ \downarrow & & \downarrow \\ BO^*(2)^{\otimes n} & \xrightarrow{(\mu^*)^{\otimes n}} & EO^*(2) * (BO^*(1)^{\otimes n}) \end{array}$$

This diagram implies the following theorem.

THEOREM. *The coaction $\mu^*\colon BO^*(2n) \to EO^*(2) * BO^*(n)$ is given on the generators $\overline{x}_m \in BO^*(2n)$ by the formula*

$$\mu^*(\overline{x}_m) = \sum_k \binom{n-k}{m-2k} \overline{e}_{m-2k} * \overline{x}_k.$$

On the remaining elements, μ^* is determined by the algebra structure.

Passing to the dual situation, we obtain the following theorem.

THEOREM. *The action $\mu_*\colon EO_*(2) * BO_*(n) \to BO_*(2n)$ is given on the primitive elements $x_k \in BO_*(n)$ by the formula*

$$\mu_*(e_i * x_k) = \binom{n-k}{i} x_{i+2k} \ + \ \textit{nonprimitive elements.}$$

It is sometimes more convenient to use other generators in BO_*. Denote the element $x_1^{\otimes m} \in BO_*(1)$ by a_m. Then BO_* is a free commutative algebra generated by the elements a_m, and the comultiplication is given on these elements by the formula

$$\nabla(a_m) = \sum_{i=0}^{m} a_i \otimes a_{m-i}.$$

Straightforward computations show that the action $\mu_*\colon EO_*(2) * BO_*(1) \to BO_*(2)$ is given by the formula

$$\mu_*(e_i * a_m) = \sum_{k \geq 0} \binom{i+k-1}{k} a_{m-k} \cdot a_{i+m+k}.$$

This formula implies that the action $\mu_*\colon EO_*(2) * BO_*(n) \to BO_*(2n)$ is given by

$$\mu_*(e_i * a_{m_1} \cdot \ldots \cdot a_{m_n}) = \sum_{\substack{i_1+\cdots+i_n=i \\ k_1,\ldots,k_n}} \binom{i_1+k_1-1}{k_1} a_{m_1-k_1} a_{i_1+m_1+k_1} \cdots \binom{i_n+k_n-1}{k_n} a_{m_n-k_n} a_{i_n+m_n+k_n}.$$

In particular, we have

$$\mu_*(e_i * a_m) = \binom{i+m-1}{m} a_{i+2m} \ + \ \text{indecomposable elements.}$$

Note that the action μ_* and the coaction μ^* are unstable, and hence they do not give an action and a coaction on the homology $H_*(BO) = \lim_n H_*(BO(n))$ and on the cohomology $H^*(BO) = \lim_n H^*(BO(n))$, respectively.

Since $BO_*(n)$ is the homology of the topological space $BO(n)$, we have the coactions

$$\tau_*: BO_*(n) \to EO^*(k) * BO_*(n)$$

and the dual actions

$$\tau^*: EO_*(k) * BO^*(n) \to BO^*(n).$$

For example, $\tau^*: EO_*(2) * BO^*(1) \to BO^*(1)$ is given by

$$\tau^*(e_i * \overline{x}_1) = \begin{cases} \overline{x}_1^2, & i = 0, \\ \overline{x}_1, & i = 1, \\ 0, & i > 1. \end{cases}$$

On the remaining elements, τ^* is determined by the algebra structure, i.e.,

$$\tau^*(e_i * \overline{x}_1^n) = \binom{n}{i} \overline{x}_1^{2n-i}.$$

Since $\tau^*: EO_*(2) * BO^*(n) \to BO^*(n)$ is compatible with the coalgebra structure, the following assertions hold.

THEOREM. *The action $\tau^*: EO_*(2) * BO^*(n) \to BO^*(n)$ is given on the generators $\overline{x}_m \in BO^*(n)$ by the formula*

$$\tau^*(e_i * \overline{x}_m) = \begin{cases} \overline{x}_{m-i}\overline{x}_m, & i \leq m, \\ 0, & i > m. \end{cases}$$

On the remaining elements, τ^* is determined by the algebra structure.

THEOREM. *The coaction $\tau_*: BO_*(n) \to EO^*(2) * BO_*(n)$ is given on the generators $a_m \in BO_*(n)$ by the formula*

$$\tau_*(a_m) = \sum_i \binom{i}{m-i} \overline{e}_{2i-m} * a_i.$$

On the remaining elements, τ_* is determined by the algebra structure on $BO_*(n)$.

Consider the homology $BU_*(n) = H_*(BU(n))$. It is a free commutative coalgebra generated by elements y_i of dimension $2i$, $1 \leq i \leq n$.

The operations $\pi_*: BU_*(n) \otimes BU_*(m) \to BU_*(n+m)$ make $BU_* = \{BU_*(n)\}$ into a free commutative Hopf algebra with a single 2-dimensional generator y_1 in $BU_*(1)$. The operation π_* is given on the elements y_i by

$$\pi_*(y_i \otimes y_j) = y_{i+j} + \text{ nonprimitive elements.}$$

The definition of the action $\mu_*: EU_*(2) * BU_*(1) \to BU_*(2)$ implies that

$$\mu_*(e_1 * 1) = 0, \quad \mu_*(e_2 * 1) = y_1.$$

On the remaining elements, μ_* is determined by the Hopf algebra structure on BU_*.

Similarly, the cohomology $BU^*(n) = H^*(BU(n))$ is a free commutative algebra generated by elements $\overline{y}_i$ of dimension $2i$, $1 \leq i \leq n$. The operations $\pi^*: BU^*(n) \to$

$BU^*(i) \otimes BU^*(n-i)$ make $BU^* = \{BU^*(n)\}$ into a free commutative Hopf algebra with a single 2-dimensional generator $\overline{y}_1$ in $BU^*(1)$.

Besides, we have the coactions $\mu^*\colon BU^*(kn) \to EU^*(k) * BU^*(n)$. In the case of $\mathbb{Z}/2$-coefficients, the definition of $\mu^*\colon BU^*(2) \to EU^*(2) * BU^*(1)$ implies that

$$\mu^*(\overline{y}_1) = \overline{e}_2 * 1, \quad \mu^*(\overline{y}_2) = \overline{e}_0 * \overline{y}_1.$$

On the remaining elements, the coaction μ^* is determined by the Hopf algebra structure on BU^*.

Let us compute the coaction $\mu^*\colon BU^*(2n) \to EU^*(2) * BU^*(n)$. Since μ^* is an algebra map, it suffices to compute it on the generators $\overline{y}_m \in BU^*(2n)$. To this end, we use the commutative diagram

$$\begin{array}{ccc} BU^*(2n) & \xrightarrow{\mu^*} & EU^*(2) * BU^*(n) \\ \downarrow & & \downarrow \\ BU^*(2)^{\otimes n} & \xrightarrow{(\mu^*)^{\otimes n}} & EU^*(2) * (BU^*(1)^{\otimes n}) \end{array}$$

which gives the following theorem.

THEOREM. *The coaction $\mu^*\colon BU^*(2n) \to EU^*(2) * BU^*(n)$ is given on the generators $\overline{y}_m$ of $BO^*(2n)$ by the formula*

$$\mu^*(\overline{y}_m) = \sum_k \binom{n-k}{m-2k} \overline{e}_{2m-4k} * \overline{y}_k.$$

On the remaining elements, μ^* is determined by the algebra structure.

Passing to the dual situation, we have the following theorem.

THEOREM. *The action $\mu_*\colon EU_*(2) * BU_*(n) \to BU_*(2n)$ is given on the primitive elements $y_k \in BU_*(n)$ by the formulas $\mu_*(e_{2i+1} * y_k) = 0$ and*

$$\mu_*(e_{2i} * y_k) = \binom{n-k}{i} y_{i+2k} + \textit{nonprimitive elements.}$$

It is sometimes more convenient to use other generators for BU_*. Denote the elements $y_1^{\otimes m} \in BO_*(1)$ by b_m. Then BU_* is a free commutative algebra generated by the elements b_m, and the comultiplication is given on these elements by

$$\nabla(b_m) = \sum_{i=0}^{m} b_i \otimes b_{m-i}.$$

Straightforward computations show that the action $\mu_*\colon EU_*(2) * BU_*(1) \to BU_*(2)$ is given by

$$\mu_*(e_{2i} * b_m) = \sum_{k \geq 0} \binom{i+k-1}{k} b_{m-k} \cdot b_{i+m+k}.$$

This formula implies that the action $\mu_*\colon EU_*(2) * BU_*(n) \to BU_*(2n)$ is given by

$$\mu_*(e_{2i} * b_{m_1} \cdot \ldots \cdot b_{m_n}) = \sum_{\substack{i_1+\cdots+i_n=i \\ k_1,\ldots,k_n}} \binom{i_1+k_1-1}{k_1} b_{m_1-k_1} b_{i_1+m_1+k_1} \cdots \binom{i_n+k_n-1}{k_n} b_{m_n-k_n} b_{i_n+m_n+k_n}.$$

In particular, we have

$$\mu_*(e_{2i} * b_m) = \binom{i+m-1}{m} b_{i+2m} + \text{indecomposable elements.}$$

Note that the action μ_* and the coaction μ^* are unstable, and hence they do not give an action and a coaction on the homology $H_*(BU) = \lim_n H_*(BU(n))$ and on the cohomology $H^*(BU) = \lim_n H^*(BU(n))$, respectively.

Since $BU_*(n)$ is the homology of the topological space $BU(n)$, we have the coactions $\tau_*\colon BU_*(n) \to EU^*(k) * BU_*(n)$ and the dual actions $\tau^*\colon EU_*(k) * BU^*(n) \to BU^*(n)$.

For example, the map $\tau^*\colon EU_*(2) * BU^*(1) \to BU^*(1)$ is given by the formula

$$\tau^*(e_i * \overline{y}_1) = \begin{cases} \overline{y}_1^2, & i = 0, \\ 0, & i = 1, \\ \overline{y}_1, & i = 2, \\ 0, & i > 1. \end{cases}$$

On the remaining elements, τ^* is determined by the algebra structure:

$$\tau^*(e_i * \overline{y}_1^n) = \binom{n}{k} \overline{y}_1^{2n-k}$$

if $i = 2k$, and $\tau^*(e_i * \overline{y}_1^n) = 0$ if $i = 2k+1$.

Since the map $\tau^*\colon EU_*(2) * BU^*(n) \to BU^*(n)$ is compatible with the coalgebra structure, the following assertions hold.

THEOREM. *The action $\tau^*\colon EU_*(2) * BU^*(n) \to BU^*(n)$ is given on the generators $\overline{y}_m$ of $BU^*(n)$ by the formula*

$$\tau^*(e_{2i} * \overline{y}_m) = \begin{cases} \overline{y}_{m-i}\overline{y}_m, & i \le m, \\ 0, & i > m. \end{cases}$$

On the remaining elements, τ^* is determined by the algebra structure.

THEOREM. *The coaction $\tau_*\colon BU_*(n) \to EU^*(2) * BU_*(n)$ is given on the generators b_m of $BU_*(n)$ by the formula*

$$\tau_*(b_m) = \sum_i \binom{i}{m-i} \overline{e}_{4i-2m} * b_i.$$

On the remaining elements, τ^* is determined by the algebra structure on $BU_*(n)$.

Now consider the homology of BU with $\mathbb{Z}/p$-coefficients. The corresponding Hopf bialgebra EU_* is generated by the element $e_2 \in EU_*(p)$ of dimension $2(p-1)$.

The definition of the action $\mu_*\colon EU_*(p) * BU_*(1) \to BU_*(p)$ implies the formula $\mu_*(e_2 * 1) = y_{p-1}$. On the remaining elements, μ_* is determined by the Hopf algebra structure on BU_*.

The coaction $\mu^*\colon BU^*(p) \to EU^*(p) * BU^*(1)$ is given by

$$\mu^*(\overline{y}_i) = \begin{cases} \overline{e}_0 * \overline{y}_0, & i = 0, \\ 0, & 0 < i < p-1, \\ \overline{e}_2 * 1, & i = p-1, \\ \overline{e}_0 * \overline{y}_1, & i = p. \end{cases}$$

On the remaining elements, the coaction μ^* is determined by the Hopf algebra structure on BU^*.

Let us compute the coaction $\mu^*\colon BU^*(pn) \to EU^*(p) * BU^*(n)$. Since μ^* is an algebra map, it suffices to compute it on the generators $\overline{y}_m \in BU^*(pn)$. To this end, we use the commutative diagram

$$\begin{array}{ccc} BU^*(pn) & \xrightarrow{\mu^*} & EU^*(p) * BU^*(n) \\ \downarrow & & \downarrow \\ BU^*(p)^{\otimes n} & \xrightarrow{(\mu^*)^{\otimes n}} & EU^*(p) * (BU^*(1)^{\otimes n}) \end{array}$$

which implies the following theorem.

THEOREM. *The coaction $\mu^*\colon BU^*(pn) \to EU^*(p) * BU^*(n)$ is given on the generators $\overline{y}_m$ of $BO^*(pn)$ by the formula*

$$\mu^*(\overline{y}_m) = \sum_{\substack{k \\ m=i(p-1)+kp}} \binom{n-k}{i} \overline{e}_{2i} * \overline{y}_k.$$

On the remaining elements, μ^* is determined by the algebra structure.

Consider the operations

$$\tau_*\colon BU_*(n) \to EU^*(k) * BU_*(n), \quad \tau^*\colon EU_*(k) * BU^*(n) \to BU^*(n)$$

for the case of $\mathbb{Z}/p$-coefficients.

For example, $\tau^*\colon EU_*(p) * BU^*(1) \to BU^*(1)$ is given by

$$\tau^*(e_{2i} * \overline{y}_1) = \begin{cases} \overline{y}_1^p, & i = 0, \\ \overline{y}_1, & i = 1, \\ 0, & i > 1. \end{cases}$$

On the remaining elements, τ^* is determined by the algebra structure,

$$\tau^*(e_{2i} * \overline{y}_1^n) = \binom{n}{i} \overline{y}_1^{pn-i(p-1)}.$$

Since the map $\tau^*\colon EU_*(p) * BU^*(n) \to BU^*(n)$ is compatible with the coalgebra structure, we obtain the following theorem.

THEOREM. *The action $\tau^*\colon EU_*(p) * BU^*(n) \to BU^*(n)$ is given on the generators $\overline{y}_m$ of $BU^*(n)$ by the formula*

$$\tau^*(e_{2i} * \overline{y}_m) = \begin{cases} \overline{y}_{m-i}^{p-1} \overline{y}_m, & i \leq m, \\ 0, & i > m. \end{cases}$$

On the remaining elements, τ^* is determined by the algebra structure.

8.2. E_∞-multiplicative spectra and Thom spectra

We now pass to spectra. Let $\mathcal{E}$ be a bioperad.

DEFINITION. A multiplicative spectrum $\mathbb{X} = \{\mathbb{X}(n)\}$ is said to be *$\mathcal{E}$-multiplicative* if we are given operations $\mu: \mathcal{E}(k) * \mathbb{X}(n) \to \mathbb{X}(kn)$ which, together with the multiplicative structure on $\mathbb{X}$, form the structure of an $\mathcal{E}$-multiplicative family. If E is an E_∞-bioperad, then the corresponding E-multiplicative spectrum $\mathbb{X} = \{\mathbb{X}(n)\}$ is said to be *E_∞-multiplicative*.

Note that the notion of an E_∞-multiplicative spectrum is analogous to the notion of an H_∞-ring prespectrum [**13**] and differs from the notion of an E_∞-ring spectrum [**13, 31**].

Let $\mathbb{S} = \{S^n\}$ be the sphere spectrum. The action of the group $O(n)$ on $\mathbb{R}^n$ induces an action of $O(n)$ on S^n. Consider the simplicial object

$$B_*(*, O(n), S^n) : S^n \leftarrow O(n) \rtimes S^n \leftarrow \dots \leftarrow O(n)^{\times m} \rtimes S^n \leftarrow \dots .$$

Its realization, denoted by $MO(n)$, is a Thom space, and the family $MO = \{MO(n)\}$ is a Thom spectrum. The Thom spectra MU and MSp are defined similarly.

In the same way as the bioperad structure on the family $GL = \{GL(n)\}$ was defined, we can define an action $\mu: O(k) \times (\mathbb{R}^n)^{\times k} \to \mathbb{R}^{kn}$ that induces an action of the bioperad $O = \{O(n)\}$ on the family of spheres $\mathbb{S} = \{S^n\}$.

Besides, there are natural maps $\pi: S^n \wedge S^m \to S^{n+m}$. Direct verification shows that the maps μ and π satisfy the required relations. Therefore, we obtain the following theorem.

THEOREM. *The sphere spectrum $\mathbb{S} = \{S^n\}$ is O-multiplicative (where $O = \{O(n)\}$).*

This structure on $\mathbb{S}$ induces maps

$$\pi: MO(n) \wedge MO(m) \to MO(n+m), \quad \mu: EO(k) * MO(n) \to MO(kn),$$

which determine the structure of an EO-multiplicative spectrum on the Thom spectrum $MO = \{MO(n)\}$. There are similar structures on the other Thom spectra. Therefore, we obtain the following theorem.

THEOREM. *The Thom spectrum MO is an EO-multiplicative spectrum, MU is an EU-multiplicative spectrum, and MSp is an ESp-multiplicative spectrum. Thus the Thom spectra are E_∞-multiplicative.*

For example, consider the Thom spectrum $MO = \{MO(n)\}$ and its homology

$$MO_*(n) = H_*(MO(n)).$$

Our goal is to compute the actions $\mu_*: EO_*(k) * MO_*(n) \to MO_*(kn)$ and the dual coactions $\mu^*: MO^*(kn) \to EO^*(k) * MO^*(n)$.

It is known that $MO_*(n)$ is a free $BO_*(n)$-comodule with a single n-dimensional generator $u_n \in S^n$, and $MO^*(n)$ is a free $BO^*(n)$-module with a single n-dimensional generator $\overline{u}_n \in S^n$. Furthermore, the family MO_*, regarded with its algebra structure, is a free Hopf BO_*-comodule with a single 1-dimensional generator $u_1 \in S^1$. Similarly, the family MO^*, regarded with its coalgebra structure, is a free Hopf BO^*-module with a single 1-dimensional generator $\overline{u}_1$.

We have the Thom isomorphisms $BO_*(n) \to MO_{*+n}(n)$, which assign to each element $x \in BO_*(n)$ the element $x \otimes u_n \in MO_{*+n}(n)$. These isomorphisms preserve the multiplication π_*. Therefore, the multiplications

$$\pi_*: MO_*(n) \otimes MO_*(m) \to MO_*(n+m)$$

are determined by the multiplications $\pi_*: BO_*(n) \otimes BO_*(m) \to BO_*(n+m)$ and the formulas $\pi_*(u_n \otimes u_m) = u_{n+m}$.

However, the Thom isomorphisms do not preserve the action μ_*. Therefore, we shall use different homomorphisms $\eta_*: BO_*(n) \to MO_*(n)$ and the dual homomorphisms $\eta^*: MO^*(n) \to BO^*(n)$. To construct them, note that $BO(n)$ is homotopy equivalent to $B(*, O(n), \mathbb{R}^n)$, and there are natural maps $\eta: B(*, O(n), \mathbb{R}^n) \to MO(n)$. These maps induce the required homomorphisms η_* and η^*.

It is readily seen that for $\eta^*: MO^*(1) \to BO^*(1)$ we have the formula

$$\eta^*(x \otimes \overline{u}_1) = x \cdot \overline{x}_1.$$

Since η^* preserves the comultiplications, in the general case we have $\eta^*(x \otimes \overline{u}_n) = x \cdot \overline{x}_n$. The corresponding formula for BO^* implies the following theorem.

THEOREM. *The coaction $\mu^*: MO^*(2n) \to EO^*(2) * MO^*(n)$ is given on the generators $\overline{x}_m \otimes \overline{u}_{2n} \in MO^*(2n)$ by the formula*

$$\mu^*(\overline{x}_m \otimes \overline{u}_{2n}) = \sum_k \binom{n-k}{m-2k} \overline{e}_{m-2k} * (\overline{x}_k \otimes \overline{u}_n).$$

In the dual situation, the homomorphism $\eta_*: BO_*(1) \to MO_*(1)$ is given by

$$\eta_*(a_m) = \begin{cases} a_{m-1} \otimes u_1, & m \geq 1, \\ 0, & i = 0. \end{cases}$$

Since η_* preserves the multiplications, in the general case for $\eta_*: BO_*(n) \to MO_*(n)$ we have the formula

$$\eta_*(a_{m_1} \dots a_{m_n}) = \begin{cases} a_{m_1-1} \dots a_{m_n-1} \otimes u_n, & m_k \geq 1, \\ 0 & \text{otherwise}. \end{cases}$$

The action $\mu_*: EO_*(2) * MO_*(n) \to MO_*(2n)$ is given by

$$\mu_*(e_i * a_{m_1} \dots a_{m_n} \otimes u_n) = \sum_{\substack{i_1+\dots+i_n=i \\ k_1,\dots,k_n}} \binom{i_1+k_1-1}{k_1} a_{m_1-k_1} a_{i_1+m_1+k_1} \dots \binom{i_n+k_n-1}{k_n} a_{m_n-k_n} a_{i_n+m_n+k_n} \otimes u_{2n}.$$

In particular, we have

$$\mu_*(e_i * a_m) = \binom{i+m-1}{m} a_{i+2m} + \text{indecomposable elements}.$$

Note that the actions $\mu_*: EO_*(k) * MO_*(n) \to MO_*(kn)$ are unstable, and hence they do not give an action on the homology $H_*(MO) = \lim_m H_{*+m}(MO(m))$ of the Thom spectrum MO.

Since $MO_*(n)$ is the homology of the space $MO(n)$, we have the coactions $\tau_*: MO_*(n) \to EO^*(k) * MO_*(n)$ and the dual actions $\tau^*: EO_*(k) * MO^*(n) \to$

$MO^*(n)$. By using the formulas for $BO_*(n)$ and the homomorphism η^*, we obtain formulas for $\tau^*\colon EO_*(2) * MO^*(n) \to MO^*(n)$, i.e.,

$$\tau^*(e_i * (\overline{x}_m \otimes \overline{u}_n) = \sum_{k\le i} \overline{x}_{m-k} \cdot \overline{x}_m \cdot \overline{x}_{n-i+k} \otimes \overline{u}_n.$$

The coaction $\tau_*\colon MO_*(n) \to EO_*(2) * MO_*(n)$ is given by

$$\tau_*(a_m \otimes u_n) = \sum_i \binom{i+1}{m-i} \overline{e}_{2i-m+n} * (a_i \otimes u_n).$$

Let us consider $H^*(MO) = \lim_n H^{*+n}(MO(n))$. It is known that $H^*(MO)$ is a free commutative algebra generated by the elements $\overline{x}_i$, $i \ge 0$. The action τ^* induces an action of the Steenrod algebra on $H^*(MO)$.

THEOREM. *The action of the Steenrod algebra on $H^*(MO)$ is given on the generators $\overline{x}_m$ of $H^*(MO)$ by the formula*

$$Sq^i(\overline{x}_m) = \sum_{k\le m} \overline{x}_k \cdot \overline{x}_{i-k} \cdot \overline{x}_m.$$

On the remaining elements, it is determined by the algebra structure on $H^*(MO)$.

Similarly, consider the Thom spectrum $MU = \{MU(n)\}$; we shall compute the action

$$\mu_*\colon EU_*(k) * MU_*(n) \to MU_*(kn)$$

and the dual coaction

$$\mu^*\colon MU^*(kn) \to EU^*(k) * MU^*(n).$$

As is known, $MU_*(2n)$ is a free $BU_*(n)$-comodule with a single $2n$-dimensional generator $u_{2n} \in S^{2n}$, and $MU^*(2n)$ is a free $BU^*(n)$-module with a single $2n$-dimensional generator $\overline{u}_{2n} \in S^{2n}$.

We have the homomorphisms $\eta_*\colon BU_*(n) \to MU_*(2n)$ and the dual homomorphisms $\eta^*\colon MU^*(2n) \to BU^*(n)$. It is readily seen that the homomorphism $\eta^*\colon MU^*(2) \to BU^*(1)$ satisfies the relation $\eta^*(y \otimes \overline{u}_2) = y \cdot \overline{y}_1$. Since η^* preserves the comultiplications, in the general case we have

$$\eta^*(y \otimes \overline{u}_{2n}) = y \cdot \overline{y}_n.$$

The following theorem is obtained by using the corresponding formula for BU^*.

THEOREM. *The coaction $\mu^*\colon MU^*(4n) \to EU^*(2) * MU^*(2n)$ is given on the generators $\overline{y}_m \otimes \overline{u}_{4n} \in MO^*(4n)$ by the formula*

$$\mu^*(\overline{y}_m \otimes \overline{u}_{4n}) = \sum_k \binom{n-k}{m-2k} \overline{e}_{2m-4k} * (\overline{y}_k \otimes \overline{u}_{2n}).$$

In the dual situation, the homomorphism $\eta_*\colon BU_*(1) \to MU_*(2)$ is given by

$$\eta_*(b_m) = \begin{cases} b_{m-1} \otimes u_2, & m \ge 1, \\ 0, & i = 0. \end{cases}$$

Since η_* preserves the multiplications, for $\eta_*\colon BU_*(n) \to MU_*(2n)$ in the general case we have

$$\eta_*(b_{m_1}\dots b_{m_n}) = \begin{cases} b_{m_1-1}\dots b_{m_n-1}\otimes u_{2n}, & m_k \geq 1,\\ 0 & \text{otherwise.}\end{cases}$$

The action $\mu_*\colon EU_*(2) * MU_*(n) \to MU_*(2n)$ is given by

$$\mu_*(e_{2i} * b_{m_1}\dots b_{m_n}\otimes u_n) = \sum_{\substack{i_1+\dots+i_n=i\\ k_1,\dots,k_n}} \binom{i_1+k_1-1}{k_1} b_{m_1-k_1} b_{i_1+m_1+k_1} \dots \binom{i_n+k_n-1}{k_n} b_{m_n-k_n} b_{i_n+m_n+k_n}\otimes u_{2n}.$$

In particular, we have

$$\mu_*(e_{2i} * (b_m \otimes u_n)) = \binom{i+m-1}{m} b_{i+2m}\otimes u_{2n} + \text{indecomposable elements.}$$

Note that the actions $\mu_*\colon EU_*(k) * MU_*(n) \to MU_*(kn)$ are unstable, and hence they do not yield an action on the homology $H_*(MU) = \lim_m H_{*+m}(MU(m))$ of the Thom spectrum MU.

Since $MU_*(n)$ is the homology of the space $MU(n)$, we have the coactions $\tau_*\colon MU_*(n) \to EU^*(k) * MU_*(n)$ and the dual actions $\tau^*\colon EU_*(k) * MU^*(n) \to MU^*(n)$.

By using the corresponding formulas for $BU_*(n)$ and the homomorphism η^*, we obtain formulas for $\tau^*\colon EU_*(2) * MU^*(n) \to MU^*(n)$, i.e.,

$$\tau^*(e_{2i} * (\overline{y}_m \otimes \overline{u}_{2n}) = \sum_{k\leq i} \overline{y}_{m-k}\cdot \overline{y}_m \cdot \overline{y}_{n-i+k}\otimes \overline{u}_{2n}.$$

Consider the cohomology $H^*(MU) = \lim_n H^{*+n}(MU(n))$ with $\mathbb{Z}/p$-coefficients, $p \geq 2$. It is known that $H^*(MU)$ is a free commutative algebra generated by the elements $\overline{y}_i$ of dimension $i \geq 0$. The action τ^* induces an action of the Steenrod algebra on $H^*(MU)$.

THEOREM. *The action of the Steenrod algebra on $H^*(MU)$ is given on the generators $\overline{y}_m \in H^*(MU)$ by the formula*

$$P^i(\overline{y}_m) = \sum_{k\leq m} \overline{y}_k^{p-1}\cdot \overline{y}_{i-k}^{p-1}\cdot \overline{y}_m.$$

On the remaining elements, it is determined by the algebra structure on $H^*(MU)$.

8.3. Generalized homology theories and the Dyer–Lashof algebra

Let us consider the homology theory $\mathbb{H}_*$ associated with a multiplicative spectrum $\mathbb{H}$. Suppose that X is a topological space; then the homology $\mathbb{H}_*(X) = \{\mathbb{H}_n(X)\}$ is defined by

$$\mathbb{H}_n(X) = \lim_i \pi_{n+i}(\mathbb{H}(i)\wedge X),$$

where the limit is taken with respect to the composite maps

$$\pi_{n+i}(\mathbb{H}(i)\wedge X) \to \pi_{n+i+1}(S\mathbb{H}(i)\wedge X) \to \pi_{n+i+1}(\mathbb{H}(i+1)\wedge X).$$

In particular, we have the following homology theories:

$$\mathbb{M}O_*,\ \mathbb{M}U_*,\ \mathbb{M}Sp_*,\ \mathbb{M}SO_*,\ \mathbb{M}SU_*.$$

Suppose now that $\mathbb{X}$ is a cospectrum, i.e., a family $\mathbb{X} = \{\mathbb{X}(m)\}$ of spaces $\mathbb{X}(m)$ together with maps $\mathbb{X}(m) \to S\mathbb{X}(m-1)$. Then the homology of $\mathbb{X}$ is defined by

$$\mathbb{H}_n(\mathbb{X}) = \lim_m \mathbb{H}_{n+m}(\mathbb{X}(m)),$$

where the limit is taken with respect to the composite maps

$$\mathbb{H}_{n+m}(\mathbb{X}(m)) \to \mathbb{H}_{n+m}(S\mathbb{X}(m-1)) \to \mathbb{H}_{n+m-1}(\mathbb{X}(m-1)).$$

It is readily seen that if X is a topological space and $\mathbb{S}X$ is the suspension cospectrum of X, then the homology of the latter is isomorphic to the homology of X.

The homology $\mathbb{H}_*(\mathbb{X})$ of a cospectrum $\mathbb{X}$ is not just a collection of abelian groups: there is an action of the algebra $\mathbb{H}_*(S^0)$ on it.

Indeed, we have the actions $\mu_*\colon \mathbb{H}_*(S^0) \otimes \mathbb{H}_*(\mathbb{X}(m)) \to \mathbb{H}_*(\mathbb{X}(m))$ that make the following diagrams commutative:

$$\begin{array}{ccc} \mathbb{H}_*(S^0) \otimes \mathbb{H}_*(\mathbb{X}(m)) & \xrightarrow{\mu_*} & \mathbb{H}_*(\mathbb{X}(m)) \\ \downarrow & & \downarrow \\ \mathbb{H}_*(S^0) \otimes \mathbb{H}_*(\mathbb{X}(m-1)) & \xrightarrow{\mu_*} & \mathbb{H}_*(\mathbb{X}(m-1)) \end{array}.$$

Passing to the limit, we obtain the required action $\mu_*\colon \mathbb{H}_*(S^0) \otimes \mathbb{H}_*(\mathbb{X}) \to \mathbb{H}_*(\mathbb{X})$.

We generalize the notion of cospectrum and introduce the notion of $\mathbb{H}$-cospectrum. Namely, we give the following definition.

DEFINITION. A family $\mathbb{X} = \{\mathbb{X}(m)\}$ of spaces $\mathbb{X}(m)$ is called an $\mathbb{H}$-*cospectrum* if a map

$$\varphi\colon \mathbb{X}(m) \to \mathbb{H}(k) \wedge \mathbb{X}(m-k)$$

is given for any m.

It is clear that an $\mathbb{S}$-cospectrum is just a cospectrum.

Let us give some nontrivial examples of $\mathbb{H}$-cospectra. To this end, consider a bioperad E and an E-multiplicative spectrum $\mathbb{H}$. For any cospectrum $\mathbb{X}$, let $E(k, \mathbb{X}(m)) = E(k) *_{\Sigma_k} \mathbb{X}(m)$, and let $E(k, \mathbb{X})$ be the family $\{E(k, \mathbb{X})(n)\}$ with

$$E(k, \mathbb{X})(km) = E(k, \mathbb{X}(m))$$

for $n = km$, and with suspensions of these spaces taken for $E(k, \mathbb{X})(n)$ for other n.

We have maps

$$\begin{aligned} E(k, \mathbb{X}(m)) &\to E(k, S\mathbb{X}(m-1)) \to E(k, \mathbb{H}(1) \wedge \mathbb{X}(m-1)) \\ &\to E(k, \mathbb{H}(1)) \wedge E(k, \mathbb{X}(m-1)) \to \mathbb{H}(k) \wedge E(k, \mathbb{X}(m-1)). \end{aligned}$$

They determine the desired $\mathbb{H}$-cospectrum structure on the family $E(k, \mathbb{X})$. In particular, for the case of the sphere cospectrum $\mathbb{S}$, we obtain the $\mathbb{H}$-cospectrum $E(k, \mathbb{S})$, which will play an important role in the construction of the Dyer–Lashof operations and the Steenrod operations.

DEFINITION. Suppose that $\mathbb{X}$ is an H-cospectrum, $\varphi\colon \mathbb{X}(m) \to \mathbb{H}(k) \wedge \mathbb{X}(m-k)$. We define the homology $\mathbb{H}_*^\varphi(\mathbb{X})$, referred to as the *twisted homology*, by setting

$$\mathbb{H}_n(\mathbb{X}) = \lim_m \mathbb{H}_{n+m}(\mathbb{X}(m)),$$

where the limit is taken with respect to the composite maps

$$\varphi_*\colon \mathbb{H}_{n+m}(\mathbb{X}(m)) \to \mathbb{H}_{n+m}(\mathbb{H}(k) \wedge \mathbb{X}(m-k)) \to \mathbb{H}_{n+m-k}(\mathbb{X}(m-k)).$$

It is clear that the homology of any spectrum regarded as an $\mathbb{H}$-cospectrum is isomorphic to the homology of the cospectrum, and for any $\mathbb{H}$-cospectrum $\mathbb{X}$ there is an action

$$\mu_*\colon \mathbb{H}_*(S^0) \otimes \mathbb{H}_*^\varphi(\mathbb{X}) \to \mathbb{H}_*^\varphi(\mathbb{X}).$$

The homology $\mathbb{H}_*^\varphi(E(k,\mathbb{X}))$ of the above cospectrum $E(k,\mathbb{X})$ provides an example of homology with nontrivial twisting. In particular, the homology

$$\mathbb{H}_n^\varphi(E(k,\mathbb{S})) = \lim_m \mathbb{H}_{n+km}(E(k,S^m))$$

is defined, where the limit is taken with respect to the homomorphism

$$\varphi_*\colon \mathbb{H}_{n+km}(E(k,S^m)) \to \mathbb{H}_{n+k(m-1)}(E(k,S^{m-1}))$$

induced by the maps $\varphi\colon E(k,S^m) \to \mathbb{H}(k) \wedge E(k,S^{m-1})$.

Note that the map φ induces an isomorphism

$$H_*(\mathbb{H} \wedge E(k,S^m)) \to H_*(\mathbb{H} \wedge E(k,S^{m-1}))$$

in ordinary homology, and hence the homomorphisms

$$\varphi_*\colon \mathbb{H}_*(E(k,S^m)) \to \mathbb{H}_*(E(k,S^{m-1}))$$

are actually isomorphisms. Thus we obtain the following result.

THEOREM. *There are isomorphisms* $\mathbb{H}_*^\varphi(E(k,\mathbb{S})) \cong \mathbb{H}_*(E(k,S^0))$.

Denote $\mathbb{H}_*^\varphi(E(k,\mathbb{S}))$ by $\mathbb{E}_*(k)$. Our goal is to construct operations on the homology $\mathbb{H}_*(X)$ of an E-multiplicative space X that are analogous to the Dyer–Lashof operations for ordinary homology.

Note that a general method for constructing the Dyer–Lashof operations for homology theories corresponding to H_∞- and E_∞-ring spectra was proposed in [**13, 31**]. To construct these operations for an H_∞-ring spectrum, a stabilization functor Z was used, which made it possible to pass from a prespectrum to an H_∞-ring spectrum. We define an analog of the Dyer–Lashof algebra and its action directly for an E_∞-multiplicative spectrum, in particular, for an H_∞-ring prespectrum. To this end we use twisted cohomology and consider a more general situation.

Let $\mathbb{X}$ be an E-multiplicative spectrum acted on by an E-multiplicative spectrum $\mathbb{H}$. We define the Dyer–Lashof operations on the homotopy groups $\pi_*(\mathbb{X}) = \lim_m \pi_{*+m}(\mathbb{X}(m))$ of the spectrum $\mathbb{X}$.

Let $x \in \pi_n(\mathbb{X})$, and let $f\colon S^{n+m} \to \mathbb{X}(m)$ be a map representing this element. It induces a map $E(k,S^{n+m}) \to E(k,\mathbb{X}(m)) \to \mathbb{X}(km)$, and hence a map

$$\mathbb{H}(j) \wedge E(k,S^{n+m}) \to \mathbb{H}(j) \wedge \mathbb{X}(km) \to \mathbb{X}(j+km).$$

The last map induces a homomorphism of homotopy groups,

$$\pi_{i+j+k(n+m)}(\mathbb{H}(j) \wedge E(k,S^{n+m})) \to \pi_{i+j+k(n+m)}(\mathbb{X}(j+km)).$$

Passing to the limit with respect to j, we obtain a homomorphism

$$\mu_f\colon \mathbb{H}_{i+k(n+m)}(E(k,S^{n+m})) \to \pi_{i+kn}(\mathbb{X}),$$

and thus an action

$$\mu_*\colon \mathbb{H}_{i+k(n+m)}(E(k,S^{n+m})) \times \pi_{n+m}(\mathbb{X}(m)) \to \pi_{i+kn}(\mathbb{X}).$$

For the suspension homomorphism $S\colon \pi_{n+m}(\mathbb{X}(m)) \to \pi_{n+m+1}(\mathbb{X}(m+1))$, we have the commutative diagram

$$\begin{array}{ccc}
\mathbb{H}_{i+k(n+m+1)}(E(k,S^{n+m+1})) \times \pi_{n+m+1}(\mathbb{X}(m+1)) & \xrightarrow{\mu_*} & \pi_{i+kn}(\mathbb{X}) \\
\uparrow{\scriptstyle 1\times S} & & \downarrow{\scriptstyle =} \\
\mathbb{H}_{i+k(n+m+1)}(E(k,S^{n+m+1})) \times \pi_{n+m}(\mathbb{X}(m)) & & \pi_{i+kn}(\mathbb{X}) \\
\downarrow{\scriptstyle \varphi_*\times 1} & & \downarrow{\scriptstyle =} \\
\mathbb{H}_{i+k(n+m)}(E(k,S^{n+m})) \times \pi_{n+m}(\mathbb{X}(m)) & \xrightarrow{\mu_*} & \pi_{i+kn}(\mathbb{X})
\end{array}$$

Passing to the limit with respect to m, we obtain the required action

$$\mu_*\colon \mathbb{H}_i^{\varphi}(E(k,\mathbb{S})) \times \pi_n(\mathbb{X}) \to \pi_{i+nk}(\mathbb{X}),$$

which is additive with respect to the first factor and not additive (in general) with respect to the second factor.

By $\mathbb{H}_*^{\varphi}(E(k,\mathbb{S})) * \pi_*(\mathbb{X})$ we denote the corresponding semitensor product, with dimension of the elements $e_i * x_n$ set equal to $i + kn$. We obtain the operations

$$\mu_*\colon \mathbb{H}_*^{\varphi}(E(k,\mathbb{S})) * \pi_*(\mathbb{X}) \to \pi_*(\mathbb{X}),$$

which preserve dimension.

In particular, if X is an E-multiplicative topological space and $\mathbb{X} = \mathbb{H} \wedge X$, then we have the Dyer–Lashof operations

$$\mu_*\colon \mathbb{E}_*(k) * \mathbb{H}_*(X) \to \mathbb{H}_*(X)$$

on the generalized homology $\mathbb{H}_*(X)$ of the E-multiplicative space X.

If in the definition of the Dyer–Lashof algebra we replace X by $E(n,S^m)$, then we obtain maps

$$\gamma_*\colon \mathbb{H}_*^{\varphi}(E(k,\mathbb{S})) * \mathbb{H}_*(E(n,S^m)) \to \mathbb{H}_*(E(kn,S^m)).$$

Passing to the limit with respect to m, we obtain operations

$$\gamma_*\colon \mathbb{H}_*^{\varphi}(E(k,\mathbb{S})) * \mathbb{H}_*^{\varphi}(E(n,\mathbb{S})) \to \mathbb{H}_*^{\varphi}(E(kn,\mathbb{S}))$$

or, what is the same, operations

$$\gamma_*\colon \mathbb{E}_*(k) * \mathbb{E}_*(n) \to \mathbb{E}_*(kn).$$

The definition of the bioperad E implies that these operations γ_* satisfy the associativity relation, and hence the family $\mathbb{E}_* = \{\mathbb{E}_*(n)\}$, $\mathbb{E}_*(n) = \mathbb{H}_*^{\varphi}(E(n,\mathbb{S}))$, can be regarded as an algebra with the operation

$$\gamma_*\colon \mathbb{E}_*(k) * \mathbb{E}_*(n) \to \mathbb{E}_*(kn).$$

If $\mathbb{X}$ is an E-multiplicative spectrum acted on by an E-multiplicative spectrum $\mathbb{H}$, then the operations

$$\mu_*: \mathbb{E}_*(k) * \pi_*(\mathbb{X}) \to \pi_*(\mathbb{X})$$

induce an action of the algebra $\mathbb{E}_*$ on the homotopy groups $\pi_*(\mathbb{X})$ of the spectrum $\mathbb{X}$.

In particular, if X is an E-multiplicative space, then the operations

$$\mu_*: \mathbb{E}_*(k) * \mathbb{H}_*(X) \to \mathbb{H}_*(X)$$

induce an action of the algebra $\mathbb{E}_*$ on the homology $\mathbb{H}_*(X)$ of the space X.

On the family $\mathbb{E}_* = \{\mathbb{E}_*(n)\}$ there is not just an algebra structure: as noted above, it possesses an $\mathbb{H}_*(S^0)$-module structure

$$\mu_*: \mathbb{H}_*(S^0) \otimes \mathbb{E}_*(n) \to \mathbb{E}_*(n).$$

We also define the action $\mu_*: \mathbb{E}_*(k) * \mathbb{H}_*(S^0) \to \mathbb{E}_*(k)$. Let $x \in \mathbb{E}_n(k)$, and let $f: S^{n+j+km} \to \mathbb{H}(j) \wedge E(k, S^m)$ be a representative of x; let $y \in \mathbb{H}_l(S^0)$, and let $g: S^{l+i} \to \mathbb{H}(i) \wedge S^0 = \mathbb{H}(i)$ be a representative of y.

For $m = l + i + p$, the composite map

$$S^{n+j+k(i+l+p)} \to \mathbb{H}(j) \wedge E(k, S^{l+i+p}) \to \mathbb{H}(j) \wedge E(k, \mathbb{H}(i) \wedge S^p) \to \mathbb{H}(j+ki) \wedge E(k, S^p)$$

determines the required element $\mu_*(x * y) \in \mathbb{E}_{n+kl}(k)$.

It is readily seen that if $\mathbb{X}$ is an E-multiplicative spectrum acted on by an E-multiplicative spectrum $\mathbb{H}$, then we have the commutative diagram

$$\begin{array}{ccc} \mathbb{E}_*(k) * \mathbb{H}_*(S^0) \times \pi_*(\mathbb{X}) & \xrightarrow{1*\mu_*} & \mathbb{E}_*(k) * \pi_*(\mathbb{X}) \\ \downarrow{\scriptstyle \mu_* \times 1} & & \downarrow{\scriptstyle \mu_*} \\ \mathbb{E}_*(k) * \pi_*(\mathbb{X}) & \xrightarrow{\mu_*} & \pi_*(\mathbb{X}) \end{array}$$

Besides, the multiplications $\pi: E(n) \times E(m) \to E(n+m)$ induce multiplications

$$\pi_*: \mathbb{E}_*(n) \otimes \mathbb{E}_*(m) \to \mathbb{E}_*(n+m).$$

The diagonal $\nabla: E(n) \to E(n) \times E(n)$ induces a map

$$E(n, S^{p+q}) \to E(n, S^p) \wedge E(n, S^q),$$

and hence, in the case where the $\mathbb{H}_*(E(n, S^m))$ are free $\mathbb{H}_*(S^0)$-modules, we have homomorphisms

$$\mathbb{H}_*(E(n, S^{p+q})) \to \mathbb{H}_*(E(n, S^p)) \otimes_{\mathbb{H}_*(S^0)} \mathbb{H}_*(E(n, S^q)).$$

It is readily seen that these homomorphisms commute with φ_*, and hence induce a comultiplication

$$\nabla_*: \mathbb{E}_*(n) \to \mathbb{E}_*(n) \otimes_{\mathbb{H}_*(S^0)} \mathbb{E}_*(n).$$

The operations γ_*, μ_*, π_*, and ∇_* satisfy all possible associativity, commutativity, and Hopf relations. We call this structure on $\mathbb{E}_* = \{\mathbb{E}_*(n)\}$ a *Hopf $\mathbb{H}_*(S^0)$-bialgebra structure.*

THEOREM. *Suppose that E is a bioperad and $\mathbb{H}$ is an E-multiplicative spectrum; then the homology $\mathbb{E}_* = \{\mathbb{E}_*(n)\}$, $\mathbb{E}_*(n) = \mathbb{H}_*^{\varphi}(E(n,\mathbb{S}))$, possesses a Hopf $\mathbb{H}_*(S^0)$-bialgebra structure. If $\mathbb{X}$ is an E-multiplicative spectrum acted on by the E-multiplicative spectrum $\mathbb{H}$, then the homotopy groups $\pi_*(\mathbb{X})$ possess the structure of an $\mathbb{H}_*(S^0)$-algebra over the Hopf $\mathbb{H}_*(S^0)$-bialgebra $\mathbb{E}_*$. In particular, if X is an E-multiplicative topological space, then the homology $\mathbb{H}_*(X)$ possesses the structure of an $\mathbb{H}_*(S^0)$-algebra over the Hopf $\mathbb{H}_*(S^0)$-bialgebra $\mathbb{E}_*$.*

In particular, we have the following actions:

$$\mu_*\colon \mathbb{E}O_* * \mathbb{M}O_* \to \mathbb{M}O_*, \qquad \mu_*\colon \mathbb{E}U_* * \mathbb{M}U_* \to \mathbb{M}U_*,$$

and so on.

Note that to define the Dyer–Lashof operations with $\mathbb{Z}/p$-coefficients we need only replace the sphere spectrum $\mathbb{S}$ by the corresponding Moore spectrum $\mathbb{S}_{(p)}$. Then we obtain the algebra

$$\mathbb{E}_*^{(p)} = \{\mathbb{E}_*^{(p)}(n) = \mathbb{H}_*^{\varphi}(E(k,\mathbb{S}_{(p)});\mathbb{Z}/p)\}$$

and the operations

$$\mu_*\colon \mathbb{E}_*^{(p)}(k) * \mathbb{H}_*(X;\mathbb{Z}/p) \to \mathbb{H}_*(X;\mathbb{Z}/p).$$

In a similar way, the Dyer–Lashof operations can be defined for ordinary homology. Namely, let $\mathbb{X}$ be an E-multiplicative spectrum acted on by an E-multiplicative spectrum $\mathbb{H}$. Consider the ordinary cohomology

$$H_*(\mathbb{X}) = \lim_{m} H_{*+m}(\mathbb{X}(m))$$

with coefficients in a field.

Let $E_*(k,S^m) = H_*(E(k,S^m))$. The maps

$$\varphi\colon E(k,S^m) \to \mathbb{H}(k) \wedge E(k,S^{m-1})$$

induce homomorphisms

$$\varphi_*\colon H_*(\mathbb{H} \wedge E(k,S^m)) \to H_*(\mathbb{H} \wedge E(k,S^{m-1})).$$

Denote the limit with respect to these homomorphism by $\mathbb{E}_*^h(k)$, i.e.,

$$\mathbb{E}_*^h(k) = \lim_{m} H_*(\mathbb{H} \wedge E(k,S^m)).$$

As noted above, there are isomorphisms

$$\mathbb{E}_*^h(k) \cong H_*(\mathbb{H} \wedge E(k,S^0)) \cong H_*(\mathbb{H}) \otimes E_*(k).$$

Our goal is to define operations

$$\mu_*^h\colon \mathbb{E}_*^h(k) * H_*(\mathbb{X}) \to H_*(\mathbb{X}).$$

To do this, we consider the chain complexes of the topological spaces and the spectra. The chain complexes $C_*(E(k))$ form an operad $C_*(E) = \{C_*(E(k))\}$. The E-multiplicative structure on the spectrum $\mathbb{X} = \{\mathbb{X}(m)\}$ induces a $C_*(E)$-multiplicative structure on $C_*(\mathbb{X}) = \{C_*(\mathbb{X}(m))\}$,

$$C_*(E(k)) \otimes_{\Sigma_k} C_*(\mathbb{X}(m))^{\otimes k} \to C_*(\mathbb{X}(km)).$$

Let $x \in H_n(\mathbb{X})$, and let $f: C_*(S^{n+m}) \to C_*(\mathbb{X}(m))$ be a map representing x. This map induces a chain map

$$C_*(E(k)) \otimes_{\Sigma_k} C_*(S^{n+m})^{\otimes k} \to C_*(E(k)) \otimes_{\Sigma_k} C_*(\mathbb{X}(m))^{\otimes k} \to C_*(\mathbb{X}(km)),$$

and hence we have a map

$$C_*(\mathbb{H}(l)) \otimes C_*(E(k)) \otimes_{\Sigma_k} C_*(S^{n+m})^{\otimes k} \to C_*(\mathbb{H}(l)) \otimes C_*(\mathbb{X}(km)) \to C_*(\mathbb{X}(l+km)).$$

Let $y \in \mathbb{E}_i^h(k)$, and let $g: C_*(S^{i+l+k(n+m)}) \to C_*(E(k)) \otimes_{\Sigma_k} C_*(S^{n+m})^{\otimes k}$ be a map representing y. Then the element $\mu_*^h(y * x) \in H_{i+kn}(\mathbb{X})$ is represented by the composite map

$$C_*(S^{i+l+k(n+m)}) \xrightarrow{g} C_*(E(k)) \otimes_{\Sigma_k} C_*(S^{n+m})^{\otimes k} \to C_*(\mathbb{X}(l+km)).$$

In particular, let X be an E-multiplicative topological space, and let $\mathbb{X} = \mathbb{H} \wedge X$. Then the homology $H_*(\mathbb{H} \wedge X)$ is referred to as the *homology of X with $\mathbb{H}$-coefficients* and is denoted by $H_*(X; \mathbb{H})$. Thus we have the operations

$$\mu_*^h: \mathbb{E}_*^h(k) * H_*(X; \mathbb{H}) \to H_*(X; \mathbb{H}).$$

If in the definition of the Dyer–Lashof operations we replace $\mathbb{X}$ by $E(n, \mathbb{S})$, then we obtain operations

$$\gamma_*^h: \mathbb{E}_*^h(k) * \mathbb{E}_*^h(n) \to \mathbb{E}_*^h(kn),$$

which give an algebra structure on $\mathbb{E}_*^h = \{\mathbb{E}_*^h(n)\}$.

On the family $\mathbb{E}_*^h = \{\mathbb{E}_*^h(n)\}$ there is not just an algebra structure. There are also an $H_*(\mathbb{H})$-module structure

$$\mu_*^h: H_*(\mathbb{H}) \otimes \mathbb{E}_*^h(n) \to \mathbb{E}_*^h(n),$$

actions

$$\mu_*^h: \mathbb{E}_*^h(k) * H_*(\mathbb{H}) \to \mathbb{E}_*^h(k),$$

and comultiplications

$$\nabla_*^h: \mathbb{E}_*^h(n) \to \mathbb{E}_*^h(n) \otimes_{H_*(\mathbb{H})} \mathbb{E}_*^h(n).$$

The definition of the Dyer–Lashof operations for the homotopy and homology groups directly implies that these operations satisfy the associativity, distributivity, and Hopf relations, and commute with the Hurewicz homomorphism, i.e., the following theorem holds.

THEOREM. *Suppose that E is a bioperad and $\mathbb{H}$ is an E-multiplicative spectrum; then the homology $\mathbb{E}_*^h = \{\mathbb{E}_*^h(n)\}$ possesses a Hopf $H_*(\mathbb{H})$-bialgebra structure. If $\mathbb{X}$ is an E-multiplicative spectrum acted on by the E-multiplicative spectrum $\mathbb{H}$, then the homology $H_*(\mathbb{X})$ possesses the structure of an $H_*(\mathbb{H})$-algebra over the Hopf $H_*(\mathbb{H})$-bialgebra $\mathbb{E}_*^h$, and the following diagrams are commutative:*

$$\begin{array}{ccc} \mathbb{E}_*(k) * \pi_*(\mathbb{X}) & \xrightarrow{\mu_*} & \pi_*(\mathbb{X}) \\ {\scriptstyle h*h}\downarrow & & \downarrow{\scriptstyle h} \\ \mathbb{E}_*^h(k) * H_*(\mathbb{X}) & \xrightarrow{\mu_*^h} & H_*(\mathbb{X}) \end{array} \qquad \begin{array}{ccc} \mathbb{E}_*(k) * \mathbb{E}_*(n) & \xrightarrow{\gamma_*} & \mathbb{E}_*(kn) \\ {\scriptstyle h*h}\downarrow & & \downarrow{\scriptstyle h} \\ \mathbb{E}_*^h(k) * \mathbb{E}_*^h(n) & \xrightarrow{\gamma_*^h} & \mathbb{E}_*^h(kn) \end{array}$$

For any topological space X, we have an isomorphism

$$\mathbb{H}_*(X) \cong H_*(X) \otimes H_*(\mathbb{H}).$$

Therefore, there is a monomorphism $i_*: H_*(X) \to \mathbb{H}_*(X)$, $i_*(x) = x \otimes 1$. In particular, we have monomorphisms $i_*: E_*(k) \to \mathbb{E}^h_*(k)$.

It follows from the definitions of the operations γ^h_* and μ^h_* that the following diagrams are commutative:

$$\begin{array}{ccc} E_*(k) * E_*(n) & \xrightarrow{\gamma_*} & E_*(kn) \\ {\scriptstyle i_* * i_*}\downarrow & & \downarrow{\scriptstyle i_*} \\ \mathbb{E}^h_*(k) * \mathbb{E}^h_*(n) & \xrightarrow{\gamma^h_*} & \mathbb{E}^h_*(kn) \end{array} \qquad \begin{array}{ccc} E_*(k) * H_*(X) & \xrightarrow{\mu_*} & H_*(X) \\ {\scriptstyle i_* * i_*}\downarrow & & \downarrow{\scriptstyle i_*} \\ \mathbb{E}^h_*(k) * H_*(X;\mathbb{H}) & \xrightarrow{\mu^h_*} & H_*(X;\mathbb{H}) \end{array}$$

Let us compute the actions

$$\mu_*: \mathbb{E}O_*(k) * \pi_*(MO) \to \pi_*(MO), \qquad \mu^h_*: \mathbb{E}O^h_*(k) * H_*(MO) \to H_*(MO).$$

Since the Hurewicz homomorphism $h: \pi_*(MO) \to H_*(MO)$ is a monomorphism and commutes with these actions, it suffices to compute only the action μ^h_*.

Consider the action

$$\mu^h_*: \mathbb{E}O^h_*(2) * H_*(MO) \to H_*(MO).$$

The homomorphisms $\varphi_*: H_*(MO) \otimes E_*(2, S^m) \to H_*(MO) \otimes E_*(2, S^{m-1})$ are given by

$$\varphi_*(a \otimes e_i * u_m) = \sum_k a \cdot a_k \otimes e_{i-k} * u_{m-1}.$$

By using the homomorphisms $E_*(2, S^m) \cong E_*(2)$, we express the above homomorphisms as

$$\varphi_*: H_*(MO) \otimes E_*(2) \to H_*(MO) \otimes E_*(2), \quad \varphi_*(a \otimes e_i) = \sum_k a \cdot a_k \otimes e_{i-k}.$$

We have the commutative diagrams

$$\begin{array}{ccc} H_*(MO) \otimes E_*(2) * MO_{n+m}(n) & \xrightarrow{1 \otimes \mu_*} & H_*(MO) \otimes MO_*(2n) \\ {\scriptstyle \varphi^n_* \otimes i_*}\downarrow & & \downarrow{\scriptstyle \pi_*} \\ H_*(MO) \otimes E_*(2) * H_m(MO) & \xrightarrow{\mu^h_*} & H_*(MO) \end{array}$$

where φ^n_* is the n-fold composition of φ_*.

Therefore, if $x \in H_{n+m}(MO(n))$, then we have

$$\mu^h_*(a \otimes e_i * i_*(x)) = \pi_*(1 \otimes \mu_*)(\varphi^{-n}_*(a \otimes e_i) * x).$$

Straightforward computations show that the iterated compositions of φ^{-1}_* are given by

$$\varphi^{-n}_*(a \otimes e_i) = a \otimes e_i + \binom{n}{1} \sum_{k>0} a a_k \otimes e_{i-k} + \binom{n+1}{2} \sum_{k_1, k_2 > 0} a a_{k_1} a_{k_2} \otimes e_{i-k_1-k_2} + \dots$$

The action $\mu^h_*: \mathbb{E}O^h_*(2) * H_*(MO) \to H_*(MO)$ is completely determined by the actions $\mu_*: EO_*(2) * MO_*(n) \to MO_*(2n)$ and the homomorphisms $\varphi^{-n}_*: H_*(MO) \otimes E_*(2) \to H_*(MO) \otimes E_*(2)$.

Since μ^h_* is compatible with the left action of $H_*(MO)$, it suffices to compute μ^h_* only on the elements $1 \otimes e_i * x \in H_*(MO) \otimes E_*(2) * H_*(MO)$. Since μ^h_* is compatible with the algebra structure on $H_*(MO)$, it suffices to compute it only on the elements $1 \otimes e_i * a_m \in H_*(MO) \otimes E_*(2) * H_*(MO)$.

The above diagram yields the following formulas:

$$\begin{aligned}\mu^h_*(1 \otimes e_i * a_m) = &\sum_{j\geq 0} \binom{i+j-1}{j} a_{m-j}a_{m+i+j}\\ &+\sum_{\substack{j\geq 0\\ k>0}} \binom{i-k+j-1}{j} a_k a_{m-j} a_{m+i-k+j}\\ &+\sum_{\substack{j\geq 0\\ k_1,k_2>0}} \binom{i-k_1-k_2+j-1}{j} a_{k_1}a_{k_2}a_{m-j}a_{m+i-k_1-k_2+j}\\ &+\dots.\end{aligned}$$

These formulas imply that the actions $\mu_*\colon \mathbb{E}O_*(2) * \pi_*(MO) \to \pi_*(MO)$ coincide with the corresponding actions studied in [**13**].

Similarly, to compute the operations $\gamma_*\colon \mathbb{E}O_*(k) * \mathbb{E}O_*(n) \to \mathbb{E}O_*(kn)$, it suffices to compute the operations $\gamma^h_*\colon \mathbb{E}O^h_*(k) * \mathbb{E}O^h_*(n) \to \mathbb{E}O^h_*(kn)$.

It follows from the corresponding commutative diagram that these operations are determined by the operations $\gamma_*\colon E_*(k) * E_*(n) \to E_*(kn)$ and the actions

$$\mu^h_*\colon \mathbb{E}^h_*(k) * H_*(MO) \to H_*(MO).$$

Consider the operations

$$\gamma^h_*\colon \mathbb{E}O^h_*(2) * \mathbb{E}O^h_*(2) \to \mathbb{E}O^h_*(4)$$

or, what is the same,

$$\gamma^h_*\colon H_*(MO) \otimes E_*(2) * H_*(MO) \otimes E_*(2) \to H_*(MO) \otimes E_*(4).$$

Then we have

$$\gamma^h_*(a \otimes e_i * x \otimes e_j) = \sum_k \mu^h_*(a \otimes e_k * x) \otimes \gamma_*(e_{i-k} * e_j).$$

8.4. Generalized cohomology theories and the Steenrod algebra

Consider the cohomology theory $\mathbb{H}^*$ associated with a spectrum $\mathbb{H}$. Suppose that X is a topological space and $\mathbb{H}$ is an E-multiplicative spectrum. Let $F(X, \mathbb{H})$ denote the spectrum with $F(X, \mathbb{H})(m)$ being the space of continuous maps $f\colon X \to \mathbb{H}(m)$. Then $F(X, \mathbb{H})$ is an E-multiplicative spectrum acted on by the spectrum $\mathbb{H}$. Therefore, operations

$$\mu_*\colon \mathbb{E}_*(k) * \pi_*(F(X, \mathbb{H})) \to \pi_*(F(X, \mathbb{H}))$$

are defined. Recall that the generalized cohomology $\mathbb{H}^*(X)$ of the space X is defined by

$$\mathbb{H}^n(X) = \pi_{-n}(F(X, \mathbb{H})).$$

Hence we have the operations

$$\mu_*\colon \mathbb{E}_i(k) * \mathbb{H}^n(X) \to \mathbb{H}^{kn-i}(X),$$

which we call the *Steenrod operations for the generalized cohomology.*

THEOREM. *Suppose that E is a bioperad and $\mathbb{H}$ is an E-multiplicative spectrum. Then the cohomology $\mathbb{H}^*(X)$ of any topological space X possesses the structure of an $\mathbb{H}^*(S^0)$-algebra over the Hopf bialgebra $\mathbb{E}_*$.*

In particular, there are Steenrod operations on cobordism theories. For example, there are Steenrod operations

$$\mu_*\colon \mathbb{E}O_* * \mathbb{M}O^*(X) \to \mathbb{M}O^*(X), \quad \mu_*\colon \mathbb{E}U_* * \mathbb{M}U^*(X) \to \mathbb{M}U^*(X)$$

analogous to the Steenrod operations in [**18**].

Note that to define the Steenrod operations for cohomology with $\mathbb{Z}/p$-coefficients, $\mathbb{H}^*(\mathbb{X};\mathbb{Z}/p)$, we need only replace the sphere spectrum $\mathbb{S}$ by the corresponding Moore spectrum $\mathbb{S}_{(p)}$. Then we obtain the operations

$$\mu_*\colon \mathbb{E}_*^{(p)}(k) * \mathbb{H}^*(\mathbb{X};\mathbb{Z}/p) \to \mathbb{H}^*(\mathbb{X};\mathbb{Z}/p).$$

In particular, there are mod-p Steenrod operations on cobordism theories. For example, we have the mod-p Steenrod operations

$$\mu_*\colon \mathbb{E}U_*^{(p)} \otimes \mathbb{M}U^*(X;\mathbb{Z}/p) \to \mathbb{M}U^*(X;\mathbb{Z}/p).$$

Steenrod operations can be defined similarly for the ordinary cohomology of a topological space X with $\mathbb{H}$-coefficients. To this end, we associate to X the spectrum $F(C_*(X), C_*(\mathbb{H}))$ in the category of chain complexes. The homology of this spectrum is referred to as the *cohomology of X with $\mathbb{H}$-coefficients* and is denoted by $H^*(X;\mathbb{H})$.

Recall that there are coactions

$$\tau_*\colon C_*(E(k)) \otimes C_*(X) \to C_*(X)^{\otimes k}$$

and actions

$$\mu_*\colon C_*(E(k)) \otimes_{\Sigma_k} C_*(\mathbb{H}(m)) \to C_*(\mathbb{H}(km)).$$

Besides, there are actions

$$C_*(\mathbb{H}(l)) \otimes F(C_*(X), C_*(\mathbb{H}(m))) \to F(C_*(X), C_*(\mathbb{H}(l+m))).$$

Hence we have the operations

$$\mu_*^h\colon \mathbb{E}_*^h(k) * H^*(X;\mathbb{H}) \to H^*(X;\mathbb{H}),$$

which are compatible with the Hurewicz homomorphism, i.e., the following diagrams are commutative:

$$\begin{array}{ccc} \mathbb{E}_*(k) * \mathbb{H}^*(X) & \xrightarrow{\mu_*} & \mathbb{H}^*(X) \\ {\scriptstyle h*h}\downarrow & & \downarrow{\scriptstyle h} \\ E_*^h(k) * H^*(X;\mathbb{H}) & \xrightarrow{\mu_*^h} & H^*(X;\mathbb{H}) \end{array}$$

These diagrams make it possible to compute the Steenrod operations. In particular, if $X = S^0$ and $\mathbb{H} = MO$, then we have $\mathbb{M}O^*(S^0) \cong MO_{-*}(S^0)$, and the formulas obtained above show that these Steenrod operations coincide with the Steenrod operations in [**18**].

Note that not only have we constructed the Steenrod operations, as was done in [**18**], but we have also defined the algebra $\mathbb{E}_*$ that performs these operations.

Consider the suspension isomorphism $S\colon \mathbb{H}^{n+1}(SX) \to \mathbb{H}^n(X)$. Suppose that an element $x \in \mathbb{H}^{n+1}(SX)$ is represented by a map $f\colon S^m SX \to \mathbb{H}(n+m+1)$. Then the element $S(x) \in \mathbb{H}^n(X)$ is represented by the same map f.

Let us define a map $\sigma\colon \mathbb{E}_{i+k-1}(k) \to \mathbb{E}_i(k)$. Suppose that an element $y \in \mathbb{E}_{i+k-1}(k)$ is represented by a map $g\colon S^{i+k-1+j+km} \to \mathbb{H}(j) \wedge E(k, S^m)$; then the element $\sigma(y) \in \mathbb{E}_i(k)$ is represented by the composite map

$$S^{i+j+k(m+1)} \xrightarrow{g\wedge 1} \mathbb{H}(j) \wedge E(k, S^m) \wedge S^1 \xrightarrow{1\wedge\nabla} \mathbb{H}(j) \wedge E(k, S^{m+1}).$$

The definitions of the operations

$$\gamma\colon \mathbb{E}_*(k) * \mathbb{E}_*(n) \to \mathbb{E}_*(kn), \quad \mu_*\colon \mathbb{E}_*(k) * \mathbb{H}^*(X) \to \mathbb{H}^*(X)$$

directly imply our next theorem.

THEOREM. *The following diagrams are commutative*:

$$\begin{array}{ccc} \mathbb{E}_{i+k-1}(k) * \mathbb{E}_{j+n-1}(n) & \xrightarrow{\gamma_*} & \mathbb{E}_{i+k-1+k(j+n-1)}(kn) \\ \Big\downarrow{\scriptstyle \sigma*\sigma} & & \Big\downarrow{\scriptstyle \sigma} \\ \mathbb{E}_i(k) * \mathbb{E}_j(n) & \xrightarrow{\gamma_*} & \mathbb{E}_{i+kj}(kn) \end{array}$$

$$\begin{array}{ccc} \mathbb{E}_{i+k-1}(k) * \mathbb{H}^{n+1}(SX) & \xrightarrow{\mu_*} & \mathbb{H}^{kn-i+1}(SX) \\ \Big\downarrow{\scriptstyle \sigma*S} & & \Big\downarrow{\scriptstyle S} \\ \mathbb{E}_i(k) * \mathbb{H}^n(X) & \xrightarrow{\mu_*} & \mathbb{H}^{kn-i}(X) \end{array}$$

Let $\mathbb{A}^m(k) = \varprojlim_n \mathbb{E}_{(k-1)n-m}(k)$, where the limit is taken with respect to the homomorphisms σ. Then the operations $\gamma_*\colon \mathbb{E}_*(k) * \mathbb{E}_*(j) \to \mathbb{E}_*(kj)$ induce operations

$$\gamma^*\colon \mathbb{A}^*(k) \otimes \mathbb{A}^*(j) \to \mathbb{A}^*(kj).$$

More precisely, the operations

$$\gamma^*\colon \mathbb{A}^m(k) \otimes \mathbb{A}^i(j) \to \mathbb{A}^{m+i}(kj)$$

are induced by the operations

$$\gamma_*\colon \mathbb{E}_{(k-1)(i+n)-m}(k) * \mathbb{E}_{(j-1)n-i}(j) \to \mathbb{E}_{(kj-1)n-i-m}(kj).$$

Similarly, the operations $\nabla_*\colon \mathbb{E}_*(n) \to \mathbb{E}_*(n) \otimes_{\mathbb{H}_*(S^0)} \mathbb{E}_*(n)$ induce operations

$$\nabla^*\colon \mathbb{A}^*(n) \to \mathbb{A}^*(n) \otimes_{\mathbb{H}^*(S^0)} \mathbb{A}^*(n).$$

If X is a topological space, then the actions

$$\mu_*\colon \mathbb{E}_*(k) * \mathbb{H}^*(X) \to \mathbb{H}^*(X)$$

induce the actions

$$\mu^*\colon \mathbb{A}^*(k) \otimes \mathbb{H}^*(X) \to \mathbb{H}^*(X).$$

More precisely, the operations

$$\mu^*\colon \mathbb{A}^m(k) \otimes \mathbb{H}^n(X) \to \mathbb{H}^{m+n}(X)$$

are induced by the operations

$$\mu_*\colon \mathbb{E}_{(k-1)n-m}(k) * \mathbb{H}^n(X) \to \mathbb{H}^{m+n}(X).$$

Therefore, the following theorem holds.

THEOREM. *The operations* γ^* *and* ∇^* *form a Hopf* $\mathbb{H}^*(S^0)$*-algebra structure on* $\mathbb{A}^* = \sum_k \mathbb{A}^*(k)$*. The cohomology* $\mathbb{H}^*(X)$ *of any topological space* X *possesses a natural structure of an* $\mathbb{H}^*(S^0)$*-algebra over the Hopf* $\mathbb{H}^*(S^0)$*-algebra* $\mathbb{A}^*$*.*

For the suspension isomorphisms we have the commutative diagrams

$$\begin{array}{ccc} \mathbb{A}^* \otimes \mathbb{H}^{*+1}(SX) & \xrightarrow{\mu^*} & \mathbb{H}^{*+1}(SX) \\ {\scriptstyle 1\otimes S}\downarrow & & \downarrow{\scriptstyle S} \\ \mathbb{A}^* \otimes \mathbb{H}^*(X) & \xrightarrow{\mu^*} & \mathbb{H}^*(X) \end{array}$$

and hence the following theorem holds.

THEOREM. *Suppose that* $\mathbb{X} = \{\mathbb{X}(m)\}$ *is a spectrum. Then the actions*

$$\mu^*(m) \colon \mathbb{A}^* \otimes \mathbb{H}^*(\mathbb{X}(m)) \to \mathbb{H}^*(\mathbb{X}(m))$$

induce the action

$$\mu^* \colon \mathbb{A}^* \otimes \mathbb{H}^*(\mathbb{X}) \to \mathbb{H}^*(\mathbb{X})$$

of the Hopf algebra $\mathbb{A}^*$ *on the cohomology* $\mathbb{H}^*(\mathbb{X})$ *of the spectrum* $\mathbb{X}$*.*

Thus the Hopf algebra $\mathbb{A}^*$ can be regarded as an analog of the Steenrod algebra for the generalized cohomology theories.

Denote the corresponding Steenrod algebras for the $\mathbb{M}O^*$-, $\mathbb{M}U^*$-, ... theories by $\mathbb{A}O^*$, $\mathbb{A}U^*$, ... For example, in the case of $\mathbb{M}O^*$-theory there is an isomorphism $\mathbb{A}O^* \cong \mathbb{M}O^* \otimes A^*$, where A^* denotes the ordinary Steenrod algebra.

Consider the Landweber–Novikov operations $\nu^* \colon \mathbb{H}^*(\mathbb{H}) \otimes \mathbb{H}^*(\mathbb{X}) \to \mathbb{H}^*(\mathbb{X})$ on the generalized cohomology of the spectrum $\mathbb{X}$; see [**26, 35**].

The action $\mu^* \colon \mathbb{A}^* \otimes \mathbb{H}^*(\mathbb{H}) \to \mathbb{H}^*(\mathbb{H})$ induces a Hopf algebra map

$$\xi^* \colon \mathbb{A}^* \to \mathbb{H}^*(\mathbb{H}), \quad \xi^*(x) = \mu^*(x \otimes 1).$$

The next theorem follows from the definitions of the Landweber–Novikov operations.

THEOREM. *The following diagrams are commutative*:

$$\begin{array}{ccc} \mathbb{A}^* \otimes \mathbb{H}^*(\mathbb{X}) & \xrightarrow{\xi^*\otimes 1} & \mathbb{H}^*(\mathbb{H}) \otimes \mathbb{H}^*(\mathbb{X}) \\ {\scriptstyle =}\downarrow & & \downarrow{\scriptstyle \nu^*} \\ \mathbb{A}^* \otimes \mathbb{H}^*(\mathbb{X}) & \xrightarrow{\mu^*} & \mathbb{H}^*(\mathbb{X}) \end{array}$$

Thus the Steenrod operations on the generalized cohomology are determined by the Landweber–Novikov operations.

PROOF. Consider a topological space X, and assume that an element $x \in \mathbb{H}^*(X)$ is represented by a map $f \colon X \to \mathbb{H}(n)$, and an element $y \in \mathbb{A}^*$ is represented by a map $g \colon S^{i+j+km} \to \mathbb{H}(j) \wedge E(k, S^m)$. Then the element $\mu^*(y \otimes x) \in \mathbb{H}^*(X)$ is represented by the map

$$S^{i+j+km} \wedge X \xrightarrow{g\wedge 1} \mathbb{H}(j) \wedge E(k, S^m) \wedge X$$
$$\to \mathbb{H}(j) \wedge E(k, S^m X) \xrightarrow{1\wedge E(k, S^m f)} \mathbb{H}(j) \wedge E(k, \mathbb{H}(n+m)) \to \mathbb{H}(j + kn + km),$$

and the element $\nu^*(\xi^*(y)\otimes x)\in \mathbb{H}^*(X)$ is represented by the map

$$S^{i+j+km}\wedge X\to \mathbb{H}(j)\wedge E(k,S^m)\wedge \mathbb{H}(n)\to \mathbb{H}(j)\wedge E(k,\mathbb{H}(n+m))\to \mathbb{H}(j+kn+km).$$

The required relation $\mu^*(y\otimes x)=\nu^*(\xi^*(y)\otimes x)$ now follows from the commutative diagram

$$\begin{array}{ccc} E(k,S^m)\wedge X & \longrightarrow & E(k,S^m\wedge X) \\ {\scriptstyle 1\wedge f}\downarrow & & \downarrow{\scriptstyle E(k,S^m f)} \\ E(k,S^m)\wedge \mathbb{H}(n) & & E(k,\mathbb{H}(n+m)) \\ \downarrow & & \downarrow{\scriptstyle =} \\ E(k,S^m\wedge \mathbb{H}(n)) & \longrightarrow & E(k,\mathbb{H}(n+m)) \end{array}$$

Since μ^* and ν^* commute with the suspension homomorphism, the required relations also holds for spectra $\mathbb{X}$ and for any $x\in \mathbb{H}^*(\mathbb{X})$, $y\in \mathbb{A}^*$. □

In particular, for the case of the spectrum MO, there is a monomorphism $A^*\to H^*(MO)$ of the ordinary Steenrod algebra into the ordinary cohomology of MO. The above computations imply that this monomorphism induces a monomorphism $\mathbb{A}O^*\to \mathbb{M}O^*(MO)$.

Note that in contrast to the Landweber–Novikov algebra $\mathbb{H}^*(\mathbb{H})$, which is defined for arbitrary (not necessarily multiplicative) spectra $\mathbb{H}$, the Steenrod algebra is only defined for E_∞-multiplicative spectra $\mathbb{H}$. Furthermore, the multiplication in the Landweber–Novikov algebra employs only the composition structure on the cohomology $\mathbb{H}^*(\mathbb{H})$, while the multiplication in the Steenrod algebra substantially uses the E_∞-multiplicative structure on $\mathbb{H}$.

Since there is a map from the Steenrod algebra to the Landweber–Novikov algebra, it follows that the action of the Steenrod algebra on the generalized cohomology of a topological space gives us only part of the cohomological operations. As for the whole set of cohomological operations, they are generated by the action of the Landweber–Novikov algebra.

Thus, in the study of questions related to applying the cohomological operations to the cohomology theories corresponding to E_∞-multiplicative spectra, it is possible to consider instead of the Landweber–Novikov algebra, which is rather large, the Steenrod algebra, which is smaller.

In particular, this Steenrod algebra can be used to construct, in analogy to the Adams spectral sequence, a spectral sequence whose E^2-term is expressed in terms of the Steenrod algebra rather than in terms of the Landweber–Novikov algebra. The operad method proposed above allows us not only to find an expression for the E^2-term of this spectral sequence, but also to define higher cohomology operations on the generalized cohomology, which yield higher differentials of the spectral sequence.

Let $\mathbb{A}_m(k)=\varprojlim_n \mathbb{E}_{m-(k-1)n}(k)$, where the limit is taken with respect to the homomorphisms σ. Then the operations $\gamma_*\colon \mathbb{E}_*(k)*\mathbb{E}_*(j)\to \mathbb{E}_*(kj)$ induce operations

$$\gamma_*\colon \mathbb{A}_*(k)\otimes \mathbb{A}_*(j)\to \mathbb{A}_*(kj).$$

More precisely, the operations

$$\gamma_*\colon \mathbb{A}_m(k)\otimes \mathbb{A}_i(j)\to \mathbb{A}_{m+i}(kj)$$

are induced by the operations

$$\gamma_*\colon \mathbb{E}_{m-(k-1)(i+n)}(k) * \mathbb{E}_{i-(j-1)n}(j) \to \mathbb{E}_{i+m-(kj-1)n}(kj).$$

Similarly, the operations $\nabla_*\colon \mathbb{E}_*(n) \to \mathbb{E}_*(n) \otimes_{\mathbb{H}_*(S^0)} \mathbb{E}_*(n)$ induce operations

$$\nabla_*\colon \mathbb{A}_*(n) \to \mathbb{A}_*(n) \otimes_{\mathbb{H}_*(S^0)} \mathbb{A}_*(n).$$

If $\mathbb{X}$ is an E-multiplicative spectrum, then the actions

$$\mu_*\colon \mathbb{E}_*(k) * \mathbb{H}_*(X) \to \mathbb{H}_*(X)$$

induce actions

$$\mu_*\colon \mathbb{A}_*(k) \otimes \mathbb{H}_*(X) \to \mathbb{H}_*(X).$$

More precisely, the operations

$$\mu_*\colon \mathbb{A}_m(k) \otimes \mathbb{H}_n(X) \to \mathbb{H}_{m+n}(X)$$

are induced by the operations

$$\mu_*\colon \mathbb{E}_{m-(k-1)n}(k) * \mathbb{H}_n(X) \to \mathbb{H}_{m+n}(X).$$

Therefore, we have the following theorem.

THEOREM. *The operations γ_* and ∇_* form a Hopf $\mathbb{H}_*(S^0)$-algebra structure on $\mathbb{A}_* = \sum_k \mathbb{A}_*(k)$. The cohomology $\mathbb{H}_*(\mathbb{X})$ of any E-multiplicative spectrum $\mathbb{X}$ possesses a natural structure of an $\mathbb{H}_*(S^0)$-algebra over the Hopf $\mathbb{H}_*(S^0)$-algebra $\mathbb{A}_*$.*

The Hopf algebra $\mathbb{A}_*$ can be regarded as an analog of the Dyer–Lashof algebra for generalized homology theories.

We denote the corresponding Dyer–Lashof algebras for the $\mathbb{M}O_*$-, $\mathbb{M}U_*$-, ... theories by $\mathbb{A}O_*$, $\mathbb{A}U_*$, ... For example, in the case of the $\mathbb{M}O_*$-theory, there is an isomorphism

$$\mathbb{A}O_* \cong \mathbb{M}O_* \otimes A_*,$$

where A_* denotes the ordinary Dyer–Lashof algebra.

CHAPTER 9

Description of the Cohomology of Groups and Algebras

9.1. Description of A_∞-algebras and A_∞-modules

Now let us consider whate it means to describe an A_∞-algebra or an A_∞-coalgebra. Ordinary algebras are described by giving indecomposable elements and relations for their products. But what is an indecomposable element in the case of A_∞-algebras? It would be a mistake to define decomposable elements of an A_∞-algebra A as elements belonging to the images of certain products $\pi_n : A^{\otimes n+2} \to A$. To give a correct definition, it is necessary to reformulate the definition of decomposable elements of algebras with Massey products for the case of A_∞-algebras.

Suppose that A is an A_∞-algebra. A sequence $(x^2, \dots, x^n)$ of elements $x^i \in A^{\otimes i}$ is called a *Massey sequence* if the following relations hold:

$$\begin{aligned}
&(\pi_0 \otimes \dots \otimes 1 - \dots + (-1)^n 1 \otimes \dots \pi_0)(x^n) = 0,\\
&(\pi_1 \otimes \dots \otimes 1 + \dots + 1 \otimes \dots \otimes \pi_1)(x^n)\\
&\qquad = (\pi_0 \otimes \dots \otimes 1 - \dots + (-1)^n 1 \otimes \dots \otimes \pi_0)(x^{n-1}), \quad \dots\\
&(\pi_{n-3} \otimes 1 + 1 \otimes \pi_{n-3})(x^n)\\
&\qquad = (\pi_{n-4} \otimes 1 + (-1)^n 1 \otimes \pi_{n-4})(x^{n-1}) + \dots + (\pi_0 \otimes 1 - 1 \otimes \pi_0)(x^3).
\end{aligned}$$

For such a Massey sequence $(x^2, \dots, x^n)$, the Massey product is defined as follows: $\mu(x^2, \dots, x^n) = \pi_0(x^2) + \dots + \pi_{n-2}(x^n)$. An element $x \in A$ is said to be *decomposable* if it is the image of some Massey product. The quotient of A by decomposable elements is called the *module of indecomposable elements* and is denoted by JA.

Let us interpret this definition in the language of bar constructions. Consider the bar construction BA on an A_∞-algebra A. The inclusion map $i : A \to BA$ induces the short exact sequence

$$0 \to A \xrightarrow{i} BA \xrightarrow{p} B^1 A = BA/A \to 0$$

and the long exact homology sequence

$$\dots \to A \xrightarrow{i_*} H_*(BA) \xrightarrow{p_*} H_*(B^1 A) \xrightarrow{\mu_*} A \to \dots.$$

The definition of Massey sequence implies that $(x^2, \dots, x^n)$ is a Massey sequence if and only if the element $x = x^2 + \dots + x^n$ is a cycle in $B^1(A)$. Moreover, the map μ_* is induced by the Massey product. Thus the element $x \in A$ is decomposable if and only if it belongs to the image of μ_*. Therefore, the module of indecomposable elements JA is isomorphic to the image of the map $i_* : A \to H_*(BA)$.

Dually, consider the cobar construction $\hat{F}K$ on a graded A_∞-coalgebra K. The projection $p : \hat{F}K \to K$ induces the short exact sequence

$$0 \to \hat{F}^1 K \xrightarrow{i} \hat{F}K \xrightarrow{p} K \to 0$$

and the long exact homology sequence

$$\cdots \to H_*(\hat{F}^1 K) \xrightarrow{i_*} H_*(\hat{F}K) \xrightarrow{p_*} K \xrightarrow{\nu} H_*(\hat{F}^1 K) \to \cdots.$$

An element $x \in K$ is said to be *primitive* if it belongs to the kernel of the homomorphism $\nu_* : K \to H_*(\hat{F}^1 K)$ or, equivalently, to the image of the homomorphism $p_* : H_*(\hat{F}K) \to K$. Denote the module of primitive elements of an A_∞-coalgebra K by PK. Note that in the case of an ordinary coalgebra K with comultiplication $\nabla : K \to K \otimes K$, primitive elements are just elements of the kernel of ∇.

Thus the description of A_∞-algebras and coalgebras requires first finding their indecomposable and primitive elements, respectively.

Now consider the question of describing relations in an A_∞-algebra. Let A be a graded A_∞-algebra. Each relation in A can be expressed in the form

$$\mu(x^2, \ldots, x^n) = \pi_0(x^2) + \cdots + \pi_{n-2}(x^n) = 0,$$

where $(x^2, \ldots, x^n)$ is a Massey sequence and $\mu(x^2, \ldots, x^n)$ is the Massey product. Finding a relation in an A_∞-algebra means finding Massey sequences whose Massey products are equal to zero. Of course, some of these relations follow from the A_∞-algebra structure; hence it suffices to find only relations that are not just consequences of the A_∞-structure.

As we have seen, Massey sequences $(x^2, \ldots, x^n)$ determine cycles in $B^1 A$. Let us show that the cycles homologous to zero determine relations following from the A_∞-structure on A.

Let $y \in A^{\otimes n+1}$, and let $d(y) = x^2 + \cdots + x^n$. Then the following relations hold:

$$x^2 = (\pi_{n-2} \otimes 1 - 1 \otimes \pi_{n-2})(y),$$

$$\cdots$$

$$x^n = (\pi_0 \otimes 1 \cdots \otimes 1 - \cdots + (-1)^n 1 \otimes \cdots \otimes 1 \otimes \pi_0)(y).$$

In this case, the Massey sequence $(x^2, \ldots, x^n)$ defines the relation

$$\sum (-1)^\epsilon \pi_i (1 \otimes \cdots \otimes \pi_{n-i-2} \otimes \cdots \otimes 1)(y) = 0,$$

which is a direct consequence of the A_∞-structure. Thus the relations we need are determined by the elements x of $H_*(B^1 A)$ such that $\mu_*(x) = 0$, where $\mu_* : H_*(B^1 A) \to A$.

On the other hand, let $A_\infty A$ be the free A_∞-algebra generated by A, and let $A^1_\infty A = A_\infty A / A$. Then finding relations in A means finding elements $x \in A^1_\infty A$ such that $\mu(x) = 0$, where $\mu : A^1_\infty A \to A$ is the map induced by the A_∞-algebra structure on A. Of course, some of these relations are just consequences of the A_∞-structure on A. To find them, consider the map

$$\gamma \times 1 - 1 \times \mu : A_\infty A^1_\infty A \to A^1_\infty A.$$

If an element x belongs to the image of this map, then it generates the relation $\mu(x) = 0$, which follows from the A_∞-structure on A. Let $A^1_\infty * A$ be the quotient of $A^1_\infty A$ by the image of the map $\gamma \times 1 - 1 \times \mu$. Then to find relations it suffices to find elements x in $A^1_\infty * A$ whose images under the map $A^1_\infty * A \to A$ vanish.

For brevity, denote the homology of the bar construction on the A_∞-algebra A by K, i.e., $K = H_*(BA)$. Further, let $\overline{A}_\infty K$ be the free A_∞-coalgebra generated by K; $\overline{A}^1_\infty K = \overline{A}_\infty K/K$. The A_∞-coalgebra structure on K induces a map $\tau\colon K \to \overline{A}^1_\infty K$. Besides, we have the duality map $\psi\colon \overline{A}^1_\infty K \to A^1_\infty A$. Consider the map $K \to A^1_\infty * A$ which is the composition of the maps τ, ψ, and the projection $A^1_\infty A \to A^1_\infty * A$. Then we have the commutative diagram

$$\begin{array}{ccccc} K & \to & H_*(B^1A) & \to & A \\ & \searrow & \downarrow & \nearrow & \\ & & A^1_\infty * A & & \end{array}$$

It follows from this diagram that to find the relations in the A_∞-algebra A, we must find the image of the map $K \to A^1_\infty * A$. Summing up, we conclude that to describe an A_∞-algebra A we must do the following.

1. Find the homology $H_*(BA)$ of the bar construction BA on A.

2. Compute the image of the homomorphism $A \to H_*(BA)$; this image is the module JA of indecomposable elements of the A_∞-algebra A.

3. Compute the image of the homomorphism $H_*(BA) \to A^1_\infty * A$; this image gives the relations in the A_∞-algebra A.

Dually, to describe an A_∞-coalgebra K, we must do the following.

1′. Find the homology $H_*(\hat{F}K)$ of the cobar construction $\hat{F}K$ on K.

2′. Compute the image of the homomorphism $H_*(\hat{F}K) \to K$; this image is the module of primitive elements of the A_∞-coalgebra K.

3′. Compute the kernel of the homomorphism $\overline{A}^1_\infty * K \to H_*(\hat{F}K)$; this kernel gives generating elements of the sought-for A_∞-coalgebra.

The above method of describing graded A_∞-algebras can be extended to graded A_∞-modules over an A_∞-algebra.

Suppose that M is a graded A_∞-module over an A_∞-algebra A. Consider the bar construction $B(A, M)$. The inclusion map $i\colon M \to B(A, M)$ induces the short exact sequence

$$0 \to M \xrightarrow{i} B(A, M) \xrightarrow{p} B^1(A, M) = B(A, M)/M \to 0$$

and the long exact homology sequence

$$\cdots \to M \xrightarrow{i_*} H_*(B(A, M)) \xrightarrow{p_*} H_*(B^1(A, M)) \xrightarrow{\mu_*} M \to \dots .$$

We say that an element $x \in M$ is *decomposable* if it is the image of some element under the map μ_*. Thus the module of indecomposable elements JM is isomorphic to the image of the homomorphism $i_*\colon M \to H_*(B(A, M))$.

Dually, for a graded A_∞-comodule L over an A_∞-coalgebra K, consider the cobar construction $F(K, L)$. The projection $p\colon \hat{F}(K, L) \to L$ induces the short exact sequence

$$0 \to F^1(K, L) \xrightarrow{i} F(K, L) \xrightarrow{p} L \to 0$$

and the long exact homology sequence

$$\cdots \to H_*(F^1(K, L)) \xrightarrow{i_*} H_*(F(K, L)) \xrightarrow{p_*} L \xrightarrow{\nu_*} H_*(\hat{F}^1(K, L)) \to \dots .$$

An element $x \in L$ is said to be *primitive* if it belongs to the kernel of the homomorphism $\nu_*\colon L \to H_*(\hat{F}^1(K,L))$ or, equivalently, to the image of the homomorphism $p_*\colon H_*(F(K,L)) \to L$. Denote the module of primitive elements of an A_∞-comodule L by PL.

Let us extend the method of describing A_∞-algebras and A_∞-coalgebras to the case of A_∞-$\vee$-algebras and A_∞-$\wedge$-coalgebras.

Let Y be a graded A_∞-$\vee$-algebra. A sequence $(y^2,\dots,y^n)$ of elements $y^i \in Y^{\vee i}$ will be called a *Massey sequence* if the following relations hold:

$$\begin{aligned}
&(\pi_0 \otimes \cdots \otimes 1 - \cdots + (-1)^n 1 \otimes \dots \pi_0)(y^n) = 0,\\
&(\pi_1 \otimes \cdots \otimes 1 + \cdots + 1 \otimes \cdots \otimes \pi_1)(y^n)\\
&\qquad = (\pi_0 \otimes \cdots \otimes 1 - \cdots + (-1)^n 1 \otimes \cdots \otimes \pi_0)(y^{n-1}),\\
&\dots\\
&(\pi_{n-3} \otimes 1 + 1 \otimes \pi_{n-3})(y^n)\\
&\qquad = (\pi_{n-4} \otimes 1 + (-1)^n 1 \otimes \pi_{n-4})(y^{n-1}) + \cdots + (\pi_0 \otimes 1 - 1 \otimes \pi_0)(y^3).
\end{aligned}$$

For such a Massey sequence $(y^2,\dots,y^n)$, the Massey product is defined as follows: $\mu(y^2,\dots,y^n) = \pi_0(y^2) + \cdots + \pi_{n-2}(y^n)$. We say that an element $y \in Y$ is *decomposable* if it is the image of some Massey product. The quotient of Y over decomposable elements is referred to as the *module of indecomposable elements* and is denoted by JY.

Let us interpret this definition in the language of bar constructions. Consider the bar construction B_cY on the A_∞-$\vee$-algebra Y. The inclusion map $i : Y \to B_cY$ induces the short exact sequence

$$0 \to Y \xrightarrow{i} B_cY \xrightarrow{p} B_c^1Y = B_cY/Y \to 0$$

and the long exact homology sequence

$$\cdots \to Y \xrightarrow{i_*} H_*(B_cY) \xrightarrow{p_*} H_*(B_c^1Y) \xrightarrow{\mu_*} Y \to \dots.$$

The definition of Massey sequences implies that a sequence $(y^2,\dots,y^n)$ is a Massey sequence if and only if the element $y = y^2 + \cdots + y^n$ is a cycle in $B_c^1(Y)$. Moreover, the map μ_* is induced by the Massey product. Thus an element $y \in Y$ is decomposable if and only if it belongs to the image of μ_*. Therefore, the module of indecomposable elements JY is isomorphic to the image of the map $i_*\colon Y \to H_*(B_cY)$.

To find the relations, we must indicate the Massey sequences $(y^2,\dots,y^n)$ whose Massey products $\mu(y^2,\dots,y^n)$ vanish. Part of these relations follows directly from the A_∞-$\vee$-algebra structure. As in the case of ordinary A_∞-algebras, it can be shown that this happens if the cycle $y^2+\cdots+y^n \in B^1Y$ is homologous to zero. Thus the relations that are not just consequences of the A_∞-$\vee$-algebra structure on Y are given by the kernel of the homomorphism $\mu_*\colon H_*(B_c^1Y) \to Y$ or, equivalently, by the image of the homomorphism $p_*\colon H_*(B_cY) \to H_*(B_c^1Y)$.

Dually, consider the cobar construction F_sX on a graded A_∞-$\wedge$-coalgebra X. The projection $p\colon F_sX \to X$ induces the short exact sequence

$$0 \to F_s^1X \xrightarrow{i} F_sX \xrightarrow{p} X \to 0$$

and the long exact homology sequence

$$\cdots \to H_*(F_s^1X) \xrightarrow{i_*} H_*(F_sX) \xrightarrow{p_*} X \xrightarrow{\nu_*} H_*(F_s^1X) \to \dots.$$

An element $x \in X$ is said to be *primitive* if it belongs to the kernel of the homomorphism $\nu_*: X \to H_*(F_s^1 X)$ or, equivalently, to the image of the homomorphism $p_*: H_*(F_s X) \to X$. Denote the module of primitive elements of an A_∞-Λ-coalgebra X by PX.

9.2. Description of the cohomology of groups

Let A be an algebra (over a field R). Then the bar construction BA on A is the chain complex

$$BA : A \leftarrow A \otimes A \leftarrow \cdots \leftarrow A^{\otimes n} \leftarrow \dots,$$

whose generators $x_1 \otimes \cdots \otimes x_n \in A^{\otimes n}$ are denoted by $[x_1, \dots, x_n]$, and whose differential is given by the formula

$$d[x_1, \dots, x_n] = \sum_{i=1}^{n-1} (-1)^i [x_1, \dots, \pi(x_i \otimes x_{i+1}), \dots, x_n],$$

where $\pi : A \otimes A \to A$ is the multiplication in A.

Apart from the structure of a chain complex, the bar construction BA possesses a coalgebra structure $\nabla: BA \to BA \otimes BA$ given by

$$\nabla[x_1, \dots, x_n] = \sum_{i=1}^{n-1} [x_1, \dots, x_i] \otimes [x_{i+1}, \dots, x_n].$$

The homology $H_*(BA)$ of the bar construction BA on A is referred to as the *homology of the algebra* A. The cohomology $H^*(BA) = H_*(\overline{BA})$, where $\overline{BA} = \mathrm{Hom}(BA, R)$, is referred to as the *cohomology of the algebra* A.

In particular, given a group G, let us consider the group ring $R(G)$ and the group algebra $A = \mathrm{Ker}\{a: R(G) \to R\}$; the homology and cohomology of the algebra A are called, respectively, the homology and cohomology of the group G.

Our aim is to describe this homology. Since the homology of groups is a special case of the homology of algebras, we consider the case of homology and cohomology of algebras.

The first problem to be solved is to elucidate the algebraic structure existing on the homology and cohomology of algebras (in addition to the usual group structure). To this end, we note that if K is a differential coalgebra, then its homology $K_* = H_*(K)$ possesses an A_∞-coalgebra structure, and hence the homology of the bar construction BA on A possesses an A_∞-coalgebra structure.

Dually, if A is a differential algebra, then its homology $A_* = H_*(A)$ possesses an A_∞-algebra structure, and hence the cohomology of the bar construction BA on A possesses an A_∞-algebra structure.

We use these structures to describe the homology and cohomology of algebras.

THEOREM. *If A is a complete algebra, then the module of primitive elements $PH_*(BA)$ of the homology of the bar construction BA is isomorphic to the module of indecomposable elements of A, i.e., $PH_*(BA) \cong JA$.*

PROOF. Indeed, by the definition of the module of primitive elements, we have

$$PH_*(BA) = \mathrm{Im}\{H_*(\hat{F} H_*(BA)) \to H_*(BA)\}.$$

By using the isomorphism $H_*(\hat{F}BA) \cong A$, we obtain the required isomorphism

$$PH_*(BA) \cong \mathrm{Im}\{A \to H_*(BA)\} \cong JA. \quad \square$$

Now let us consider a coalgebra K, and let us find the module $JH_*(FK)$ of indecomposable elements of the cohomology of its cobar construction FK.

THEOREM. *If K is a coalgebra of finite type, then the module $JH_*(FK)$ of indecomposable elements of the homology $H_*(FK)$ is isomorphic to the module of primitive elements PK of the coalgebra K.*

PROOF. Indeed, by the definition of the module of indecomposable elements of an A_∞-algebra, we have

$$JH_*(FK) = \mathrm{Im}\{H_*(FK) \to H_*(BH_*(FK))\}.$$

By using the isomorphism $H_*(BH_*(FK)) \cong K$, we obtain the required isomorphism

$$JH_*(FK) \cong \mathrm{Im}\{H_*(FK) \to K\} \cong PK. \quad \square$$

Note that if the homology $A_* = H_*(BA)$ of an algebra A is a finitely generated graded module, then the cohomology $A^* = H^*(BA)$ is an A_∞-algebra, and the module of indecomposable elements JA^* of this A_∞-algebra is isomorphic to the module $\overline{PA_*}$ that is dual to the module of primitive elements PA_* of the A_∞-coalgebra A_*. Besides, if A is a complete algebra, then the module of indecomposable elements of the cohomology of A is isomorphic to the module of one-dimensional cohomology, $JH^*(BA) \cong H^1(BA)$.

We now consider a coalgebra K of finite type and the cohomology of its cobar construction $A_* = H_*(FK)$. As shown above, the module of indecomposable elements of A_* is isomorphic to the module $PK = H_1(FK)$ of primitive elements of K. Thus to describe the homology of the cobar construction, it remains to find the relations between the operations in $H_*(FK)$. Denote the iteration of the comultiplication $\nabla: K \to K \otimes K$ by $\nabla(n): K \to K^{\otimes n}$ and the projection onto the module of primitive elements by $p: K \to PK$.

THEOREM. *If K is a coalgebra of finite type, then the relations in the A_∞-algebra $A_* = H_*(FK)$ can be written in the form*

$$\sum_{n \geq 0} \pi_n(p \otimes \cdots \otimes p)\nabla(n+2)(x) = 0, \quad x \in K.$$

Note that for arbitrary elements $x_1, \dots, x_n \in PK = H_1(FK)$, the elements $\pi_{n-2}(x_1 \otimes \cdots \otimes x_n)$ belong to $H_2(FK)$. Therefore, all relations described in the theorem are relations in $H_2(FK)$, while the relations in the homology groups of higher dimensions are given by the A_∞-algebra structure on $H_*(FK)$. Hence we obtain the following result.

COROLLARY. *Suppose that K is a coalgebra of finite type. Then the homology $H_*(FK)$ is determined by the homology groups $H_1(FK)$, $H_2(FK)$ and the operations $\pi_n : H_1(FK)^{\otimes n+2} \to H_2(FK)$.*

Dually, consider the homology $K_* = H_*(BA)$ of a complete algebra A. As shown above, the module of primitive elements of K_* is isomorphic to the module

$JA = H_1(BA)$ of indecomposable elements of A. Hence to describe the homology of the bar construction, it remains to find the generating elements of $\overline{A}^1_\infty JA$. Denote the inclusion map of the indecomposable elements of A into the algebra A itself by $q : JA \to A$, and the iteration of the multiplication π: $A \otimes A \to A$ by $\pi(n)$: $A^{\otimes n} \to A$.

THEOREM. *If A is a complete algebra, then the generating elements of the A_∞-coalgebra $H_*(BA)$ can be written in the form*

$$\sum_{n \geq 0} \nabla_n(x_1 \otimes \cdots \otimes x_{n+2}),$$

where the elements $x_1, \ldots, x_{n+2} \in JA$ satisfy the relation

$$\sum_{n \geq 0} \pi(n+2)(x_1 \otimes \cdots \otimes x_{n+2}) = 0.$$

COROLLARY. *If A is a complete algebra, then its homology $H_*(BA)$ is determined by the homology groups $H_1(BA)$, $H_2(BA)$ and the operations*

$$\nabla_n : H_2(BA) \to H_1(BA)^{\otimes n+2}.$$

THEOREM. *Suppose that A is a complete algebra and the homology $A_* = H_*(BA)$ is a finitely generated graded module. Then the relations in the cohomology $A^* = H^*(BA)$, which is regarded as an A_∞-algebra, can be written in the form*

$$\sum_{n \geq 0} \pi_n(\overline{q \otimes \cdots \otimes q})\overline{\pi}(n+2)(x) = 0, \quad x \in \overline{A}.$$

Note that the assumption that the algebra is complete can be replaced by the assumption that it is weakly complete. We say that an algebra A is *weakly complete* if the corresponding map $A \to \hat{A}$ induces an isomorphism in homology. It is clear that the above theorems concerning generating elements and relations hold in this case as well.

We now consider the question as to what groups lead to complete or weakly complete group algebras.

Let C_p denote a cyclic group of prime order p. The corresponding group algebra A with coefficients in $\mathbb{Z}/p$ is generated by the elements $T_1, \ldots, T_{p-1}$, and the multiplication is defined on them by the formula $\pi(T_i \otimes T_j) = T_{i+j} - T_i - T_j$, where $i + j$ is taken modulo p and $T_0 = 0$. Direct verification shows that the algebra A has a finite filtration, and hence it is complete. It has a unique indecomposable element, which is generated by the element T_1 of A; denote this indecomposable element by u. The dual element $\overline{u}$ generates the only primitive element in the coalgebra $\overline{A}$. The formulas for multiplication in A imply that the relations in the cohomology of C_p, regarded as an A_∞-algebra, have the form

$$\pi_0(\overline{u} \otimes \overline{u}) = 0, \ldots, \pi_{p-3}(\overline{u} \otimes \cdots \otimes \overline{u}) = 0.$$

For a prime $q \neq p$, the cyclic group C_q leads to a noncomplete group algebra with coefficients in $\mathbb{Z}/p$. However, this algebra is weakly complete. Similarly, the cyclic group C_∞ leads to a weakly complete group algebra with coefficients in $\mathbb{Z}/p$.

Note that if a group G leads to a weakly complete algebra over the field $\mathbb{Z}/p$ and F is an abelian group, then the central extension G by F also leads to a weakly complete algebra. Therefore, nilpotent groups lead to weakly complete algebras over $\mathbb{Z}/p$, and hence the above method can be used to describe their cohomology.

An example of an algebra which is not weakly complete is provided by the symmetric group Σ_p with coefficients in $\mathbb{Z}/p$ for $p > 2$.

If G is a finite group whose group algebra over $\mathbb{Z}/p$ is not weakly complete, then to describe the cohomology $H^*(G;\mathbb{Z}/p)$ we can use the facts that this cohomology is isomorphic to the module of G-invariant elements in the cohomology $H^*(H;\mathbb{Z}/p)$ of a Sylow p-subgroup $H \subset G$, and that p-groups are nilpotent. In particular, if H is a normal subgroup of G, then we can use the isomorphism $H^*(G;\mathbb{Z}/p) \cong H^*(H;\mathbb{Z}/p)^{G/H}$.

For example, in the case of the symmetric group Σ_p, the Sylow p-group is the cyclic group C_p. The inclusion $C_p \subset \Sigma_p$ induces a monomorphism in cohomology

$$H^*(\Sigma_p;\mathbb{Z}/p) \to H^*(C_p;\mathbb{Z}/p),$$

whose image is the set of elements which are invariant under the automorphism

$$ad^*_\gamma\colon H^*(C_p;\mathbb{Z}/p) \to H^*(C_p;\mathbb{Z}/p), \quad \gamma \in \Sigma_p;$$

see [**47**].

On the one-dimensional cohomology, this automorphism is given by the formula $ad^1_\gamma(\overline{u}) = 2\overline{u}$. Consider the automorphism $ad_\gamma\colon M\widetilde{A}_\infty\{\overline{u}\} \to M\widetilde{A}_\infty\{\overline{u}\}$ defined on the generators by the recursive formula

$$ad_\gamma(\pi_i(x_1 \otimes \cdots \otimes x_{i+2})) = \pi_i(ad_\gamma(x_1) \otimes \cdots \otimes ad_\gamma(x_{i+2})).$$

The elements invariant under this automorphism form a module isomorphic to $H^*(\Sigma_p;\mathbb{Z}/p)$.

Now let us consider an A_∞-module M over an A_∞-algebra A. Then the homology of the bar construction $B(A,M)$ is an A_∞-comodule over the A_∞-coalgebra $H_*(B(A))$. Let us find the primitive elements of this homology.

As in the case of A_∞-algebras, it can be shown that the following theorem holds.

THEOREM. *Suppose that M is an A_∞-module over an A_∞-algebra A. Then the module of primitive elements $PH_*(B(A,M))$ in the homology of the bar construction $B(A,M)$ is isomorphic to the module of indecomposable elements in M, i.e., $PH_*(B(A,M)) \cong JM$.*

Dually, let us consider an A_∞-comodule L over an A_∞-coalgebra K. Then the homology of the cobar construction $F(K,L)$ is an A_∞-module over the A_∞-algebra $H_*(F(K))$. Let us find the indecomposable elements of this homology.

As in the case of A_∞-coalgebras, it can be shown that the following theorem holds.

THEOREM. *Suppose that L is an A_∞-comodule over an A_∞-coalgebra K. Then the module of indecomposable elements $JH_*(F(K,L))$ in the homology of the cobar construction $F(K,L)$ is isomorphic to the module of primitive elements of L, i.e., $JH_*(F(K,L)) \cong PL$.*

In a similar way, by using the $\wedge$-bar construction and the $\vee$-cobar construction, we can define the notion of decomposable elements and find the relations for a commutative graded A_∞-algebra, and define the notion of primitive elements for a commutative graded A_∞-coalgebra.

THEOREM. *If Y is a graded A_∞-$\vee$-algebra, then the module of primitive elements $PH_*(B_cY)$ of the homology of the bar construction B_cY is isomorphic to the module of indecomposable elements of the A_∞-$\vee$-algebra Y,*

$$PH_*(B_cY) \cong JY.$$

PROOF. By the definition of the module of primitive elements, we have

$$PH_*(B_cY) = \operatorname{Im}\{H_*(F_\vee H_*(B_cY)) \to H_*(B_cY)\}.$$

By using the isomorphism $H_*(F_\vee H_*(B_cY)) \cong Y$, we obtain the isomorphism

$$PH_*(B_cY) \cong \operatorname{Im}\{Y \to H_*(B_cY)\}.$$

It is readily seen that this image coincides with the image of the homomorphism $Y \to H_*(B_cY)$, which by definition gives the module of indecomposable elements of the A_∞-$\vee$-algebra Y. □

Dually, we have the following theorem.

THEOREM. *If X is an arbitrary A_∞-$\wedge$-coalgebra, then the module of indecomposable elements $JH_*(FX)$ is isomorphic to the module of primitive elements of the A_∞-$\wedge$-coalgebra X.*

9.3. Description of the cohomology of Banach algebras

In this section we consider the question of how the methods of describing the cohomology of an algebra that are based on the use of Stasheff's A_∞-structures can be applied to describing the cohomology of a Banach algebra.

For simplicity we shall only consider the case of the cohomology of algebras with trivial coefficients. First we recall the necessary information about homological algebra in the category of Banach spaces; see [**53**].

A *Banach complex* is a family $X = \{X_n\}_{n\in\mathbb{Z}}$ of Banach spaces X_n with a family $d = \{d_n\}$ of continuous operators $d_n: X_n \to X_{n-1}$ (called the *differentials*) that satisfy the relations $d_{n-1} \circ d_n = 0$. As a rule, we shall consider Banach complexes $X = \{X_n\}$ with either $X_n = 0$ for $n < 0$ or $X_n = 0$ for $n > 0$.

A Banach complex X is said to be *admissible* if there exists a family of continuous operators $s = \{s_n\}$, $s_n: X_{n-1} \to X_n$, satisfying the relations $d_n \circ s_n \circ d_n = d_n$.

The spaces $Z_n(X) = \operatorname{Ker}\{d_n: X_n \to X_{n-1}\}$ are called the *spaces of n-dimensional cycles*. The spaces $B_n(X) = \operatorname{Im}\{d_{n+1}: X_{n+1} \to X_n\}$ are called the *spaces of n-dimensional boundaries*. It is clear that $B_n(X) \subset Z_n(X)$. The quotient space $H_n(X) = Z_n(X)/B_n(X)$ is called the *n-dimensional homology of the Banach complex X*. The family $H_*(X) = \{H_n(X)\}$ is called the *homology of X*. If X is an admissible Banach complex, then the map $d_{n+1} \circ s_n: Z_n(X) \to B_n(X)$ is an inverse of the inclusion map $i_n: B_n(X) \to Z_n(X)$. Therefore, in this case the homology $H_*(X)$ consists of Banach spaces $H_n(X)$. The short exact sequences

$$0 \to B_n(X) \xrightarrow{i_n} Z_n(X) \xrightarrow{p_n} H_n(X) \to 0$$

split, i.e., there exist continuous operators $q_n: H_n(X) \to Z_n(X)$ satisfying the relations

$$p_n \circ q_n = \operatorname{Id}: H_n(X) \to H_n(X), \quad \operatorname{Id} - q_n \circ p_n = d_{n+1} \circ s_n: Z_n(X) \to B_n(X).$$

For a Banach complex X, denote by $\overline{X}$ the dual complex $\overline{X} = \{\overline{X}_n\}$, where $\overline{X}_n$ is the Banach space dual to X_{-n}. The differentials $d_n: X_n \to X_{n-1}$ induce differentials $\overline{d}_n: \overline{X}_{-n+1} \to \overline{X}_{-n}$. Clearly, if X is an admissible Banach complex, so is $\overline{X}$. The homology of $\overline{X}$ is called the *cohomology of* X and is denoted by $H^*(X) = \{H^n(X)\}$, $H^n(X) = H_{-n}(\overline{X})$.

A *map of Banach complexes* $f: X \to Y$ *of dimension* m is a family $f = \{f_n\}$ of continuous operators $f_n: X_n \to Y_{n+m}$.

Clearly, such maps form a linear space; it is denoted by $\mathrm{Hom}_m(X, Y)$.

We define linear operators $d: \mathrm{Hom}_m(X, Y) \to \mathrm{Hom}_{m-1}(X, Y)$, called the *differentials*, by setting

$$d(f)_n = d_{n+m} \circ f_n + (-1)^m f_{n-1} \circ d_n: X_n \to Y_{n+m-1}.$$

It is readily seen that $d \circ d = 0$, and thus the family $\mathrm{Hom}(X, Y) = \{\mathrm{Hom}_m(X, Y)\}$ forms a complex in the category of linear spaces.

A *chain map of Banach complexes* is a map f of dimension 0 satisfying the relation $d(f) = 0$ or, equivalently, $d_n \circ f_n = f_{n-1} \circ d_n$.

A chain map of Banach complexes $f: X \to Y$ induces maps in homology

$$H_*(f) = \{H_n(f)\}: H_*(X) \to H_*(Y), \quad H_n(f): H_n(X) \to H_n(Y),$$

and in cohomology

$$H^*(f) = \{H^n(f)\}: H^*(Y) \to H^*(X), \quad H^n(f): H^n(Y) \to H^n(X).$$

Two chain maps $f, g: X \to Y$ of Banach complexes are said to be *homotopic* (and we write $f \simeq g$) if there exists a map $h: X \to Y$ of dimension 1 satisfying the relation $d(h) = g - f$ or, equivalently, $d_{n+1} \circ h_n + h_{n-1} \circ d_n = g_n - f_n$. Such a map h is called a *homotopy between* f *and* g. It is readily seen that the homotopy relation is an equivalence relation.

Banach complexes X and Y are said to be *homotopy equivalent* (and we write $X \simeq Y$) if there exist chain maps $f: X \to Y$ and $g: Y \to X$ such that $g \circ f \simeq \mathrm{Id}_X$ and $f \circ g \simeq \mathrm{Id}_Y$. A Banach complex is said to be *contractible* if it is homotopy equivalent to the zero complex.

It is readily seen that if $f \simeq g: X \to Y$, then $\overline{f} \simeq \overline{g}: \overline{Y} \to \overline{X}$, and if $X \simeq Y$, then $\overline{X} \simeq \overline{Y}$. Furthermore, homotopic maps induce the same homomorphisms in homology (and in cohomology), and homotopy equivalent Banach complexes have isomorphic homology (and cohomology).

THEOREM. *If X is an admissible Banach complex, then its homology $H_*(X)$, regarded as a Banach complex with zero differential, is homotopy equivalent to X, i.e., $H_*(X) \simeq X$.*

PROOF. Consider the map $\mathrm{Id} - s \circ d: X \to Z(X)$. Denote its composition with the projection $p: Z(X) \to H_*(X)$ by $\eta: X \to H_*(X)$. The map $q: H_*(X) \to Z(X)$ considered above gives a map $\xi: H_*(X) \to X$. It is readily seen that $\eta \circ \xi = \mathrm{Id}: H_*(X) \to H_*(X)$. Let us show that the maps $s_n: X_n \to X_{n+1}$ constitute the required homotopy between the maps Id and $\xi \circ \eta$. Indeed, we have $\mathrm{Id} - \xi_n \circ \eta_n = \mathrm{Id} - q_n \circ p_n(\mathrm{Id} - s_n \circ d_n) = d_{n+1} \circ s_n + s_n \circ d_n$. □

In particular, this theorem implies the following result.

COROLLARY. *If X is an admissible Banach complex, then $H^*(X) \cong \overline{H_*(X)}$.*

PROOF. Indeed, the homotopy equivalence $X \simeq H_*(X)$ implies the homotopy equivalence $\overline{X} \simeq \overline{H_*(X)}$, and hence we have the required isomorphism. $\square$

Let X be a Banach complex. We define the Banach complex SX, which is called the *suspension of* X, by setting $(SX)_{n+1} = X_n$. The elements of SX will be denoted by $[x]$, where $x \in X$. If $x \in X$ has dimension n, then $[x]$ has dimension $n+1$. The differential in SX is defined by the formula $d[x] = -[d(x)]$.

It is clear that if X is an admissible Banach complex, so is the suspension SX, and the homology of the suspension is isomorphic to the suspension of the homology, i.e., $H_*(SX) \cong SH_*(X)$.

Similarly, the desuspension $S^{-1}X$ of a Banach complex X is defined by setting $(S^{-1}X)_{n-1} = X_n$. Elements of the desuspension will be denoted by $[x]$, where $x \in X$, whenever the context allows this without ambiguity. If $x \in X$ has dimension n, then $[x]$ has dimension $n-1$. The differential on SX is defined by the formula $d[x] = -[d(x)]$.

Clearly, the functors of suspension and desuspension are mutually inverse.

The *tensor product* of Banach complexes X' and X'' is the Banach complex $X = X' \otimes X''$, where

$$(X' \otimes X'')_n = \sum_{p+q=n} X'_p \otimes X''_q,$$

and the differential is defined by

$$d_n(x'_p \otimes x''_q) = d'_p(x'_p) \otimes x''_q + (-1)^p x'_p \otimes d''_q(x''_q), \quad p+q=n.$$

Given two maps $f'\colon X' \to Y'$ and $f''\colon X'' \to Y''$ of dimensions m' and m'', respectively, we define a map $f = f' \otimes f''\colon X' \otimes X'' \to Y' \otimes Y''$ of dimension $m = m' + m''$ by setting, on the generators $x'_p \otimes x''_q \in X'_p \otimes X''_q$,

$$(f' \otimes f'')(x'_p \otimes x''_q) = (-1)^{m''p} f'_p(x'_p) \otimes f''_q(x''_q).$$

The formula for the differential in the tensor product of Banach complexes can be rewritten as $d = d' \otimes 1 + 1 \otimes d''$.

THEOREM. *The tensor product of admissible Banach complexes is an admissible Banach complex.*

PROOF. Let (X', d', s') and (X'', d'', s'') be admissible Banach complexes. We define a map $s\colon X' \otimes X'' \to X' \otimes X''$ by setting

$$s = s' \otimes 1 + 1 \otimes s'' - (d' \circ s' + s' \circ d') \otimes s'' - s' \otimes (d'' \circ s'' + s'' \circ d'').$$

Straightforward computations give the required relation $d \circ s \circ d = d$. $\square$

Suppose that maps $f'\colon X' \to Y'$ and $f''\colon X'' \to Y''$ are homotopic to maps $g'\colon X' \to Y'$ and $g''\colon X'' \to Y''$, respectively; then the map $f' \otimes f''\colon X' \otimes X'' \to Y' \otimes Y''$ is homotopic to $g' \otimes g''\colon X' \otimes X'' \to Y' \otimes Y''$. The required homotopy $h\colon X' \otimes X'' \to Y' \otimes Y''$ is given by the formula $h = h' \otimes g'' + f' \otimes h''$, where h' and h'' are the homotopies between f', g' and f'', g'', respectively.

In particular, homotopy equivalent chain complexes $X' \simeq Y'$, $X'' \simeq Y''$ lead to homotopy equivalent tensor products $X' \otimes X'' \simeq Y' \otimes Y''$.

COROLLARY. *If X and Y are admissible Banach complexes, then there is an isomorphism $H_*(X \otimes Y) \cong H_*(X) \otimes H_*(Y)$.*

Let A be a Banach algebra, i.e., a Banach space endowed with an associative multiplication $\pi\colon A \otimes A \to A$. Consider the *bar construction* BA on A, where

$$(BA)_n = \begin{cases} A^{\otimes n}, & n \geq 1, \\ 0, & n \leq 0. \end{cases}$$

The generators of dimension n in $A^{\otimes n}$ are denoted by $[x_1, \dots, x_n]$, $x_i \in A$. The differential d is defined by

$$d_n[x_1, \dots, x_n] = \sum_{i=1}^{n-1} (-1)^{i+1}[x_1, \dots, \pi(x_i \otimes x_{i+1}), \dots, x_n].$$

A Banach algebra A is said to be *admissible* if the complex BA is admissible.

Clearly, any finite-dimensional Banach algebra is admissible. As an example of an infinite-dimensional admissible Banach algebra, consider the algebra $A = l_1$ of absolutely convergent series $\sum_{n=1}^{\infty} x_n$ with the multiplication

$$\sum_{n=1}^{\infty} x_n \cdot \sum_{n=1}^{\infty} y_n = \sum_{n=2}^{\infty} \left(\sum_{k=1}^{n-1} x_k \cdot y_{n-k} \right).$$

Let e_i be the series with 1 in the ith place and zeros elsewhere. We define maps $s\colon A^{\otimes n} \to A^{\otimes n+1}$ by setting

$$s(e_{i_1} \otimes \dots \otimes e_{i_n}) = \begin{cases} e_1 \otimes e_{i_1 - 1} \otimes \dots \otimes e_{i_n}, & i_1 > 1, \\ 0, & i_1 = 1. \end{cases}$$

On the remaining elements, s is defined by linearity and continuity.

Straightforward computations show that $d \circ s \circ d = d$, which means that A is an admissible algebra.

Among the properties of admissible Banach algebras, we note the following ones.

THEOREM. *The tensor product of admissible Banach algebras is an admissible Banach algebra.*

PROOF. Let A', A'' be admissible algebras, and let

$$s'\colon A'^{\otimes n} \to A'^{\otimes n+1}, \quad s''\colon A''^{\otimes n} \to A''^{\otimes n+1}$$

be the corresponding maps of the bar constructions. Then the required map $s\colon A^{\otimes n} \to A^{\otimes n+1}$ for the algebra $A = A' \otimes A''$ is defined to be the tensor product of s' and s''. □

THEOREM. *The direct sum of admissible Banach algebras is an admissible Banach algebra.*

PROOF. Let A', A'' be admissible algebras, and let

$$s'\colon A'^{\otimes n} \to A'^{\otimes n+1}, \quad s''\colon A''^{\otimes n} \to A''^{\otimes n+1}$$

be the corresponding maps of the bar constructions.

Consider the Banach algebra $A = A' \oplus A''$. The generators of $A^{\otimes n}$ are of the form $a_1 \otimes \cdots \otimes a_n$, where each a_i belongs either to A' or to A''. We partition this expression into blocks b_i consisting of elements of the same algebra, and rewrite it as $b_1 \otimes \cdots \otimes b_m$, where the blocks $b_1, \ldots, b_m$ are tensor products of elements of A' or A'', and neighboring blocks contain elements of different algebras. Hence we have either $b_i \in BA'$ or $b_i \in BA''$.

We define the map $s\colon A^{\otimes n} \to A^{\otimes n+1}$ by setting

$$\begin{aligned} s(b_1 \otimes \cdots \otimes b_m) &= s(b_1) \otimes b_2 \otimes \cdots \otimes b_m \\ &\quad + (-1)^{n_1}(b_1 - d \circ s(b_1)) \otimes s(b_2) \otimes \cdots \otimes b_m + \ldots \\ &\quad + (-1)^{n_1+\cdots+n_{m-1}}(b_1 - d \circ s(b_1)) \otimes \cdots \otimes (b_{m-1} - d \circ s(b_{m-1})) \otimes s(b_m), \end{aligned}$$

where n_i is the dimension of b_i.

Straightforward computations show that the required relations hold. □

The homology (cohomology) of the bar construction BA on an algebra A is referred to as the *homology* (*cohomology*) of A and is denoted by $H_*(A)$ ($H^*(A)$). Thus, by definition, we have $H_*(A) = H_*(BA)$ ($H^*(A) = H^*(BA)$).

The problem of homological triviality (amenability) of a Banach algebra has been fairly well studied. In particular, if an algebra A with identity possesses a diagonal, i.e., an element $d \in A \otimes A$ such that $\pi(d)$ is the identity in A and $a \cdot d = d \cdot a$ for any $a \in A$, then the homology of A is trivial.

An example of such an algebra is the algebra $\mathcal{M}_n$ of $n \times n$ matrices. Let e_{kl} be the matrix with 1 in the (k,l) place and zeros elsewhere. Then the following element is a diagonal:

$$d_n = \frac{1}{n} \sum_{k,l=1}^{n} e_{kl} \otimes e_{lk}.$$

A weaker condition on A, which also implies its homological triviality, is the existence of an approximate diagonal, i.e., a net $d_\nu \in A \otimes A$ such that $\pi(d_\nu)$ is an approximate identity in A and $\lim_\nu (a \cdot d_\nu - d_\nu \cdot a) = 0$ for any $a \in A$.

An example of such an algebra is the algebra $\mathcal{K}(H)$ of compact operators in a separable Hilbert space H.

Another example of a homologically trivial algebra is the algebra $L_1(G)$ on a locally compact amenable group G.

We consider the following examples of Banach and polynormed algebras with nontrivial cohomology:

1. $A = \mathbb{C}$ with trivial multiplication. In this case $(BA)_n \cong \mathbb{C}$ and the differential is identically zero. Hence the homology of this algebra in each dimension $n \geq 1$ is isomorphic to $\mathbb{C}$.

2. $A = l_1$, the Banach algebra of absolutely convergent series. It is known that the one-dimensional homology of this algebra is isomorphic to $\mathbb{C}$, and the remaining homology groups are zero.

3. $A = l_1^{(m)}$, the subalgebra of l_1 consisting of the series whose first m terms are zero. It follows directly from the definition of homology that the one-dimensional homology of this algebra is isomorphic to $\mathbb{C}^{\times m+1}$.

4. $A = l_1^m = l_1/l_1^{(m)}$. It is readily seen that the one-dimensional homology of this algebra is isomorphic to $\mathbb{C}$.

5. $A = \mathcal{O}_{z_0}(\mathcal{D})$, the polynormed algebra of analytic functions on a domain $\mathcal{D} \subset \mathbb{C}$ that vanish at a point $z_0 \in \mathcal{D}$. It is known that the one-dimensional homology of this algebra is isomorphic to $\mathbb{C}$, and the remaining homology groups are zero.

6. $A = \mathcal{O}_{z_0}^{(m)}(\mathcal{D})$, the polynormed algebra of analytic functions on a domain $\mathcal{D} \subset \mathbb{C}$ that vanish at a point z_0 together with all their derivatives up to order m. It follows directly from the definition of homology that the one-dimensional homology of this algebra is isomorphic to $\mathbb{C}^{\times m+1}$.

A Banach complex X is called a *differential Banach algebra* if there is a map $\pi\colon X \otimes X \to X$ (called *multiplication*) satisfying the associativity relation $\pi(\pi \otimes 1) = \pi(1 \otimes \pi)$.

Dually, a Banach complex Y is called a *differential Banach coalgebra* if there is a map $\nabla\colon Y \to Y \otimes Y$ (called *comultiplication*) that satisfies the relation $(\nabla \otimes 1)\nabla = (1 \otimes \nabla)\nabla$.

It is clear that if Y is a differential Banach coalgebra, then the dual complex $\overline{Y}$ is a differential Banach algebra with the multiplication defined as the composite map $\overline{Y} \otimes \overline{Y} \xrightarrow{i} \overline{Y \otimes Y} \xrightarrow{\overline{\nabla}} \overline{Y}$.

A *graded Banach algebra* (*coalgebra*) is a differential Banach algebra (coalgebra) with zero differential.

It is readily seen that if X is an admissible Banach algebra, then its homology $H_*(X)$ is an admissible Banach algebra with respect to the multiplication $\pi_*\colon H_*(X) \otimes H_*(X) \to H_*(X)$ given by the formula $\pi_* = \eta \circ \pi \circ (\xi \otimes \xi)$, where $\xi\colon H_*(X) \to X$, $\eta\colon X \to H_*(X)$, $\pi\colon X \otimes X \to X$.

Dually, the homology of an admissible differential Banach coalgebra is a graded Banach algebra. In particular, the homology $H_*(Y)$ of an admissible Banach coalgebra Y is a graded Banach coalgebra, and its cohomology $H^*(Y)$ is a graded Banach algebra.

Given a differential graded algebra X, we define a differential coalgebra BX, called the *bar construction on* X. To this end, we consider the tensor algebra $TX = \sum_{n \geq 1} X^{\otimes n}$. Denote the elements $x_1 \otimes \cdots \otimes x_n \in X^{\otimes n}$ by $[x_1, \dots, x_n]$, and let the dimension of such an element be equal to the sum of the dimensions of the x_i plus n. The differential is defined by

$$\begin{aligned} d[x_1, \dots, x_n] = &-\sum_{i=1}^{n} (-1)^{\epsilon_{i-1}} [x_1, \dots, d(x_i), \dots, x_n] \\ &+ \sum_{i=1}^{n-1} (-1)^{\epsilon_i} [x_1, \dots, \pi(x_i \otimes x_{i+1}), \dots, x_n], \end{aligned}$$

where $\epsilon_i = \dim(x_1) + \cdots + \dim(x_i) + i$.

Endowed with the above grading and differential, the tensor algebra TX is called the *bar construction on* X and is denoted by BX. Apart from the structure of a chain complex, BX carries a comultiplication $\nabla\colon BX \to BX \otimes BX$, which makes it into a differential coalgebra,

$$\nabla[x_1, \dots, x_n] = \sum_{i=1}^{n-1} [x_1, \dots, x_i] \otimes [x_{i+1}, \dots, x_n].$$

Dually, for a differential Banach coalgebra Y, we define a differential algebra FY, called the *cobar construction on* Y. We take the tensor algebra TY, denote its elements of the form $y_1 \otimes \cdots \otimes y_n$ by $[y_1, \ldots, y_n]$, and define their dimension to be the sum of the dimensions of the y_i minus n. The differential is given by

$$\begin{aligned} d[y_1, \ldots, y_n] = &- \sum_{i=1}^{n} (-1)^{\epsilon_{i-1}} [y_1, \ldots, d(y_i), \ldots, y_n] \\ &+ \sum_{i=1}^{n-1} (-1)^{\epsilon_i} [y_1, \ldots, x_i', x_i'', \ldots, x_n], \end{aligned}$$

where $\nabla(y_i) = \sum y_i' \otimes y_i''$.

The multiplication $\pi\colon FY \otimes FY \to FY$ is defined by

$$\pi([y_1, \ldots, y_n] \otimes [y_{n+1}, \ldots, y_{n+m}]) = [y_1, \ldots, y_{n+m}].$$

As in the case of ordinary differential algebras, the following assertion can be proved.

THEOREM. *For any differential Banach algebra* X, *there is a homotopy equivalence* $FBX \simeq X$.

Given a differential Banach algebra X, consider the decreasing filtration

$$X = X^1 \supset X^2 = X \cdot X \supset \cdots \supset X^n \supset \ldots,$$

and set $\hat{X} = \varprojlim X/X^n$.

We have a natural map of algebras $X \to \hat{X}$. An algebra X is said to be *complete* if this map is an isomorphism.

The following theorem is proved as in the case of ordinary differential algebras.

THEOREM. *If* X *is a complete algebra, then there is a homotopy equivalence of algebras* $\hat{F}BX \simeq X$.

We now pass to the dual situation. Given a Banach coalgebra Y, we define a filtration $Y^1 \subset Y^2 \subset \cdots \subset Y^n \subset \cdots \subset Y$ by setting

$$Y^1 = PY = \{y \in Y : \nabla(y) = 0\}, \quad Y^2 = \{y \in Y : \nabla(y) \in Y^1 \otimes Y^1\},$$

and so on.

Let $\check{Y}$ denote the direct limit of Y^n, $\check{Y} = \varinjlim Y^n$. It is clear that $\check{Y}$ is a subcoalgebra of Y. We say that Y is a *coalgebra of finite type* if $\check{Y} = Y$.

The following theorem is dual to the previous one.

THEOREM. *If* Y *is a coalgebra of finite type, then there is a homotopy equivalence of coalgebras* $BFY \simeq Y$.

A Banach complex X is called a *differential* A_∞*-algebra* if it is endowed with operations $\pi_n\colon X^{\otimes n+2} \to X$ of dimension $n \geq 0$ that satisfy the relations

$$d(\pi_{n+1}) = \sum_{i=0}^{n} (-1)^{\epsilon} \pi_i (1 \otimes \cdots \otimes \pi_{n-i} \otimes \cdots \otimes 1),$$

where $\epsilon = nk + ik + n + k$.

A *graded* A_∞*-algebra* is a differential A_∞-algebra with zero differential.

Dually, a Banach complex Y is called a *differential A_∞-coalgebra* if it is endowed with operations $\nabla_n : Y \to Y^{\otimes n+2}$ of dimension $n \geq 0$ that satisfy the relations

$$d(\nabla_{n+1}) = \sum_{i=0}^{n} (-1)^{\epsilon}(1 \otimes \cdots \otimes \nabla_{n-i} \otimes \cdots \otimes 1)\nabla_i,$$

where $\epsilon = nk + ik + ni + i + k + 1$.

Let X be a Banach complex. Consider the suspension SX and the tensor algebra TSX. Then providing X with an A_∞-algebra structure is equivalent to defining a differential on TSX that makes it into a differential coalgebra; the latter is called the *bar construction on* X and is denoted by BX. It is readily seen that if the higher operations π_i for $i \geq 1$ are trivial, i.e., if X is an ordinary differential algebra, then the bar construction BX coincides with the ordinary bar construction BX on the algebra X.

By duality, given a Banach complex Y, consider the desuspension $S^{-1}Y$ and the tensor algebra $\hat{T}S^{-1}Y$. Then providing Y with a differential A_∞-coalgebra structure is equivalent to defining a differential on $\hat{T}S^{-1}Y$ that makes it into a differential algebra; the latter is called the *cobar construction on* Y and is denoted by $\hat{F}Y$. If Y is an ordinary differential Banach coalgebra, i.e., if the higher operations ∇_i for $i \geq 1$ are trivial, then this cobar construction coincides with the ordinary cobar construction $\hat{F}Y$ on Y.

Let X', X'' be differential A_∞-algebras. A *map of A_∞-algebras* $f\colon X' \to X''$ is a family of maps $f_i\colon X'^{\otimes i+1} \to X''$ of dimension i such that the corresponding maps $f(i+1)\colon (SX')^{\otimes i+1} \to SX''$ satisfy the relations

$$\begin{aligned} d(f(n)) = &\sum_{i=1}^{n-1} f(i)(1 \otimes \cdots \otimes \pi'(n-i+1) \otimes \cdots \otimes 1) \\ &- \sum_{n_1+\cdots+n_i=n} \pi''(i)(f(n_1) \otimes \cdots \otimes f(n_i)). \end{aligned}$$

Giving such operations is equivalent to giving a map of differential Banach coalgebras $B(f)\colon BX' \to BX''$.

By using the products of families $f \cup \pi'$ and $\pi'' \cup f$, the relations for a map of A_∞-algebras can be rewritten in the form $d(f) = f \cup \pi' - \pi'' \cup f$.

Maps $f, g\colon X' \to X''$ of differential A_∞-algebras are said to be *homotopic* (and we write $f \simeq g$) if there is a family $h = \{h(i)\}$ such that the relation $d(h) = g - f - \pi'' \cup h - h \cup \pi'$ holds. The family h is called a *homotopy between f and g*. It is easy to check that if f and g are homotopic, then the corresponding maps of the bar constructions $Bf, Bg\colon BX' \to BX''$ are also homotopic. Differential A_∞-algebras X', X'' are said to be *homotopy equivalent* (and we write $X' \simeq X''$) if there exist maps $f\colon X' \to X''$ and $g\colon X'' \to X'$ such that $g \circ f \simeq \mathrm{Id}_{X'}$ and $f \circ g \simeq \mathrm{Id}_{X''}$.

The notions of a map of differential A_∞-coalgebras, homotopy relation, and homotopy equivalence of A_∞-coalgebras are defined in a dual way.

The following theorem is proved as in the case of ordinary differential algebras.

THEOREM. *If X is an admissible differential Banach algebra, then its homology $X_* = H_*(X)$ has a graded A_∞-algebra structure, and there is a homotopy equivalence of A_∞-algebras $X \simeq X_*$ inducing a homotopy equivalence of differential coalgebras $BX \simeq BX_*$.*

Dually, we have the following theorem.

THEOREM. *If Y is an admissible differential Banach coalgebra, then its homology $Y_* = H_*(Y)$ has a graded A_∞-coalgebra structure, and there is a homotopy equivalence of A_∞-coalgebras $Y \simeq Y_*$ inducing a homotopy equivalence of differential algebras $\hat{F}Y \simeq \hat{F}Y_*$.*

In particular, the homology $H_*(A)$ of an admissible Banach algebra A has an A_∞-coalgebra structure, and the cohomology $H^*(A)$ has an A_∞-algebra structure.

Now let us consider the question of describing Banach A_∞-algebras. As in the case of ordinary differential algebras, to describe a Banach A_∞-algebra A, we must do the following.

1. Find the homology $H_*(BA)$ of the bar construction BA on A.

2. Compute the image of the homomorphism $A \to H_*(BA)$; this image is the module JA of indecomposable elements of the A_∞-algebra A.

3. Compute the image of the homomorphism $H_*(BA) \to A^1_\infty * A$; this image gives the relations in the A_∞-algebra A.

Dually, to describe an A_∞-coalgebra K, we must do the following.

1′. Find the homology $H_*(\hat{F}K)$ of the cobar construction $\hat{F}K$ on K.

2′. Compute the image of the homomorphism $H_*(\hat{F}K) \to K$; this image is the module of primitive elements of the A_∞-coalgebra K.

3′. Compute the kernel of the homomorphism $\overline{A}^1_\infty * K \to H_*(\hat{F}K)$; this kernel gives generating elements of the sought-for A_∞-coalgebra.

To describe A_∞-algebras and coalgebras, we must first find the indecomposable and primitive elements, respectively.

Let us consider an algebra A, and let us find the primitive elements of the homology $H_*(BA)$ of its bar construction. The following theorem holds.

THEOREM. *Suppose that A is a complete algebra. Then the space of primitive elements $PH_*(BA)$ of the homology of the bar construction BA is isomorphic to the space of indecomposable elements of A, i.e., $PH_*(BA) \cong JA$.*

Dually, we have the following theorem.

THEOREM. *Suppose that K is a coalgebra of finite type. Then the space $JH_*(FK)$ of indecomposable elements of the homology $H_*(FK)$ is isomorphic to the space PK of primitive elements of K.*

Note that if the homology $A_* = H_*(BA)$ of A is a graded space with the approximation property, then the cohomology $A^* = H^*(BA)$ is an A_∞-algebra, and the space JA^* of indecomposable elements of this A_∞-algebra is isomorphic to the space $\overline{PA_*}$ dual to the space PA_* of primitive elements of the A_∞-coalgebra A_*. Moreover, if the algebra A is complete, then the space of indecomposable elements of the cohomology of A is isomorphic to the space of one-dimensional cohomology, i.e., $JH^*(BA) \cong H^1(BA)$.

Now consider a coalgebra K of finite type and the homology of its cobar construction $A_* = H_*(FK)$. As shown above, the space of indecomposable elements of A_* is isomorphic to the space $PK = H_1(FK)$ of primitive elements of K. Thus, to describe the homology of the cobar construction, it remains to find the relations between operations on $H_*(FK)$.

Denote the iteration of the comultiplication $\nabla\colon K \to K \otimes K$ by $\nabla(n)\colon K \to K^{\otimes n}$, and the projection onto the space of primitive elements by $p\colon K \to PK$.

THEOREM. *If K is a coalgebra of finite type, then the relations in the A_∞-algebra $A_* = H_*(FK)$ can be written in the form*

$$\sum_{n\geq 0} \pi_n(p\otimes\cdots\otimes p)\nabla(n+2)(x) = 0, \quad x\in K.$$

Note that for any elements $x_1,\ldots,x_n \in PK = H_1(FK)$, the elements $\pi_{n-2}(x_1\otimes\cdots\otimes x_n)$ belong to $H_2(FK)$. Hence all relations are relations in $H_2(FK)$, while the relations in the homology of higher dimension are given by the A_∞-algebra structure on $H_*(FK)$. Therefore, we obtain the following corollary.

COROLLARY. *If K is a coalgebra of finite type, then the homology $H_*(FK)$ is determined by the homology groups $H_1(FK)$, $H_2(FK)$ and the operations $\pi_n : H_1(FK)^{\otimes n+2} \to H_2(FK)$.*

Dually, for a complete algebra A, consider the homology $K_* = H_*(BA)$. As shown above, the space of primitive elements of K_* is isomorphic to the space $JA = H_1(BA)$ of indecomposable elements of A. Thus, to describe the homology of the bar construction, it remains to find generating elements in $\overline{A}^{1}_{\infty} JA$. Denote the inclusion map of the indecomposable elements of A into the algebra A itself by $q : JA \to A$, and the iteration of the multiplication $\pi : A\otimes A \to A$ by $\pi(n) : A^{\otimes n} \to A$.

THEOREM. *If A is a complete algebra, then the generating elements of the A_∞-coalgebra $H_*(BA)$ can be written in the form*

$$\sum_{n\geq 0} \nabla_n(x_1\otimes\cdots\otimes x_{n+2}),$$

where the elements $x_1,\ldots,x_{n+2}\in JA$ satisfy the relation

$$\sum_{n\geq 0} \pi(n+2)(x_1\otimes\cdots\otimes x_{n+2}) = 0.$$

COROLLARY. *If A is a complete algebra, then the homology $H_*(BA)$ is determined by the homology groups $H_1(BA)$, $H_2(BA)$ and the operations*

$$\nabla_n : H_2(BA) \to H_1(BA)^{\otimes n+2}.$$

THEOREM. *Suppose that A is a complete algebra and the homology $A_* = H_*(BA)$ is a graded space with the approximation property. Then the relations in the cohomology $A^* = H^*(BA)$, regarded as an A_∞-algebra, can be written in the form*

$$\sum_{n\geq 0} \pi_n(\overline{q\otimes\cdots\otimes q})\overline{\pi}(n+2)(x) = 0, \quad x\in \overline{A}.$$

Note that the assumption that the algebra is complete can be replaced by the assumption that it is weakly complete. An algebra A is said to be *weakly complete* if the corresponding map $A\to \hat{A}$ induces an isomorphism in homology. It is clear that the above theorems concerning generating elements and relations hold in this case as well.

Now we give some examples of describing the homology of algebras.

1. $A = l_1^m$. It is clear that this algebra is complete. The cohomology $H^*(A)$ has one generator $u_1 \in H^1(A)$, which corresponds to the element $e_1 \in l_1^m$; the relations are $\pi_{n-2}(u_1 \otimes \cdots \otimes u_1) = 0$, $2 \le n \le m$.

This description implies that the one-dimensional cohomology $H^1(A)$ is isomorphic to $\mathbb{C}$. The two-dimensional cohomology is isomorphic to $\mathbb{C}$ and is generated by $u_2 = \pi_{m-1}(u_1 \otimes \cdots \otimes u_1)$. In even dimension, the cohomology is isomorphic to $\mathbb{C}$ and is generated by

$$u_2^n = u_2 \cdot \ldots \cdot u_2 \in H^{2n}(A).$$

In odd dimension, the cohomology is isomorphic to $\mathbb{C}$ and is generated by

$$u_2^n \cdot u_1 \in H^{2n+1}(A).$$

2. $A = l_1$. This algebra is not complete. The corresponding algebra $\hat{l}_1$ consists of all formal power series with complex coefficients. The cohomology $H^*(\hat{l}_1)$ has a single generator $u_1 \in H^1(\hat{l}_1)$, which corresponds to the element $e_1 \in l_1$; the relations are as follows:

$$\pi_{n-2}(u_1 \otimes \cdots \otimes u_1) = 0, \quad n \ge 2.$$

This description implies that the one-dimensional cohomology $H^1(\hat{l}_1)$ is isomorphic to $\mathbb{C}$, and the higher homology groups are zero. On the other hand, by using the isomorphism $H^*(A) \cong Ext_A(\mathbb{C}, \mathbb{C})$, we readily see that $H^1(l_1) \cong \mathbb{C}$ and $H^n(l_1) = 0$, $n \ge 2$. Thus the map $l_1 \to \hat{l}_1$ induces an isomorphism in cohomology, and hence l_1 is a weakly complete algebra.

3. $A = l_1^{(m)}$. Since the sequence $0 \to l_1^{(m)} \to l_1 \to l_1^m \to 0$ is exact and the algebras l_1, l_1^m are weakly complete, it follows that $l_1^{(m)}$ is also weakly complete.

The cohomology group $H^1(A)$ is generated by the elements $u_{m+1}, \ldots, u_{2m+1}$ corresponding to $e_{m+1}, \ldots, e_{2m+1} \in A$. The relations are as follows:

$$\begin{aligned} u_{m+1}u_{m+1} &= 0, \\ u_{m+1}u_{m+2} + u_{m+2}u_{m+1} &= 0, \\ &\cdots \\ u_{m+1}u_{2m+1} + \cdots + u_{2m+1}u_{m+1} &= 0, \\ \pi_1(u_{m+1} \otimes u_{m+1} \otimes u_{m+1}) + u_{m+2}u_{2m+1} + \cdots + u_{2m+1}u_{m+2} &= 0, \\ &\cdots \end{aligned}$$

4. $A = \mathcal{O}_{z_0}(\mathcal{D})$. As above, it can be shown that this algebra is weakly complete. The cohomology $H^*(A)$ has one generator $u \in H^1(A)$, which corresponds to the element $z - z_0 \in A$; the relations are $\pi_{n-2}(u \otimes \cdots \otimes u) = 0$, $n \ge 2$. This description implies that the one-dimensional cohomology $H^1(A)$ is isomorphic to $\mathbb{C}$, and all higher-dimensional cohomology groups are zero.

9.4. Description of the E_2-term of the Adams spectral sequence for stable homotopy groups of topological spaces

Let K be the Milnor coalgebra (which is dual to the Steenrod algebra). It is a Hopf algebra. If $R = \mathbb{Z}/2$, then K is a polynomial algebra with generators ξ_i

of dimension $2^i - 1$, and the comultiplication $\nabla\colon K \to K \otimes K$ is defined on the generators ξ_i by

$$\nabla(\xi_i) = \sum_{k=0}^{i} \xi_{i-k}^{2^k} \otimes \xi_k.$$

If $R = \mathbb{Z}/p$, then it is a free commutative algebra with generators ξ_i of dimension $2(p^i - 1)$ and τ_i of dimension $2p^i - 1$. The comultiplication $\nabla\colon K \to K \otimes K$ is defined on the generators by

$$\nabla(\xi_i) = \sum_{k=0}^{i} \xi_{i-k}^{p^k} \otimes \xi_k, \quad \nabla(\tau_i) = \tau_i \otimes 1 + \sum_{k=0}^{i} \xi_{i-k}^{p^k} \otimes \tau_k.$$

Recall that the homology of the cobar construction FK on the Milnor coalgebra K is isomorphic to the second term of the Adams spectral sequence for stable homotopy groups of the spheres. It follows directly from the definition of the comultiplication in the Milnor coalgebra that its primitive elements are the elements $\xi_1^{2^i}$, $i \geq 0$, if $R = \mathbb{Z}/2$, and the elements $\xi_1^{p^i}$ and τ_0 if $R = \mathbb{Z}/p$. Denote the corresponding indecomposable elements of $H_*(FK)$ by h_i and g_0 (the latter corresponds to τ_0). The general theorem on describing the structure of the homology of algebras implies the following theorem.

THEOREM. *If $R = \mathbb{Z}/2$, then the module of indecomposable elements of the A_∞-algebra $E_2 = H_*(FK)$ is generated by the elements h_i of dimension $2^i - 1$. If $R = \mathbb{Z}/p$, then the corresponding module of indecomposable elements is generated by the elements h_i of dimension $(2p-2)^i - 1$ and the element g_0 of dimension* 0.

This theorem is a reformulation of the corresponding result from [**22**].

Now let us pass to finding relations for the second term of the Adams spectral sequence. Let $R = \mathbb{Z}/2$. Denote by $p\colon K \to QA_*$ the projection given by the formulas $p(\xi_1^{2^i}) = h_i$ and $p(x) = 0$ in all other cases. To find the relations, we must compute the map $\psi\colon K \to \mathcal{A}_\infty QA_*$. It is readily seen that

$$\psi(\xi_i) = \pi_{i-2} \times (h_{i-1} \otimes \cdots \otimes h_0),$$

and hence we have the relations

$$\pi_{i-2}(h_{i-1} \otimes \cdots \otimes h_0) = 0,$$

which will be referred to as the *basic relations.*

Similarly, in the case $R = \mathbb{Z}/p$, the elements ξ_i and τ_i generate the basic relations

$$\pi_{i-2}(h_{i-1} \otimes \cdots \otimes h_0) = 0, \quad \pi_{i-1}(h_{i-1} \otimes \cdots \otimes h_0 \otimes g_0) = 0.$$

To find the remaining relations, we must compute the values of ψ on the products of the elements ξ_i. To this end, we make use of the Hopf relation in the coalgebra K. Namely, we define the $\cup$-products in QA_* by setting $h_i \cup h_i = h_{i+1}$ and $h_i \cup h_j = 0$ in all other cases. Similarly, we set $h_i \cup h_i \cup h_{i+1} = h_i \cup h_{i+1} \cup h_i = h_{i+1} \cup h_i \cup h_i = h_{i+2}$ and $h_i \cup h_j \cup h_k = 0$ in all other cases. And so on.

By using these products, we define the products of the elements

$$\pi_{n-2} \times (x_1 \otimes \cdots \otimes x_n)$$

in $\mathcal{A}_\infty QA_*$ by setting

$$\begin{aligned}&\pi_{n-2} \times (x_1 \otimes \dots \otimes x_n) \cup \pi_{m-2} \times (x_{n+1} \otimes \dots \otimes x_{n+m})\\ &\quad = \sum \pi_{n+m-2} \times (x_{i_1} \otimes \dots \otimes x_{i_{n+m}})\\ &\quad\quad + \sum \pi_{n+m-3} \times (x_{i_1} \otimes \dots \otimes x_{i_k} \cup x_{j_k} \otimes \dots \otimes x_{i_{n+m}}) + \dots,\end{aligned}$$

where the summation is over all (n, m)-shuffles of $(1, \dots, n+m)$, $1 \leq i_1 < i_2 < \dots \leq n$, $n+1 \leq j_1 < j_2 \dots \leq n+m$.

Similarly, we define the products

$$\pi_{n-2} \times (x_1 \otimes \dots \otimes x_n) \cup \pi_{m-2} \times (x_{n+1} \otimes \dots \otimes x_{n+m}) \cup \pi_{k-2} \times (x_{n+m+1} \otimes \dots \otimes x_{n+m+k}),$$

and so on.

Taking the products of the basic relations, we obtain the relations

$$\begin{aligned}&\pi_{n-2}(x_1 \otimes \dots \otimes x_n) \cup \pi_{m-2}(x_{n+1} \otimes \dots \otimes x_{n+m}) = 0,\\ &\pi_{n-2}(x_1 \otimes \dots \otimes x_n) \cup \pi_{m-2}(x_{n+1} \otimes \dots \otimes x_{n+m})\\ &\quad\quad \cup \pi_{k-2}(x_{n+m+1} \otimes \dots \otimes x_{n+m+k}) = 0,\end{aligned}$$

and so on. These relations are in a sense commutativity relations. Commutative A_∞-algebras will be called *S_∞-algebras.*

THEOREM. *If $R = \mathbb{Z}/2$, then the E_2-term of the Adams spectral sequence for stable homotopy groups of the spheres, regarded as an S_∞-algebra, is generated by the elements h_i of dimension $2^i - 1$ and the relations*

$$\pi_{i-2}(h_{i-1} \otimes \dots \otimes h_0) = 0.$$

If $R = \mathbb{Z}/p$, then the E_2-term, regarded as an S_∞-algebra, is generated by the elements h_i of dimension $(2p-2)^i - 1$ and g_0 of dimension 0 and the relations

$$\pi_{i-2}(h_{i-1} \otimes \dots \otimes h_0) = 0, \quad \pi_{i-1}(h_{i-1} \otimes \dots \otimes h_0 \otimes g_0) = 0.$$

This method makes it possible to describe the second term of the Adams spectral sequence not only for the stable homotopy groups of the spheres, but for the stable homotopy groups of topological spaces as well.

Indeed, if X is a topological space, then the E_2-term of the corresponding Adams spectral sequence is isomorphic to the homology of the cobar construction $F(K, H_*(X))$, where $H_*(X)$ is regarded as a comodule over the Milnor coalgebra. Hence the E_2-term can be described as an A_∞-module over the A_∞-algebra $H_*(FK)$. Its generators are the primitive elements of $H_*(X)$ with respect to the coaction $\tau\colon H_*(X) \to K \otimes H_*(X)$.

9.5. Description of the E_2-term of the Adams spectral sequence for unstable homotopy groups of topological spaces

In the previous section we discussed a method for describing the E_2-term of the Adams spectral sequence for stable homotopy groups of topological spaces. It was shown that the E_2-term possesses an A_∞-structure; besides, its generators and relations were found. In this section we consider the question of describing the E_2-term of the Bousfield–Kan spectral sequence for unstable homotopy groups of topological spaces [**9**].

Let K be the Milnor coalgebra and L an unstable K-comodule. Consider the cobar construction

$$F(K,L) : L \to K \times L \to K \times K \times L \to \cdots \to K^{\times n} \times L \to \dots.$$

As usual, elements of this cobar construction will be denoted by $[x_1, \dots, x_n]x$, where $x_i \in K$, $x \in L$. The differential is defined on these elements by

$$d([x_1, \dots, x_n]x) = \sum_{i=1}^{n} [x_1, \dots, x_i', x_i'', \dots, x_n]x + [x_1, \dots, x_n, x']x'',$$

where $\nabla(x_i) = \sum x_i' \times x_i''$, $\tau(x) = \sum x' \times x''$.

If X is a trivial K-comodule, then the corresponding cobar construction is denoted by $F(K)(X)$.

Suppose that X is a topological space; then its homology $X_* = H_*(X)$ (with coefficients in $\mathbb{Z}/2$) is an unstable K-comodule, and if the space X is good in the Massey–Peterson sense, then the homology of the cobar construction $F(K, X_*)$ is isomorphic to the E_2-term of the Bousfield–Kan spectral sequence for the unstable homotopy groups of X. In particular, if X is the n-dimensional sphere S^n, then the cohomology $F(K)(S^n)$ gives the E_2-term of the spectral sequence for the unstable homotopy groups of the n-dimensional sphere.

For the homology $H_*F(K, S^n)$ of the cobar construction, we have the composition product

$$\gamma\colon H_kF(K, S^m) \otimes H_mF(K, S^n) \to H_kF(K, S^n)$$

and the composition action

$$\mu\colon H_kF(K, S^m) \otimes H_mF(K, L) \to H_kF(K, L).$$

Moreover, as in the stable case, there are higher composition products, which define the structure of an A_∞-algebra on $H_*F(K, S^*)$ and the structure of an A_∞-module over the A_∞-algebra $H_*F(K, S^*)$ on $H_*F(K, L)$. However, in the unstable case these operations are not enough to describe the homology of the cobar construction $F(K, L)$ and hence the E_2-term of the spectral sequence for the unstable homotopy groups of topological spaces.

Our goal is to find an A_∞-structure that suffices to describe the homology of the cobar construction $F(K, L)$, and hence to describe the E_2-term of the spectral sequence for the unstable homotopy groups of topological spaces.

To this end, note that the correspondence $X \longmapsto F(K)(X)$ defines a functor $F(K)$ from the category of graded modules to the category of differential modules. However, its domain of definition cannot be extended to the category of differential modules, primarily due to the fact that for differential modules X the product $K \times X$ cannot be naturally defined. To get around this difficulty, we consider bigraded differential modules.

DEFINITION. A *bigraded differential module* is a bigraded module $X = \sum X_{n,m}$ endowed with a differential that has total degree -1 and does not lower the second grading, i.e., has the form

$$d_{n,m}\colon X_{n,m} \to X_{n-1,m} + X_{n-2,m+1} + \dots.$$

For bigraded differential modules, we can naturally define the notions of tensor product,

$$(X \otimes Y)_{n,m} = \sum_{\substack{i+j=n \\ p+q=m}} X_{i,p} \otimes Y_{j,q};$$

suspension, $(SX)_{n,m} = X_{n-1,m}$; desuspension, $(S^{-1}X)_{n,m} = X_{n+1,m}$; chain map, chain homotopy, chain equivalence, homology, etc.

Let us regard the Milnor coalgebra K as a bigraded differential module, where the first grading is zero, and the second coincides with the usual grading of the Milnor coalgebra.

For a bigraded differential module X, by $K \times X$ we denote the bigraded differential module which is the submodule of $K \otimes X$ generated by the elements $x \otimes y_{n,m}$ such that $e(x) \leq m$.

If L is a bigraded differential K-comodule, then the cobar construction $F(K, L)$ coincides as a bigraded module with $\sum_n S^{-n}(K^{\times n} \times L)$, and the differential is given on the elements $[x_1, \ldots, x_n]x$ by

$$\begin{aligned} d([x_1, \ldots, x_n]x) = &\sum_{i=1}^{n} [x_1, \ldots, x_i', x_i'', \ldots, x_n]x \\ &+ [x_1, \ldots, x_n, x']x'' + [x_1, \ldots, x_n]d(x), \end{aligned}$$

where $\nabla(x_i) = \sum x_i' \times x_i''$, $\tau(x) = \sum x' \times x''$.

Then $F(K, L)$ is also a bigraded differential module. If X is a trivial K-comodule, then the corresponding cobar construction is denoted by $F(K)(X)$. The correspondence $X \longmapsto F(K)(X)$ defines a functor on the category of bigraded differential modules.

THEOREM. *The functor $F(K)$ is a monad on the category of bigraded differential modules. If L is a bigraded differential K-comodule, then $F(K, L)$ is an algebra over the monad $F(K)$.*

PROOF. It is readily seen that the monad structure $\gamma\colon F(K) \circ F(K) \to F(K)$ is given by

$$\gamma([x_1, \ldots, x_n]([x_{n+1}, \ldots, x_{n+m}]x)) = [x_1, \ldots, x_{n+m}]x.$$

By a similar formula, $F(K, L)$ is given the structure of an algebra over the monad $F(K)$, $\mu\colon F(K)(F(K, L)) \to F(K, L)$. □

Note that giving the last map is equivalent to giving a family of maps

$$F_{p,q}(K, S^m) \otimes F_{n,m}(K, L) \to F_{n+p,q}(K, L)$$

permutable with the differential.

For an arbitrary monad A on the category of bigraded differential modules, consider the bar construction $B(A)$. If X is a bigraded module, then $B(A)(X)$ coincides, as a bigraded module, with $\sum_n S^n(A^{\circ n}(X))$. If a bigraded module M is an algebra over A, then the bar construction $B(A, M)$ is defined; as a bigraded module, it coincides with $\sum_n S^n(A^{\circ n}(M))$.

THEOREM. *Let K be the Milnor coalgebra, and let L be an unstable K-comodule. Then there are chain equivalences*

$$B(F(K)) \simeq K, \quad B(F(K), F(K, L)) \simeq L.$$

The required chain homotopy is induced by the standard chain homotopy for ordinary cobar constructions.

Thus in the category of bigraded differential modules we have all the assertions that are necessary for applying to this case the technique developed above for describing the homology of A_∞-algebras and A_∞-modules over an A_∞-algebra. Namely, we have the following theorem.

THEOREM. *The homology A_* of the monad $A = F(K)$ is an A_∞-monad, and the homology M_* of the bigraded differential module $F(K, L)$ is an A_∞-algebra over the A_∞-monad A_*.*

To define the indecomposable elements in A_* and the relations between their products, we consider the short exact sequence

$$0 \to A_* \xrightarrow{i} BA_* \xrightarrow{p} B^1A_* = BA_*/A_* \to 0$$

and the long exact homology sequence

$$\dots \xrightarrow{i_*} K \cong H_*(BA_*) \xrightarrow{p_*} H_*(B^1A_*) \xrightarrow{\mu_*} A_* \xrightarrow{i_*} K \cong H_*(BA_*) \xrightarrow{p_*} \dots$$

Then the decomposable elements in A_* are the elements of the image of μ_*, and the module of indecomposable elements JA_* is isomorphic to the image of the homomorphism $i_*\colon A_* \to K$.

It is readily seen that this image in K contains the elements of the form $\xi_1^{2^i} \in K$. Denote the corresponding indecomposable elements by h_i. Thus, for a graded module X_*, the indecomposable elements in $A_*(X_*)$ are the elements of the form $h_i \otimes x_n$ with $2^i \le n$.

The relations are given by the kernel of the homomorphism $\mu_*\colon H_*(B^1A_*) \to A_*$ or, equivalently, by the image of the homomorphism $p_*\colon K \cong H_*(BA_*) \to H_*(B^1A_*)$. To find this image, we define the projection $p\colon K \to JA_*$ onto the indecomposable elements by setting $p(\xi_1^{2^i}) = h_i$ and $p(x) = 0$ in all other cases. Then the required relations in A_* are generated by the elements $x \in K$ and have the form

$$\sum_{i\ge 2} \gamma_{i-2}(p(x^i) \otimes \dots \otimes p(x^1)) = 0,$$

where $\nabla(i)(x) = \sum x^i \otimes \dots \otimes x^1$; $\nabla(i)$ is the iteration of the comultiplication $\nabla\colon K \to K \otimes K$.

Quite similarly, to define the indecomposable elements for an A_∞-algebra M_* over the A_∞-monad A_*, we must consider the short exact sequence

$$0 \to M_* \xrightarrow{i_*} B(A_*, M_*) \xrightarrow{p_*} B^1(A_*, M_*) = B(A_*, M_*)/M_* \to 0$$

and the corresponding long exact homology sequence

$$\dots \xrightarrow{p_*} H_*(B^1(A_*, M_*)) \xrightarrow{\mu_*} M_* \xrightarrow{i_*} L \cong H_*(B(A_*, M_*)) \xrightarrow{p_*} \dots$$

Then the decomposable elements in M_* are the elements of the image of the homomorphism μ_*, and the module JM_* of indecomposable elements is isomorphic to the image of the homomorphism i_*.

The relations are given by the kernel of the homomorphism

$$\mu_*: H_*(B^1(A_*, M_*)) \to M_*$$

or, equivalently, by the image of the homomorphism

$$p_*: L \cong H_*(B(A_*, M_*)) \to H_*(B^1(A_*, M_*)).$$

Let X be a topological space, and let X_* be its homology; then the bigraded module $M_{n,m}(X_*)$ is isomorphic to the term $E_2^{n,m}(X)$ of the spectral sequence for the homotopy groups of X. The A_∞-monad structure on A_* and the structure of an A_∞-algebra over the A_∞-monad on M_*, when applied to the E_2-term, yield the operations

$$\begin{aligned} \gamma_{i-2}&: E_2^{n_i,m_i}(S^{m_{i-1}}) \otimes \cdots \otimes E_2^{n_1,m_1}(S^m) \to E_2^{n_1+\cdots+n_i,m_i}(S^m), \qquad i \geq 2; \\ \mu_{i-2}&: E_2^{n_i,m_i}(S^{m_{i-1}}) \otimes \cdots \otimes E_2^{n_1,m_1}(L) \to E_2^{n_1+\cdots+n_i,m_i}(L), \qquad i \geq 2. \end{aligned}$$

The above discussion implies the following results.

THEOREM. *Let us regard the term $E_2(S^*)$ of the spectral sequence for the unstable homotopy groups of the spheres as an A_∞-monad. Then its indecomposable elements are the elements of the form $h_i \otimes u^n$ with $2^i \leq n$.*

If the suspension homomorphism, which is defined for spheres, is included into the set of operations under consideration, then only the elements of the form $h_i \otimes u^{2^i}$ are indecomposable.

Furthermore, if the $\cup_1$-product, $h_i \cup_1 h_i = h_{i+1}$, is additionally included into the set of operations under consideration, then only the element of the form $h_0 \otimes u^1$ remains indecomposable in the E_2-term. The remaining elements can be obtained from this one by applying the $\cup_1$-product, the suspension homomorphism, and the Massey products.

THEOREM. *Let us regard the term $E_2(S^*)$ of the spectral sequence for the unstable homotopy groups of the spheres as an A_∞-monad. Then the relations in it are generated by the elements $x \in K$ and have the form*

$$\sum_{i \geq 2} \gamma_{i-2}(p(x^i) \otimes \cdots \otimes p(x^1))(u^n) = 0, \qquad e(x) \leq n.$$

In particular, the elements $\xi_i \in K$ give rise to the generating relations, which have the form

$$\gamma_{i-2}(h_{i-1} \otimes \cdots \otimes h_0)(u^n) = 0.$$

The remaining relations can be obtained as products of the generating relations.

If the suspension is included into the set of operations under consideration, then the only generating relations are those of the form

$$\gamma_{i-2}(h_{i-1} \otimes \cdots \otimes h_0)(u^1) = 0.$$

The above theorems provide a complete description of the E_2-term of the spectral sequence for the homotopy groups of the spheres.

The $E_2(X)$-term of the spectral sequence for the homotopy groups of a good topological space X can be described in a similar way. In this case the generators are the primitive elements of the homology $H_*(X)$ with respect to the coaction

$$\tau: H_*(X) \to K \times H_*(X).$$

In the case of an arbitrary space X, the A_∞-$\vee$-algebra structure should be additionally included into the set of operations on the term $E_2(X)$ of the Bousfield–Kan spectral sequence. In this case the generators are the primitive elements of the homology $H_*(X)$ with respect to the coaction

$$\tau \colon H_*(X) \to K \times H_*(X)$$

and the comultiplication

$$\nabla \colon H_*(X) \to H_*(X) \otimes H_*(X).$$

CHAPTER 10

Homology Operations and Differentials in the Adams Spectral Sequence

10.1. Functional homology operations

Functional homology operations were defined by Steenrod and Peterson [**34**] as partial and multivalued maps. Here we define everywhere defined and single-valued functional homology operations, which are unique up to an equivalence relation that we introduce.

Let *Chain* be the category of chain complexes over a field R, and let *Mod* be the category of graded modules over R. Consider the homology functor $H_*\colon Chain \to Mod$ that assigns the graded homology module $X_* = H_*(X)$ to each chain complex X and the induced homomorphism in homology $f_*\colon X^1_* \to X^2_*$ to each chain map $f\colon X^1 \to X^2$.

For each chain complex X, let us choose a homomorphism $\xi\colon X_* \to X$ of selecting representatives in homology classes. In general, the correspondence $X \longmapsto \xi$ is not a transformation of functors $\xi\colon H_* \to \mathrm{Id}$, since for a given chain map $f\colon X^1 \to X^2$, the diagram

$$\begin{array}{ccc} X^1_* & \xrightarrow{f_*} & X^2_* \\ {\scriptstyle \xi^1}\downarrow & & \downarrow{\scriptstyle \xi^2} \\ X^1 & \xrightarrow{f} & X^2 \end{array}$$

is not necessarily commutative, but is only homotopy commutative, i.e., there is a map $h\colon X^1_* \to X^2$ of dimension 1 such that $d(h) = \xi \circ f_* - f \circ \xi$. For each chain map $f\colon X^1 \to X^2$, let us fix this homotopy $h = h(f)\colon X^1_* \to X^2$.

Now suppose that a pair of chain maps is given, $f^1\colon X^1 \to X^2$ and $f^2\colon X^2 \to X^3$. Consider the map $H(f^2, f^1)\colon X^1_* \to X^3$ given by

$$H(f^2, f^1) = f^2 \circ h(f^1) + h(f^2) \circ f^1_* - h(f^2 \circ f^1).$$

Straightforward computations show that $dH(f^2, f^1) = 0$, and hence the map $H(f^2, f^1)$ induces a map

$$H_*(f^2, f^1)\colon X^1_* \to X^3_*$$

of dimension 1. These maps will be called *secondary functional homology operations*. Straightforward computations show that for any triple of chain maps

$$f^1\colon X^1 \to X^2, \quad f^2\colon X^2 \to X^3, \quad f^3\colon X^3 \to X^4,$$

the following relation holds:

$$H_*(f^3 \circ f^2, f^1) - H_*(f^3, f^2 \circ f^1) = f^3_* \circ H_*(f^2, f^1) - H_*(f^3, f^2) \circ f^1_*.$$

Of course, the functional homology operations $H_*(f^2, f^1)$ constructed above depend on the choice of the maps $\xi: X_* \to X$ and the homotopies $h(f): X_*^1 \to X^2$. Suppose that $\xi', \xi'': X_* \to X$ are arbitrary homomorphisms of choosing representatives in homology classes, and for each chain map $f: X^1 \to X^2$, the corresponding homotopies $h', h'': X_*^1 \to X^2$ are given. Let $g: X_* \to X$ be a chain homotopy between ξ' and ξ''. Then direct computations show that the map $c(f) = h'' - h' + f \circ g - g \circ f_*$ satisfies the relation $dc(f) = 0$, and hence it induces a map $c_*(f): X_*^1 \to X_*^2$. Further, the functional homology operations $H'_*(f^2, f^1)$, $H''_*(f^2, f^1)$ corresponding to the maps ξ', h' and ξ'', h'' satisfy the relation

$$H''_*(f^2, f^1) - H'_*(f^2, f^1) = f_*^2 \circ c_*(f^1) + c_*(f^2) \circ f_*^1 - c_*(f^2 \circ f^1).$$

Let us introduce an equivalence relation on the set of secondary homology operations by setting $H'_* \sim H''_*$ if there exist maps $c_*(f): X_*^1 \to X_*^2$ raising the dimension by one and satisfying the above relation. Up to this equivalence, the secondary functional homological operations are independent of the choice of the maps ξ and h. Our next goal is to define functional homology operations $H_*(f^n, \dots, f^1): X_*^1 \to X_*^{n+1}$ for a collection of chain maps $f^1: X^1 \to X^2, \dots, f^n: X^n \to X^{n+1}$. With this aim in mind, let us consider a more general situation.

DEFINITION. A *chain precategory* $\mathcal{X}$ consists of a family of objects $\{X\}$ and a family of chain complexes $\mathrm{Hom}(X^1, X^2)$ (whose elements are called *morphisms from* X^1 *to* X^2) given for each pair of objects $X^1, X^2 \in \mathcal{X}$.

DEFINITION. A *prefunctor* $f: \mathcal{X} \to \mathcal{Y}$ *from a chain precategory* $\mathcal{X}$ *to a chain precategory* $\mathcal{Y}$ is a correspondence assigning an object $F(X) \in \mathcal{Y}$ to each object $X \in \mathcal{X}$, and for each pair of objects $X^1, X^2 \in \mathcal{X}$, a homomorphism $\mathrm{Hom}(X^1, X^2) \to \mathrm{Hom}(F(X^1), F(X^2))$ which is, in general, not permutable with the differential, and which assigns a morphism $F(f): F(X^1) \to F(X^2)$ of $\mathcal{Y}$ to a morphism $f: X^1 \to X^2$ of $\mathcal{X}$.

In a similar way we define a transformation of prefunctors $\alpha: F \to G$.

For chain precategories $\mathcal{X}'$ and $\mathcal{X}''$ with the same family of objects, we define the chain precategory $\mathcal{X}' \times \mathcal{X}''$ by setting

$$\mathrm{Hom}_{\mathcal{X}' \times \mathcal{X}''}(X^1, X^2) = \sum_{X \in \mathcal{X}'} \mathrm{Hom}_{\mathcal{X}'}(X, X^2) \otimes \mathrm{Hom}_{\mathcal{X}''}(X^1, X).$$

It is clear that a precategory $\mathcal{X}$ is a category if it is endowed with an algebra structure with respect to the product $\times$ defined above, and a prefunctor $F: \mathcal{X} \to \mathcal{Y}$ is a functor if it is a map of algebras.

The analogy of the notions of category and functor with the notions of algebra and map of algebras allows us to extend to categories the main constructions and methods of E. Brown's theory of twisting cochains and N. A. Berikashvili's theory of predifferentials. Namely, given a chain category $\mathcal{X}$, we define a chain precategory $B\mathcal{X}$, which we call the *bar construction on* $\mathcal{X}$, and which is a coalgebra with respect to the product $\times$. The objects in $B\mathcal{X}$ are the same as in $\mathcal{X}$, and the morphisms are generated by sequences of morphisms of the form

$$[f^n, \dots, f^1], \quad f^1: X^1 \to X^2, \dots, f^n: X^n \to X^{n+1}.$$

For the dimension of the morphism $[f^n, \ldots, f^1]$ we take the sum of the dimensions of the morphisms f^i plus n, and the differential is given by the formula

$$\begin{aligned} d[f^n, \ldots, f^1] = &\sum_{i=1}^{n-1} (-1)^{\epsilon_i} [f^n, \ldots, f^{i+1} \circ f^i, , \ldots, f^1] \\ &- \sum_{i=1}^{n-1} (-1)^{\eta_i} [f^n, \ldots, d(f^i), \ldots, f^1], \end{aligned}$$

where $\epsilon_i = \deg[f^n, \ldots, f^{i+1} \circ f^i]$, $\eta_i = \deg[f^n, \ldots, f^{i+1}]$.

Suppose that $\mathcal{X}$ and $\mathcal{Y}$ are chain categories; a *twisting prefunctor* $\Phi\colon B\mathcal{X} \to \mathcal{Y}$ is a prefunctor that lowers the dimension by one and satisfies the relation $d\Phi + \Phi \cup \Phi = 0$.

For any chain category $\mathcal{X}$, we have a twisting prefunctor $J\colon B\mathcal{X} \to \mathcal{X}$ that assigns to each morphism $[f^n, \ldots, f^1]$ in $B\mathcal{X}$ the morphism f^1 in $\mathcal{X}$ if $n = 1$ and 0 otherwise.

Note that since the precategory $B\mathcal{X}$ is a coalgebra, we can define in a standard way the cobar construction $FB\mathcal{X}$ on it, which is an algebra and hence a category.

Any twisting prefunctor $\Phi\colon B\mathcal{X} \to \mathcal{Y}$ defines a functor $F\Phi\colon FB\mathcal{X} \to \mathcal{Y}$, and in the case $\Phi = J\colon B\mathcal{X} \to \mathcal{X}$ we obtain the functor $FJ\colon FB\mathcal{X} \to \mathcal{X}$. The properties of the bar and cobar constructions show that this functor, which acts as the identity on objects, is a chain equivalence on morphisms.

Let $\Phi, \Psi\colon B\mathcal{X} \to \mathcal{Y}$ be twisting prefunctors. A *transformation* $\xi\colon \Phi \to \Psi$ *of twisting prefunctors* is a transformation of prefunctors that preserves dimensions of the morphisms and satisfies the relation

$$d(\xi) = \xi \cup \Phi - \Psi \cup \xi.$$

Two transformations of twisting prefunctors $\xi', \xi''\colon \Phi \to \Psi$ are said to be *homotopic* (we write $\xi' \simeq \xi''$) if there is a transformation of prefunctors $h\colon \Phi \to \Psi$ that raises the dimension by one and satisfies the relation

$$d(h) = \xi'' - \xi' - \Psi \cup h - h \cup \Phi.$$

Two twisting prefunctors $\Phi, \Psi\colon B\mathcal{X} \to \mathcal{Y}$ are said to be *homotopy equivalent* (we write $\Phi \sim \Psi$) if there are transformations of twisting prefunctors $\xi\colon \Phi \to \Psi$, $\eta\colon \Psi \to \Phi$ such that their compositions $\eta \circ \xi$ and $\xi \circ \eta$ are homotopic to the identity transformations.

THEOREM. *The homology functor $H_*\colon BChain \to Mod$ can be extended to a twisting prefunctor $H_*\colon BChain \to Mod$ so that it is homotopy equivalent to the standard twisting prefunctor $J\colon BChain \to Chain$.*

PROOF. For each chain complex X, let us fix chain maps $\xi\colon X_* \to X$, $\eta\colon X \to X_*$ and a chain homotopy $h\colon X \to X$ that satisfy the relations

$$\eta \circ \xi = \mathrm{Id}, \quad d(h) = \xi \circ \eta - \mathrm{Id}, \quad h \circ \xi = 0, \quad \eta \circ h = 0, \quad h \circ h = 0.$$

We define the prefunctor $H_*\colon BChain \to Mod$ by setting

$$H_*[f^n, \ldots, f^1] = (-1)^n \eta \circ f^n \circ h \circ \cdots \circ f^1 \circ \xi.$$

Straightforward computations show that $dH_* + H_* \cup H_* = 0$, and hence H_* is a twisting prefunctor.

We define the transformations of prefunctors $\xi\colon H_* \to J$ and $\eta\colon J \to H_*$ by setting

$$\xi[f^n,\dots,f^1] = (-1)^{n+1} h \circ f^n \circ \dots \circ f^1 \circ \xi,$$
$$\eta[f^n,\dots,f^1] = \eta \circ f^n \circ h \circ \dots \circ f^1 \circ h.$$

Straightforward computations show that

$$d\xi = \xi \cup H_* - J \cup \xi, \quad d\eta = \eta \cup J - H_* \cup \eta.$$

Let us show that $\eta \circ \xi = \mathrm{Id}$. Indeed, $\eta \circ \xi(X) = \mathrm{Id}_X$ and

$$\begin{aligned}\eta \circ \xi[f^n,\dots,f^1] &= (\eta \cup \xi)[f^n,\dots,f^1]\\ &= \sum_{i=1}^{n-1} \eta[f^n,\dots,f^{i+1}] \circ \xi[f^i,\dots,f^1] = 0.\end{aligned}$$

Let us show that $\xi \circ \eta \simeq \mathrm{Id}$. To this end, we define the transformation $h\colon J \to J$ by setting

$$h[f^n,\dots,f^1] = h \circ f^n \circ \dots \circ f^1 \circ h.$$

Straightforward computations show that $d(h) = \xi \circ \eta - \mathrm{Id} - J \cup h - h \cup J$, and hence h is a homotopy between $\xi \circ \eta$ and Id. This completes the proof. □

The maps $H_*[f^n,\dots,f^1]\colon X^1_* \to X^{n+1}_*$ raising the dimension by n are called *n-ary functional homology operations*. By writing out the formula $dH_* + H_* \cup H_* = 0$ in full, the relations between the functional homology operations can be expressed in the form

$$\begin{aligned}&\sum_{i=1}^{n-1} (-1)^{n-i+1} H_*[f^n,\dots,f^{i+1} \circ f^i,\dots,f^1]\\ &\qquad = \sum_{i=1}^{n-1} (-1)^{n-i} H_*[f^n,\dots,f^{i+1}] \circ H_*[f^i,\dots,f^1].\end{aligned}$$

Note that giving functional homology operations is equivalent, by this formula, to giving a family of chain maps

$$H_*[f^n,\dots,f^1]\colon I^n \otimes X^1_* \to X^{n+1}_*$$

satisfying the relations

$$H_*[f^n,\dots,f^1]\delta^i_0 = H_*[f^n,\dots,f^{i+1} \circ f^i,\dots,f^1],$$
$$H_*[f^n,\dots,f^1]\delta^i_1 = H_*[f^n,\dots,f^{i+1}] \circ H_*[f^i,\dots,f^1],$$

where I is the normalized chain complex of the standard 1-dimensional simplex, $I^n = (I^1)^{\otimes n}$, and $\delta^i_\epsilon\colon I^{n-2} \to I^{n-1}$ are the coface operators

$$\delta^i_\epsilon(u_{n-2} \otimes \dots \otimes u_1) = u_{n-2} \otimes \dots \otimes \epsilon \otimes \dots \otimes u_1, \quad \epsilon = 0, 1.$$

Another possible language for describing functional homology operations is the language of Stasheff's A_∞-structures. Namely, by analogy with the notion of A_∞-algebra, we can define the notions of A_∞-functor, A_∞-transformation of functors, homotopy equivalence of A_∞-functors, etc. Reformulating the previous theorem, we obtain the following assertion.

THEOREM. *The homology functor* $H_*: BChain \to Mod$ *can be extended to an* A_∞*-functor so that it is homotopy equivalent to the identity functor* $\mathrm{Id}: Chain \to Chain$.

Now consider the question of how the functional homology operations behave with regard to the tensor product of morphisms, which is defined in the category of chain complexes. Suppose that $(f^n, \dots, f^1)$, $(g^n, \dots, g^1)$ are two collections of chain maps $f^i: X^i \to X^{i+1}$, $g^i: Y^i \to Y^{i+1}$, $1 \le i \le n$. Consider the collection $(f^n \otimes g^n, \dots, f^1 \otimes g^1)$.

THEOREM. *If the collections of chain maps* $(f^n, \dots, f^1)$, $(g^n, \dots, g^1)$ *are as above, then the maps*

$$\xi^i: X^i_* \otimes Y^i_* \to X^i \otimes Y^i, \quad \eta^i: X^i \otimes Y^i \to X^i_* \otimes Y^i_*$$

and the homotopy $h^i: X^i \otimes Y^i \to X^i \otimes Y^i$ *can be chosen so that*

$$\begin{aligned} H_*[f^n \otimes g^n, \dots, f^1 \otimes g^1] = \sum H_*[f^{n_k}, \dots, f^{n_{k-1}+1}] \circ \dots \circ H_*[f^{n_1}, \dots, f^1] \\ \otimes H_*[g^{n_k} \circ \dots \circ g^{n_{k-1}+1}, \dots, g^{n_1} \circ \dots \circ g^1], \end{aligned}$$

$1 \le n_1 < \dots < n_k \le n$.

PROOF. Consider the maps

$$\begin{aligned} \xi(X^i): X^i_* \to X^i, \quad \eta(X^i): X^i \to X^i_*, \quad h(X^i): X^i \to X^i; \\ \xi(Y^i): Y^i_* \to Y^i, \quad \eta(Y^i): Y^i \to Y^i_* \quad h(Y^i): Y^i \to Y^i. \end{aligned}$$

For the maps ξ^i, η^i, h^i we take, respectively,

$$\xi(X^i) \otimes \xi(Y^i), \quad \eta(X^i) \otimes \eta(Y^i), \quad \xi(X^i) \circ \eta(X^i) \otimes h(X^i) + h(X^i) \otimes \mathrm{Id}(Y^i).$$

Straightforward computations show that the required relation holds. □

Let us give some examples in which the functional homology operations appear.

Example 1. Suppose that $X^1 \xrightarrow{d^1} X^2 \to \dots \to X^n \xrightarrow{d^n} X^{n+1} \to \dots$ is a bicomplex, i.e., the X^n $(n \ge 1)$ are chain complexes, the $d^n: X^n \to X^{n+1}$ are chain maps, and $d^{n+1} \circ d^n = 0$. The functional homology operations

$$H_*[d^{n+m}, \dots, d^m]: X^m_* \to X^{n+m+1}_*$$

define Berikashvili's predifferentials (see [**6**]) in the spectral sequence for the homology of the total complex $TotX$. In this case, the relations for the functional homology operations mean that these operations define a differential in the total complex $TotX_*$. Denote the corresponding complex by $\widetilde{TotX}_*$. The transformation $\xi: H_* \to J$ of twisting prefunctors defines a map of chain complexes $\widetilde{TotX}_* \to TotX$, which is a chain equivalence if the convergence conditions for the spectral sequence hold. Thus, in this case, giving the homology X^n_* of the complexes X^n and the functional homology operations $H_*[d^{n+m}, \dots, d^m]$ determines the homology of the total complex.

Example 2. Let K be a differential coalgebra, and let $\nabla\colon K \to K \otimes K$ be its comultiplication. Consider the cobar construction FK, which is the total complex for the bicomplex

$$K \xrightarrow{\nabla} K^{\otimes 2} \to \cdots \to K^{\otimes n} \xrightarrow{\nabla(n)} K^{\otimes n+1} \to \dots,$$

where $\nabla(n) = \nabla \otimes 1 \otimes \cdots \otimes 1 - \cdots + (-1)^{n-1} 1 \otimes \cdots \otimes 1 \otimes \nabla$. Then the functional homology operations

$$H_*[\nabla(n+m), \dots, \nabla(m)]\colon K_*^{\otimes m} \to K_*^{\otimes n+m+1}$$

define a differential in FK_*, and the resulting complex $\widetilde{F}K_*$ is chain equivalent to the complex FK, provided that the convergence conditions for the spectral sequence hold.

By using the formula for the functional homology operations on the tensor product of maps, we obtain

$$H_*[\nabla(n+m-1), \dots \nabla(m)] = \sum_{i=1}^{m} (-1)^{(i+1)n} \operatorname{Id}^{\otimes m-i} \otimes H_*[\nabla(n), \dots, \nabla] \otimes \operatorname{Id}^{\otimes i-1},$$

which means that the differential in FK_* defined by these operations is compatible with the standard algebra structure, and hence it defines an A_∞-coalgebra structure on K_*; furthermore, the corresponding differential algebra FK_* is chain equivalent to FK.

Dually, for a differential algebra A with multiplication $\pi\colon A \otimes A \to A$, the functional homology operations

$$H_*[\pi, \dots, \pi(n)]\colon A_*^{\otimes n} \to A_*$$

define an A_∞-algebra structure on the homology A_* of A. The corresponding differential coalgebra BA_* is chain equivalent to the bar construction BA.

Example 3. Let L be a comodule over a coalgebra K, and let $\tau\colon L \to K \otimes L$ be the coaction. We set

$$\tau(n) = \nabla \otimes \cdots \otimes 1 - \cdots + (-1)^{n-1} 1 \otimes \cdots \otimes \tau\colon K^{\otimes n-1} \otimes L \to K^{\otimes n} \otimes L,$$

and consider the cobar construction $F(K, L)$, which is the total complex of the bicomplex

$$L \xrightarrow{\tau} K \otimes L \to \cdots \to K^{\otimes n-1} \otimes L \xrightarrow{\tau(n)} K^{\otimes n} \otimes L \to \dots$$

Then the functional homology operations

$$H_*[\tau(n+m), \dots, \tau(m)]\colon K_*^{\otimes m-1} \otimes L_* \to K_*^{\otimes n+m} \otimes L_*$$

define a differential in $F(K_*, L_*)$, and the FK_*-module $F(K_*, L_*)$ is chain equivalent to the $F(K)$-module $F(K, L)$ (in case the convergence conditions for the spectral sequence hold).

Dually, for a module M over an algebra A, the functional homology operations

$$H_*[\mu, \dots, \mu(n)]\colon A_*^{\otimes n-1} \otimes M_* \to M_*$$

define on the homology M_* of M the structure of an A_∞-module over the A_∞-algebra A_*. The corresponding BA_*-comodule $B(A_*, M_*)$ is chain equivalent to the BA-comodule $B(A, M)$.

Example 4. Let X be a simplicial set, and consider the augmented cosimplicial object

$$R^*X\colon\ RX \xrightarrow{\delta^0} R^2X \to \cdots \to R^nX \xrightarrow{\delta^0,\dots,\delta^{n-1}} R^{n+1}X \to \dots,$$

which was used by Bousfield and Kan to construct the spectral sequence for the homotopy groups of X with coefficients in R. By the definition of the higher differentials of this spectral sequence, they can be expressed in terms of the functional homology operations

$$H_*[\delta,\dots,\delta]\colon H_*(R^{n-1}X,R) \to H_*(R^{n+m-1}X,R).$$

Therefore, the following assertion holds.

THEOREM. *The E_1-term of the Bousfield–Kan spectral sequence can be endowed with a differential that makes it into a complex whose homology is isomorphic to the E_∞-term of this spectral sequence.*

As was shown in Chapter 7, instead of the augmented cosimplicial object of the Bousfield–Kan spectral sequence one can take the augmented cosimplicial object

$$F^*(C,RX)\colon\ RX \to CRX \to \cdots \to C^{n-1}RX \to C^nRX \to \dots$$

in the category of simplicial sets, or the augmented cosimplicial object

$$F^*(E,C_*(X;R))\colon\ C_*(X;R) \to \overline{E}C_*(X;R) \to \cdots \to \overline{E}^nC_*(X;R) \to \dots$$

in the category of chain complexes.

Example 5. Let E be an operad, and consider the comonad $\overline{E}$ and the corresponding cobar construction

$$F\overline{E}\colon\ \overline{E} \xrightarrow{\overline{\gamma}} \overline{E}\circ\overline{E} \to \cdots \to \overline{E}^{\circ n} \xrightarrow{\overline{\gamma}(n)} \overline{E}^{\circ n+1} \to \dots,$$

where $\overline{\gamma}(n) = \overline{\gamma}\overline{E}^{\circ n-1} - \cdots + (-1)^{n-1}\overline{E}^{\circ n-1}\overline{\gamma}$. Then the functional homology operations

$$H_*[\overline{\gamma}(n),\dots,\overline{\gamma}]\colon \overline{E}_* \to \overline{E}_*^{\circ n+1}$$

define an A_∞-comonad structure on $\overline{E}_*$. Passing from $\overline{E}_*$ to the Milnor coalgebra $\mathcal{K}$, we obtain an A_∞-coalgebra structure on the Milnor coalgebra.

Example 6. Let X be a coalgebra over an operad E. Consider the cobar construction

$$F(E,X)\colon\ X \xrightarrow{\tau} \overline{E}(X) \to \cdots \to \overline{E}^{n-1}(X) \xrightarrow{\tau(n)} \overline{E}^n(X) \to \dots,$$

where $\tau(n) = \overline{E}^{\circ n-1}\tau - \cdots + (-1)^{n-1}\overline{\gamma}\overline{E}^{\circ n-1}$. Then the functional homology operations

$$H_*[\tau(n),\dots,\tau]\colon X_* \to \overline{E}_*^n(X_*)$$

define on X_* the structure of an A_∞-coalgebra over the A_∞-comonad $\overline{E}_*$ and determine the higher differentials of the spectral sequence for the homotopy groups of the E-coalgebra X. Therefore, the following assertion holds.

THEOREM. *If X is the chain complex of a simplicial set or a topological space, then the homology of the corresponding cobar construction $F(\overline{E}_*, X_*)$ is isomorphic to the E_∞-term of the Bousfield–Kan spectral sequence for the homotopy groups of this space.*

Recall that the suspension SX of an E-coalgebra X is an SE-coalgebra, and the following diagrams are commutative:

$$\begin{array}{ccc} \overline{E} & \xrightarrow{\overline{\gamma}} & \overline{E}\circ\overline{E} \\ \downarrow & & \downarrow \\ \overline{SE} & \xrightarrow{S\overline{\gamma}} & \overline{SE}\circ\overline{SE} \end{array} \qquad \begin{array}{ccc} SX & \xrightarrow{\tau} & \overline{SE}(SX) \\ =\downarrow & & \downarrow \\ SX & \xrightarrow{S\tau} & S(\overline{E}(X)) \end{array}$$

Besides, it follows from the expressions for the homology $\overline{E}_*$ of the comonad $\overline{E}$ that the maps $\xi\colon \overline{E}_* \to \overline{E}$, $\eta\colon \overline{E} \to \overline{E}_*$, and $h\colon \overline{E} \to \overline{E}$ can be chosen so that they commute with the suspension homomorphism $\overline{SE} \to \overline{E}$. Therefore, the functional homology operations constructed above commute with the suspension homomorphism, and hence the following assertions hold.

THEOREM. *The A_∞-comonad structure on $\overline{E}_*$ commutes with suspension, and hence is stable. In passing to the Milnor coalgebra, it induces an A_∞-coalgebra structure on the stable Milnor coalgebra K.*

THEOREM. *The functional homology operations that give the higher differentials of the Bousfield–Kan spectral sequence commute with suspension, and hence are stable. They induce the differentials of the Adams spectral sequence for the stable homotopy groups of topological spaces.*

10.2. A_∞-cosimplicial objects

Recall that a family of objects $X^* = \{X^n\}_{n\geq 0}$ of a certain category is called a *precosimplicial object* (see Chapter 2) if there are given *coface operators*

$$\delta^i\colon X^n \to X^{n+1}, \quad 1 \leq i \leq n+1,$$

and *codegeneracy operators*

$$\sigma^i\colon X^n \to X^{n-1}, \quad 0 \leq i \leq n-1,$$

satisfying the following relations:

$$\begin{aligned} \delta^j\delta^i &= \delta^i\delta^{j-1}, \quad i<j, \\ \sigma^j\sigma^i &= \sigma^i\sigma^{j+1}, \quad i \leq j, \\ \sigma^j\delta^i &= \begin{cases} \delta^i\sigma^{j-1}, & i<j, \\ \mathrm{Id}, & i=j, \quad i=j+1, \\ \delta^{i-1}\sigma^j, & i>j+1. \end{cases} \end{aligned}$$

A *map $f^*\colon X \to Y$ of precosimplicial objects* is a family $f^* = \{f^n\}_{n\geq 0}$ of maps $f^n\colon X^n \to Y^n$ that commute with the coface and codegeneracy operators, i.e.,

$$\begin{aligned} \delta^i f_n &= f^{n+1}\delta^i, \\ \sigma^i f^n &= f_{n-1}\sigma^i. \end{aligned}$$

By I^n we denote the unit cube, $I^n = \{(t_1, \dots, t_n) | \ 0 \le t_i \le 1\}$. Let

$$u_i^\epsilon : I^n \to I^{n+1}, \quad 1 \le i \le n+1, \quad \epsilon = 0, 1,$$
$$v_i : I^n \to I^{n-1}, \quad 0 \le i \le n,$$

be maps defined by

$$u_i^\epsilon(t_1, \dots, t_n) = (t_1, \dots t_{i-1}, \epsilon, t_i, \dots, t_n);$$
$$v_i(t_1, \dots, t_n) = \begin{cases} (t_2, \dots, t_n), & i = 0, \\ (t_1, \dots, t_i * t_{i+1}, \dots, t_n), & 1 \le i \le n-1, \\ (t_1, \dots, t_{n-1}), & i = n, \end{cases}$$

where $t_i * t_{i+1} = t_i + t_{i+1} - t_i \cdot t_{i+1}$.

A precosimplicial object $X^* = \{X^n\}$ in the category of topological space is called an *A_∞-cosimplicial object* or just an *A_∞-cosimplicial space* if there are given coface operators $\delta_m^0 : X^n \times I^m \to X^{n+m+1}$ satisfying the following relations:

$$\begin{aligned} \sigma^0 \delta_0^0 &= \mathrm{Id}; \\ \delta_m^0 (1 \times u_i^0) &= \delta^i \delta_{m-1}^0, \quad 1 \le i \le m; \\ \delta_m^0 (1 \times u_i^1) &= \delta_{i-1}^0 (\delta_{m-1}^0 \times 1), \quad 1 \le i \le m; \\ \delta_{m-1}^0 (1 \times v_i) &= \sigma^i \delta_m^0, \quad 0 \le i \le m, \quad m \ge 1; \\ \delta_m^0 (\delta^i \times 1) &= \delta^{i+m+1} \delta_m^0, \quad i \ge 1; \\ \delta_m^0 (\sigma^i \times 1) &= \sigma^{i+m+1} \delta_m^0, \quad i \ge 0. \end{aligned}$$

The A_∞-cosimplicial objects form a category whose morphisms are the maps $f^* : X^* \to Y^*$ preserving the A_∞-structure.

Let $X^* = \{X^n\}$ be an A_∞-cosimplicial space, and let $Y^* = \{Y^n\}$ be a cosimplicial space. An *A_∞-map from X^* to Y^** is a map $f^* = \{f^n\}$, $f^n : X^n \to Y^n$, of precosimplicial objects together with a family of maps $f_m^n : X^n \times I^m \to Y^{n+m}$ satisfying the following relations:

$$\begin{aligned} f_0^n &= f^n; \\ f_m^n (1 \times u_i^0) &= \delta^{i-1} f_{m-1}^n, \quad 1 \le i \le m; \\ f_m^n (1 \times u_i^1) &= f_i^{n+m-i-1} (\delta_{m-i}^0 \times 1), \quad 1 \le i \le m; \\ f_{n-1}^m (d_i \times 1) &= d_{i-m} f_n^m, \quad i > m; \\ f_{m-1}^n (1 \times v_i) &= \sigma^i f_m^n, \quad 0 \le i < m; \\ f_m^n (\delta^i \times 1) &= \delta^{i+m} f_m^{n-1}, \quad i \ge 1; \\ f_{m-1}^n (\sigma^i \times 1) &= \sigma^{i+m} f_n^m, \quad i \ge 0. \end{aligned}$$

An A_∞-map is called an *A_∞-homotopy equivalence* if the corresponding maps f_n are homotopy equivalences.

Now consider the category of chain complexes. In this case, for the objects I^n we must take the normalized chain complexes of the n-dimensional cubes. The definitions of an A_∞-cosimplicial object and of an A_∞-map can be reformulated in the following form.

A precosimplicial object $X^* = \{X^n\}$ in the category of chain complexes is called an *A_∞-cosimplicial object* if we are given maps $\delta_m^0 : X^n \to X^{n+m+1}$ that raise the dimension by m and satisfy the relations

$$\begin{aligned} d(\delta^0_m) &= \sum_{i=1}^{m}(-1)^{i-1}(\delta^i\delta^0_{m-1} - \delta^0_{i-1}\delta^0_{m-i}); \\ \delta^0_m\delta^i &= \delta^{i+m+1}\delta^0_m, \quad i \geq 1; \\ \sigma^i\delta^0_m &= \begin{cases} 0, & 0 \leq i \leq m, \quad m \geq 1, \\ \mathrm{Id}, & i = m = 0, \\ \delta^0_m\sigma_{i-m-1}, & i > m. \end{cases} \end{aligned}$$

An A_∞*-map of an* A_∞*-cosimplicial object* X^* *to a cosimplicial object* Y^* *in the category of chain complexes* is a map of precosimplicial objects $f^* = \{f^n\}$, $f^n\colon X^n \to Y^n$, together with a family of maps $f^n_m\colon X^n \to Y^{n+m}$ that raise the dimension by m and satisfy the relations

$$\begin{aligned} f^n_0 &= f^n; \\ d(f^n_m) &= \sum_{i=1}^{m}(-1)^{i-1}(\delta^{i-1}f^n_{m-1} - f^{n+m-i-1}_i\delta^0_{m-i}); \\ f^{n+1}_m\delta^i &= \delta^{i+m}f^n_m, \qquad i \geq 1; \\ \sigma^i f^n_m &= \begin{cases} 0, & i < m, \\ f^{n+1}_m\sigma^{i-m}, & i \geq m. \end{cases} \end{aligned}$$

Let us extend the perturbation theory of chain complexes [**21**] to A_∞-cosimplicial objects.

We say that a chain complex $\widetilde{X}$ is a *deformation retract* of a chain complex X if there exist chain maps $\xi\colon \widetilde{X} \to X$, $\eta\colon X \to \widetilde{X}$ and a chain homotopy $h\colon X \to X$ such that

$$\eta\xi = \mathrm{Id}, \quad d(h) = \mathrm{Id} - \xi\eta.$$

It turns out that in this case we can additionally assume that the following relations hold [**21**]:

$$h\xi = 0, \quad \eta h = 0, \quad hh = 0.$$

A precosimplicial object $\widetilde{X}^*$ in the category of chain complexes will be called a *deformation retract of a precosimplicial object* X^* if for each n the chain complex $\widetilde{X}^n$ is a deformation retract of the chain complex X^n, and the corresponding chain maps and chain homotopies commute with the precosimplicial structure.

The following theorem is a cosimplicial analog of the main lemma of the perturbation theory of chain complexes [**21**].

THEOREM. *Let* $X^* = \{X^n\}$ *be a cosimplicial object in the category of chain complexes, and let a precosimplicial object* $\widetilde{X}^* = \{\widetilde{X}^n\}$ *be a deformation retract of* X^* *regarded as a precosimplicial object. Then* $\widetilde{X}^*$ *possesses the structure of an* A_∞*-cosimplicial object, and there is an* A_∞*-cosimplicial homotopy equivalence between* X^* *and* $\widetilde{X}^*$.

PROOF. Let $\xi^*\colon \widetilde{X}^* \to X^*$, $\eta^*\colon X^* \to \widetilde{X}^*$, and $h^*\colon X^* \to X^*$ be the corresponding maps. We define the operators $\delta^0_m\colon \widetilde{X}^n \to \widetilde{X}^{n+m+1}$ by setting

$$\delta^0_m = \eta\delta^0 h^{n+m} \ldots \delta^0 h^{n+1}\delta_0\xi.$$

Straightforward computations show that the required relations hold.

Further, we define the maps $\xi^n_m\colon \widetilde{X}^n \to X^{n+m}$ by setting

$$\xi^n_m = h^{n+m}\delta^0 \dots h^{n+1}\delta^0\xi.$$

They give an A_∞-cosimplicial homotopy equivalence between $\widetilde{X}^*$ and X^*. □

THEOREM. *Let $f^*\colon X^* \to Y^*$ be a map of cosimplicial objects in the category of chain complexes, and let precosimplicial objects $\widetilde{X}^*$ and $\widetilde{Y}^*$ be deformation retracts of X^* and Y^* regarded as precosimplicial objects. On each of $\widetilde{X}^*$ and $\widetilde{Y}^*$, consider the structure of an A_∞-cosimplicial object defined above. Then there exists an A_∞-map $\widetilde{f}^*\colon \widetilde{X}^* \to \widetilde{Y}^*$ such that the following diagram is homotopy commutative:*

$$\begin{array}{ccc} X^* & \xrightarrow{f^*} & Y^* \\ {\scriptstyle \xi^*}\uparrow & & \uparrow{\scriptstyle \xi^*} \\ \widetilde{X}^* & \xrightarrow{\widetilde{f}^*} & \widetilde{Y}^* \end{array}$$

The proof is similar to the proof of the previous theorem.

Now consider the augmented cosimplicial object

$$F^*(E,X)\colon\ X \to \overline{E}(X) \to \dots \to \overline{E}^{n-1}(X) \to \overline{E}^n(X) \to \dots$$

constructed for an E-coalgebra X. Its deformation retract is the augmented cosimplicial object

$$F^*(\overline{E}_*,X_*)\colon\ X_* \to \overline{E}_*(X_*) \to \dots \to \overline{E}_*^{n-1}(X_*) \to \overline{E}_*^n(X_*) \to \dots .$$

Then, by the above theorem, the functional homology operations that give the higher differentials of the Bousfield–Kan spectral sequence can be chosen so that they form the structure of an A_∞-cosimplicial object on $\widetilde{F}^*(\overline{E}_*,X_*)$.

10.3. Functorial homology operations

By $\Delta^* = \{\Delta^n\}$ we denote the cosimplicial object in the category of chain complexes consisting of the chain complexes of the standard n-dimensional simplices. Further, let F be a functor on the category of chain complexes, and let transformations $\Delta^n \otimes F(X) \to F(\Delta^n \otimes X)$ be given that commute with the cofaces and codegeneracy operators. Such a functor F will be called a *chain functor*.

The functors $\underline{E}$ and $\overline{E}$ provide examples of chain functors we are interested in. Indeed, let us define maps $\Delta^n \otimes \underline{E}(X) \to \underline{E}(\Delta^n \otimes X)$ or, equivalently, families of maps

$$\Delta^n \otimes \underline{E}(j;X) \to \underline{E}(j;\Delta^n \otimes X)$$

as the compositions

$$\begin{aligned} \Delta^n \otimes \underline{E}(j;X) = \Delta^n \otimes E(j) \otimes_{\Sigma_j} X^{\otimes j} &\xrightarrow{1\otimes\nabla\otimes 1} \Delta^n \otimes E(j) \otimes E(j) \otimes_{\Sigma_j} X^{\otimes j} \\ \xrightarrow{\tau\otimes 1\otimes 1} \Delta^{n\otimes j} \otimes E(j) \otimes_{\Sigma_j} X^{\otimes j} &\to E(j) \otimes_{\Sigma_j} (\Delta^n \otimes X)^{\otimes j} = \underline{E}(j;\Delta^n \otimes X), \end{aligned}$$

where $\tau\colon \Delta^n \otimes E(j) \to \Delta^{n\otimes j}$ is the E-coalgebra structure on the complex Δ^n. Direct verification shows that the required relations hold.

Similarly, we define maps $\Delta^n \otimes \overline{E}(j;X) \to \overline{E}(j;\Delta^n \otimes X)$ or, equivalently, maps

$$E(j) \otimes \Delta^n \otimes \operatorname{Hom}_{\Sigma_j}(E(j); X^{\otimes j}) \to (\Delta^n \otimes X)^{\otimes j}$$

as the compositions

$$E(j) \otimes \Delta^n \otimes \mathrm{Hom}_{\Sigma_j}(E(j); X^{\otimes j}) \xrightarrow{\nabla \otimes 1 \otimes 1} E(j) \otimes E(j) \otimes \Delta^n \otimes \mathrm{Hom}_{\Sigma_j}(E(j); X^{\otimes j})$$
$$\to E(j) \otimes \Delta^{n \otimes j} \otimes \mathrm{Hom}_{\Sigma_j}(E(j); X^{\otimes j}) \to (\Delta^n \otimes X)^{\otimes j}.$$

Direct verification shows that the required relations hold.

Let F be a chain functor, and F_* the functor that assigns the graded module of homology $F_*(X) = H_*(F(X))$ to a chain complex X. Then F_* is not just a functor but an A_∞-functor, i.e., there are functional operations assigning maps

$$F_*(f^n, \dots, f^1) = H_*(F(f^n), \dots, F(f^1)) \colon F_*(X^1) \to F_*(X^{n+1})$$

of dimension $n-1$ to sequences of maps of chain complexes $f^1 \colon X^1 \to X^2, \dots,$ $f^n \colon X^n \to X^{n+1}$.

In particular, the functors $\underline{E}_*$ and $\overline{E}_*$ are A_∞-functors.

THEOREM. *Suppose that F is a chain functor and $f^1 \colon X^1 \to X^2, \dots, f^n \colon X^n \to X^{n+1}$ is a sequence of maps of chain complexes. Then*

$$H_*(F(f^n), \dots, F(f^1)) = \sum (-1)^\epsilon F_*(H_*(f^n, \dots, f^{n_m+1}), \dots, H_*(f^{n_1}, \dots, f^1)),$$

where the summation is over all positive integers m and all $n_1, \dots, n_m$ such that $1 \le n_1 < \dots < n_m < n$.

PROOF. Indeed, given a chain complex X, for the representative map $F_*(X_*) \to F(X)$ we can take the composition of the maps

$$\xi(F) \colon F_*(X_*) \to F(X_*), \quad F(\xi) \colon F(X_*) \to F(X).$$

For the projection $F(X) \to F_*(X_*)$ we can take the composition of the maps

$$F(\eta) \colon F(X) \to F(X_*), \quad \eta(F) \colon F(X_*) \to F_*(X_*).$$

For the homotopy $F(X) \to F(X)$ we can take the sum $F(\xi) \circ h(F) \circ F(\eta) + F(h)$.

Substituting these maps into the formula for functional homology operations, we obtain the required identities. □

Our goal is to compute the functional homology operations for the functors $\underline{E}_*$ and $\overline{E}_*$. This means that for each sequence

$$f^1 : X^1 \to X^2, \dots, f^n : X^n \to X^{n+1}$$

of maps of chain complexes, we must compute the maps

$$\underline{E}_*(f^n, \dots, f^1) \colon \underline{E}_*(X^1) \to \underline{E}_*(X^{n+1}), \quad \overline{E}_*(f^n, \dots, f^1) \colon \overline{E}_*(X^1) \to \overline{E}_*(X^{n+1}).$$

First, let us consider the functor $\underline{E}(2; -)$ that assigns to a complex X the complex

$$\overline{E}(2; X) = E(2) \otimes_{\Sigma_2} X \otimes X,$$

where $E(2)$ is the Σ_2-free acyclic complex with generators e_i of dimension i and with differential given by

$$d(e_i) = e_{i-1} + e_{i-1}T, \quad T \in \Sigma_2.$$

Here we assume the ground ring to be the field $\mathbb{Z}/2$.

As noted above, the homology $\underline{E}_*(2;-)$ of this functor is not just a functor but also an A_∞-functor, i.e., a map

$$\underline{E}_*(2; f^n, \dots, f^1): \underline{E}_*(2; X^1) \to \underline{E}_*(2; X^{n+1})$$

is assigned to each sequence of maps of chain complexes

$$f^1 : X^1 \to X^2, \dots, f^n : X^n \to X^{n+1}.$$

Let us compute these functional operations.

Note that there is an isomorphism $\underline{E}_*(2, X) \cong \underline{E}_*(2, X_*)$ for any chain complex X. Furthermore, if X_* is a graded module, then $\underline{E}_*(2; X_*)$ is the direct sum of the quotient module $X_* \cdot X_*$ of the tensor product $X_* \otimes X_*$ by the permutations of the factors, and the module generated by the elements of the form $e_i \times y_n$, $i \geq 1$, of dimension $i + 2n$. The elements $y_n \cdot y_n \in X_* \cdot X_*$ will also be denoted by $e_0 \times y_n$.

Let $\xi: X_* \to X$, $\eta: X \to X_*$, $h: X \to X$ be maps that give a chain equivalence between X and X_*. By

$$E(\xi): E(2, X_*) \to E(2, X),\ \ E(\eta): E(2, X) \to E(2, X_*),\ \ E(h): E(2, X) \to E(2, X)$$

we denote the maps defined by the formulas

$$\begin{gathered} E(\xi)(e_i \otimes y_1 \otimes y_2) = e_i \otimes \xi(y_1) \otimes \xi(y_2), \\ E(\eta)(e_i \otimes x_1 \otimes x_2) = e_i \otimes \eta(x_1) \otimes \eta(x_2), \\ E(h)(e_i \otimes x_1 \otimes x_2) = e_i \otimes (x_1 \otimes h(x_2) + h(x_1) \otimes \xi\eta(x_2)) + e_{i-1} \otimes h(x_1) \otimes h(x_2). \end{gathered}$$

It is clear that they give a chain equivalence $\underline{E}(2, X) \simeq \underline{E}(2, X_*)$.

Let us define the following maps:

$$\begin{gathered} \xi(E): \underline{E}_*(2; X_*) \to \underline{E}(2; X_*), \quad \eta(E): \underline{E}(2; X_*) \to \underline{E}_*(2; X_*), \\ h(E): \underline{E}(2; X_*) \to \underline{E}(2; X_*). \end{gathered}$$

To do this, we chose a basis $\{y\}$ in X_*, which is somehow ordered (we mean that a basis is chosen in the module $H_n(X)$ for each n).

We define the map $\xi(E)$ by setting

$$\xi(E)(e_i \times y) = e_i \otimes y \otimes y; \quad \xi(E)(y_1 \cdot y_2) = e_0 \otimes (y_1 \otimes y_2), \quad y_1 \leq y_2.$$

We define the map $\eta(E)$ by setting

$$\eta(E)(e_i \otimes y_1 \otimes y_2) = \begin{cases} e_i \times y_1, & y_1 = y_2, \\ y_1 \cdot y_2, & y_1 < y_2,\ i = 0, \\ 0 & \text{otherwise} \end{cases}$$

The map $h(E)$ is defined by setting

$$h(E)(e_i \otimes y_1 \otimes y_2) = \begin{cases} e_{i+1} \otimes y_2 \otimes y_1, & y_1 > y_2, \\ 0 & \text{otherwise.} \end{cases}$$

Straightforward computations show that the required relations hold.

The maps

$$\begin{gathered} E(\xi) \circ \xi(E): E_*(2, X_*) \to E(2, X), \quad \eta(E) \circ E(\eta): E(2, X) \to E_*(2, X_*), \\ E(\xi) \circ h(E) \circ E(\eta) + E(h): E(2, X) \to E(2, X) \end{gathered}$$

give a chain equivalence between $E(2, X)$ and $E_*(2, X_*)$.

The general formula for the functional homology operators for a chain functor implies the following formula for the functor $\underline{E}(2, -)$:

$$\underline{E}_*(2, f^n, \dots, f^1) = \sum \underline{E}_*(2, H_*(f^n, \dots, f^{n_m+1}), \dots, H_*(f^{n_1}, \dots, f^1)),$$

where the summation is over all positive integers m and all $n_1, \dots, n_m$ such that $1 \leq n_1 < \dots < n_m < n$.

Let X be a graded module in which an ordered basis $\{x_i\}$ is chosen; we define maps $p: X \otimes X \to X$, $q: X \to X \otimes X$, and $r: X \otimes X \to X \otimes X$ by setting $q(x_i) = x_i \otimes x_i$ and

$$p(x_i \otimes x_j) = \begin{cases} x_i, & i = j, \\ 0, & i \neq j, \end{cases} \qquad r(x_i \otimes x_j) = \begin{cases} x_j \otimes x_i, & i > j, \\ 0, & i \leq j. \end{cases}$$

Suppose that $f^1: X^1 \to X^2, \dots, f^n: X^n \to X^{n+1}$ is a sequence of maps of graded modules, in each of which an ordered basis is chosen; we define a map $(f^n, \dots, f^1): X^1 \to X^{n+1}$ by setting

$$(f^n, \dots, f^1) = p \circ (f^n)^{\otimes 2} \circ r \circ (f^{n-1})^{\otimes 2} \circ \dots \circ r \circ (f^1)^{\otimes 2} \circ q.$$

The definition of the homology operations directly implies the following theorem.

THEOREM. *Suppose that $f^1: X^1 \to X^2, \dots, f^n: X^n \to X^{n+1}$ is a sequence of maps of graded modules. Then*

$$\underline{E}_*(2; f^n, \dots, f^1)(e_i \times x) = e_{i+n-1} \times (f^n, \dots, f^1)(x).$$

Similarly, for the functor $\overline{E}_*(2; -)$, the following theorem holds.

THEOREM. *Suppose that $f^1: X^1 \to X^2, \dots, f^n: X^n \to X^{n+1}$ is a sequence of maps of graded modules. Then*

$$\overline{E}_*(2; f^n, \dots, f^1)(\overline{e}_i \times x) = \begin{cases} \overline{e}_{i-n+1} \times (f^n, \dots, f^1)(x), & i \geq n-1, \\ 0 & \textit{otherwise}. \end{cases}$$

To compute the functional homology operations for the entire functor $\underline{E}_*$, we can use the monad structure $\underline{\gamma}_*: \underline{E}_* \circ \underline{E}_* \to \underline{E}_*$ and the formula

$$\underline{E}_*(f^n, \dots, f^1) \circ \gamma_* = \sum \underline{\gamma}_* \circ \underline{E}_*(\underline{E}_*(f^n, \dots, f^{n_m+1}), \dots, \underline{E}_*(f^{n_1}, \dots, f^1)),$$

where the summation is over all positive integers m and all $n_1, \dots, n_m$ such that $1 \leq n_1 < \dots < n_m < n$.

Passing to the Dyer–Lashof algebra $\mathcal{R}$, we obtain operations

$$\mathcal{R}(f^n, \dots, f^1): \mathcal{R} \otimes X^1 \to \mathcal{R} \otimes X^{n+1},$$

which are computed on the generators Q^i by

$$\mathcal{R}(f^n, \dots, f^1)(Q^i \otimes x) = Q^{i+n-1} \otimes (f^n, \dots, f^1)(x).$$

To compute the functional homology operations for the entire functor $\overline{E}_*$, we can use the monad structure $\overline{\gamma}_*: \overline{E}_* \to \overline{E}_* \circ \overline{E}_*$ and the formula

$$\overline{\gamma}_* \circ \overline{E}_*(f^n, \dots, f^1) = \sum \overline{E}_*(\overline{E}_*(f^n, \dots, f^{n_m+1}), \dots, \overline{E}_*(f^{n_1}, \dots, f^1)) \circ \overline{\gamma}_*,$$

where the summation is over all positive integers m and all $n_1, \dots, n_m$ such that $1 \leq n_1 < \dots < n_m < n$.

Passing to the Milnor coalgebra $\mathcal{K}$, we obtain operations

$$\mathcal{K}(f^n, \dots, f^1): \mathcal{K} \otimes X^1 \to \mathcal{K} \otimes X^{n+1},$$

which are computed by

$$\mathcal{K}(f^n, \dots, f^1)(y \otimes x) = y \cdot \xi_1^{n-1} \otimes (f^n, \dots, f^1)(x).$$

Let K be a Hopf algebra with multiplication $\pi: K \otimes K \to K$ and comultiplication $\nabla: K \to K \otimes K$. Denote by$\nabla^{\otimes 2}: K^{\otimes 2} \to K^{\otimes 2} \otimes K^{\otimes 2}$ the comultiplication in the tensor product $K^{\otimes 2}$; $\nabla^{\otimes 2} = (1 \otimes T \otimes 1)(\nabla \otimes \nabla)$.

Consider the operation

$$\Psi^1 = (\pi \otimes \pi, \nabla^{\otimes 2}): K^{\otimes 2} \to K^{\otimes 2}.$$

It follows directly from the definition that this operation is given by the formula

$$\Psi^1(x \otimes y) = \sum_{x' > y'} y'x' \otimes y''x'',$$

where $\nabla(x) = \sum x' \otimes x''$, $\nabla(y) = \sum y' \otimes y''$.

We are primarily interested in the case $x = y$. The corresponding operation will also be denoted by $\Psi^1: K \to K \otimes K$. Clearly, it is given by the formula

$$\Psi^1(x) = \sum_{x' > x_1} x_1 x' \otimes x_2 x'',$$

where $\nabla(x) = \sum x' \otimes x'' = \sum x_1 \otimes x_2$.

In particular, for any primitive element $x \in K$, we have

$$\Psi^1(x) = x \otimes x, \quad x \in PK.$$

Similarly, for the operation

$$\Psi^2 = (\pi \otimes \pi \otimes \pi, \nabla^{\otimes 2} \otimes 1^{\otimes 2}, \nabla^{\otimes 2}): K^{\otimes 2} \to K^{\otimes 3}$$

we have the following formula:

$$\Psi^2(x \otimes y) = \sum_{\substack{x' > y' \\ (y')' > (x')'}} (x')'(y')' \otimes (x')''(y')'' \otimes x''y'',$$

where $\nabla(x') = \sum (x')' \otimes (x')''$, $\nabla(y') = \sum (y')' \otimes (y')''$.

In the case $x = y$, the corresponding operation will also be denoted by

$$\Psi^2: K \to K^{\otimes 3}.$$

Clearly, it is given by the formula

$$\Psi^2(x) = \sum_{\substack{x' > x_1 \\ (x_1)_1 > (x')'}} (x_1)_1 (x')' \otimes (x_1)_2 (x')'' \otimes x_2 x'',$$

where $\nabla(x') = \sum (x')' \otimes (x')''$, $\nabla(x_1) = \sum (x_1)_1 \otimes (x_1)_2$.

Note that if $x \in K$ is a primitive element, then $\Psi^2(x) = 0$.

The operations $\Psi^m: K \to K^{\otimes m+1}$ for a Hopf algebra K are defined and computed in a similar way.

For example, consider the Milnor coalgebra $\mathcal{K}$. The comultiplication ∇ is given on the elements ξ_1^n by

$$\nabla(\xi_1^n) = \sum_{i=0}^{n} \binom{n}{i} \xi_1^i \otimes \xi_1^{n-i}.$$

This directly implies that

$$\Psi^1(\xi_1^n) = \sum_{i<j} \binom{n}{i}\binom{n}{j} \xi_1^{i+j} \otimes \xi_1^{2n-i-j}.$$

In particular, for primitive elements $\xi_1^{2^i} \in \mathcal{K}$, we have $\Psi^1(\xi_1^{2^i}) = \xi_1^{2^i} \otimes \xi_1^{2^i}$.

Similarly, for the operation Ψ^2, we have

$$\Psi^2(\xi_1^n) = \sum_{\substack{i<j\\k>l}} \binom{n}{i}\binom{n}{j}\binom{i}{k}\binom{j}{k} \xi_1^{k+l} \otimes \xi_1^{i+j-k-l} \otimes \xi_1^{2n-i-j}.$$

In particular, for the elements $\xi_1^{2^i+1} \in \mathcal{K}$, the following formula holds:

$$\begin{aligned}\Psi^2(\xi_1^{2^i+1}) &= \xi_1 \otimes (\xi_1^{2^i} \otimes \xi_1^{2^i+1} + \xi_1^{2^i+1} \otimes \xi_1^{2^i}) \\ &\quad + (\xi_1^{2^i} \otimes \xi_1^{2^i+1} + \xi_1^{2^i+1} \otimes \xi_1^{2^i}) \otimes \xi_1.\end{aligned}$$

10.4. Differentials in the Adams spectral sequence

Consider the cobar construction on the operad E,

$$F\overline{E} : E \xrightarrow{\overline{\gamma}} \overline{E}\overline{E} \to \cdots \to \overline{E}^n \to \dots,$$

and the commutative diagram

$$\begin{array}{ccccccc} \underline{E}\overline{E} & \xrightarrow{E\overline{\gamma}} & \underline{E}\overline{E}^2 & \longrightarrow \dots \longrightarrow & \underline{E}\overline{E}^n & \longrightarrow & \dots \\ {\scriptstyle T}\downarrow & & {\scriptstyle T}\downarrow & & {\scriptstyle T}\downarrow & & \\ \overline{E}\underline{E} & \xrightarrow{\overline{\gamma}} & \overline{E}^2\underline{E} & \longrightarrow \dots \longrightarrow & \overline{E}^n\underline{E} & \longrightarrow & \dots \end{array}$$

This diagram induces the diagram

$$\begin{array}{ccccccc} \underline{E}_*\overline{E}_* & \xrightarrow{E_*\overline{\gamma}_*} & \underline{E}_*\overline{E}_*^2 & \longrightarrow \dots \longrightarrow & \underline{E}_*\overline{E}_*^n & \longrightarrow & \dots \\ {\scriptstyle T_*}\downarrow & & {\scriptstyle T_*}\downarrow & & {\scriptstyle T_*}\downarrow & & \\ \overline{E}_*\underline{E}_* & \xrightarrow{\overline{\gamma}_*E_*} & \overline{E}_*^2\underline{E}_* & \longrightarrow \dots \longrightarrow & \overline{E}_*^n\underline{E}_* & \longrightarrow & \dots \end{array}$$

in which, apart from the maps indicated by arrows, there are also higher operations.

Passing to the Dyer–Lashof algebra and the Milnor coalgebra, we obtain the diagram

$$\begin{array}{ccccccc} \mathcal{R}\otimes\mathcal{K} & \xrightarrow{1\otimes\nabla} & \mathcal{R}\otimes\mathcal{K}^{\otimes 2} & \longrightarrow \dots \longrightarrow & \mathcal{R}\otimes\mathcal{K}^{\otimes n} & \longrightarrow & \dots \\ {\scriptstyle T}\downarrow & & {\scriptstyle T}\downarrow & & {\scriptstyle T}\downarrow & & \\ \mathcal{K}\otimes\mathcal{R} & \xrightarrow{\nabla\otimes 1} & \mathcal{K}^{\otimes 2}\otimes\mathcal{R} & \longrightarrow \dots \longrightarrow & \mathcal{K}^{\otimes n}\otimes\mathcal{R} & \longrightarrow & \dots \end{array}$$

in which, apart from the maps indicated by arrows, there are also higher operations.

This diagram defines an action of the Dyer–Lashof algebra $\mathcal{R}$ on the cobar construction $F\mathcal{K}$. This action can be used to compute the differentials in the Adams spectral sequence. To do this, we must first compute the action itself.

THEOREM. *The homology operation $T^1: \mathcal{R} \otimes \mathcal{K} \to \mathcal{K}^{\otimes 2} \otimes \mathcal{R}$ corresponding to the commutative diagram*

$$\begin{array}{ccc} \underline{E}\overline{E} & \xrightarrow{T} & \overline{E}\underline{E} \\ {\scriptstyle \underline{E}\overline{\gamma}}\downarrow & & \downarrow{\scriptstyle \overline{\gamma}\underline{E}} \\ \underline{E}\overline{E}^2 & \xrightarrow{\overline{E}T\circ T\overline{E}} & \overline{E}^2\underline{E} \end{array}$$

is given by $T^1(Q^{i+n} \otimes x_n) = \Psi^1(x_n) \otimes Q^{i+1}$.

PROOF. Indeed, for the chain homotopy h in the diagram

$$\begin{array}{ccc} \underline{E}_*\overline{E}_* & \xrightarrow{E_*\overline{\gamma}_*} & \underline{E}_*\overline{E}_*^2 \\ {\scriptstyle \xi}\downarrow & & \downarrow{\scriptstyle \xi} \\ \underline{E}\overline{E}_* & \xrightarrow{E_*\overline{\gamma}_*} & \underline{E}\overline{E}_*^2 \end{array}$$

we have the formula

$$h(e_i \times x_n) = \sum_{x' < x_1} e_{i+1} \otimes (x' \otimes x'') \otimes (x_1 \otimes x_2).$$

Substituting it into the general formula for functional homology operations and using the formula for the permutation map, we obtain the required formula for T^1. $\square$

In particular, for the primitive elements $\xi_1^{2^k} \in \mathcal{K}$, we have

$$T^1(Q^{i+2^k} \otimes \xi_1^{2^k}) = \xi_1^{2^k} \otimes \xi_1^{2^k} \otimes Q^{i+1}.$$

It is similarly shown that an analog of this formula holds in the general case as well. Namely, the following theorem holds.

THEOREM. *The operations $T^m: \mathcal{R} \otimes \mathcal{K} \to \mathcal{K}^{\otimes m+1} \otimes \mathcal{R}$ are given by*

$$T^m(Q^{i+n} \otimes x_n) = \Psi^m(x_n) \otimes Q^{i+m}.$$

Let us extend the action of the Dyer–Lashof algebra on the cobar construction $F\mathcal{K}$ by setting, for $x_n \in \mathcal{K}$,

$$Q^{n+i}[x_n] = \begin{cases} [Q^{n+i}(x_n)], & i \geq 0, \\ \Psi^{-i}(x_n), & i < 0. \end{cases}$$

On the remaining elements of the cobar construction $F\mathcal{K}$, the action of the Dyer–Lashof algebra is defined by the permutation map

$$\begin{aligned} &Q^{n+m+i}(x_n \times y_m) \\ &\quad = \sum_{k \geq 0} Q^{n+i-k}(x_n \xi_0^k) \times Q^{m+k}(y_m) + \sum_{k \geq 1} Q^{n+i+k}(x_n) \times Q^{m-k}(y_m) \\ &\qquad + \sum_{k \geq 1} Q^{n+i}(x_n)\xi_1^k \times Q^{m-k}(y_m) + \sum_{k \geq 1} Q^{n-k}(x_n) \times Q^{m+i+k}(y_m). \end{aligned}$$

Since the elements of the Milnor coalgebra can be obtained by applying the action of elements of the Dyer–Lashof algebra to elements of the Milnor coalgebra of smaller dimension, the above formulas make it possible, in principle, to compute the higher differentials of the Adams spectral sequence. For example, for the elements $h_i = [\xi_1^{2^i}]$, $i \geq 2$, we have $\xi_1^{2^i} = Q^{2^{i-1}}(\xi_1^{2^{i-1}})$.

The second differential is defined on these elements by the values of the secondary operation $\delta_1^0 \colon \mathcal{K} \to \mathcal{K}^{\times 3}$, for which we have the formulas

$$\delta_1^0(\xi_1^{2^i}) = \delta_1^0(Q^{2^{i-1}}(\xi_1^{2^{i-1}})) = Q^{2^{i-1}}(\xi_0^{2^i} \times \xi_1^{2^{i-1}}) = \xi_1 \xi_0^{2^{i+1}} \times \xi_1^{2^{i-1}} \xi_0^{2^i} \times \xi_1^{2^{i-1}} \xi_0^{2^{i-1}}.$$

Hence the second differential on these elements is given by $d^2(h_i) = h_0 h_{i-1}^2$, $i \geq 2$.

Taking into account the relations $h_0 h_1 = 0$, $h_0 h_2^2 = 0$, and $h_0 h_{i-1}^2 \neq 0$ for $i > 3$ in the E_2-term of the Adams spectral sequence, we see that among the elements h_i only h_0, h_1, h_2, and h_3 are cycles. By dimensional arguments, all higher differentials on them are zero, and hence these elements determine elements of the stable homotopy groups of the spheres in dimensions 0, 1, 3, and 7, respectively.

10.5. Description of the homotopy groups of topological spaces

The problem of describing and computing the homotopy groups $\pi_m(S^n)$ of the spheres S^n is among the oldest problems of algebraic topology. It is not hard to show that $\pi_m(S^n) = 0$ if $m < n$, and $\pi_n(S^n) \cong \mathbb{Z}$. The first nontrivial computations of the homotopy groups of the spheres were done by Hopf [**23**] in 1935. He constructed a homomorphism $H \colon \pi_{2n-1}(S^n) \to \mathbb{Z}$ and proved that for $n = 2k$, its image contains $2\mathbb{Z}$, and for $n = 2, 4, 8$, in the corresponding homotopy groups there exist elements h_1, h_2, h_3 such that $H(h_1) = H(h_2) = H(h_3) = 1$. These elements are called the *Hopf elements.*

The next step in computing the homotopy groups of the spheres was done by Freudenthal [**19**] in 1937. He defined the suspension homomorphism

$$S \colon \pi_m(S^n) \to \pi_{m+1}(S^{n+1})$$

which assigns the homotopy classes $[Sf]$ of the suspensions $Sf \colon S^{m+1} = SS^m \to SS^n = S^{n+1}$ to homotopy classes $[f]$ of maps $f \colon S^m \to S^n$. By using this homomorphism, Freudenthal proved that there are isomorphisms

$$\pi_{n+1}(S^n) \cong \mathbb{Z}/2, \quad n \geq 3.$$

The generating elements η_n of these groups satisfy the relations $\eta_{n+1} = S\eta_n$ and $\eta_3 = Sh_1$. Thus all generators of $\pi_{n+1}(S^n)$ can be obtained by applying the suspension homomorphism to the Hopf element h_1.

It was also shown that the suspension homomorphism is an isomorphism for $m < 2n - 1$ and an epimorphism for $m = 2n - 1$. Hence the homotopy groups $\pi_{n+k}(S^n)$ for sufficiently large n (i.e., for $n > k + 1$) do not depend on k, which makes it possible to define the so-called *stable homotopy groups of the spheres*

$$\sigma_k = \varinjlim \pi_{n+k}(S^n).$$

In 1950 Whitehead [**50**] computed the groups $\pi_{n+2}(S^n)$,

$$\pi_{n+2}(S^n) \cong \mathbb{Z}/2, \quad n \geq 2.$$

In doing this he used the composition product

$$\circ\colon \pi_m(S^n) \times \pi_k(S^m) \to \pi_k(S^n)$$

assigning the homotopy class $[f_1 \circ f_2]$ of the composition $f_1 \circ f_2\colon S^k \to S^n$ to the homotopy classes $[f_1], [f_2]$ of maps $f_1\colon S^m \to S^n$, $f_2\colon S^k \to S^m$. Taking the compositions $\eta_n \circ \eta_{n+1}$ of the Freudenthal elements, Whitehead obtained generators for $\pi_{n+2}(S^n)$.

It is notable that all these and other computations of that time used the Hopf elements. New elements were obtained by applying to them either the suspension homomorphism of Freudenthal or the composition product of Whitehead. The following question arose: Can all elements of the homotopy groups of the spheres be obtained in this way? The answer turned out to be negative.

To find new elements in the homotopy groups of the sphere, i.e., elements that cannot be obtained as above, in 1962 Toda [**49**] introduced new operations

$$\pi_{m_1}(S^n) \times \cdots \times \pi_{m_i}(S^{m_{i-1}}) \to \pi_{m_i+i-2}(S^n),$$

called the *higher composition products*, which are partial and multivalued maps. With the help of these operations, Toda computed the homotopy groups of the spheres $\pi_{n+k}(S^n)$ for $k \leq 19$.

Let us recall the definitions of the first Toda operation:

$$\langle\ ,\ ,\ \rangle\colon \pi_m(S^n) \times \pi_k(S^m) \times \pi_l(S^k) \to \pi_{l+1}(S^n),$$

called the *secondary composition product*. It is defined for the triples $(\alpha_1, \alpha_2, \alpha_3)$ of elements $\alpha_1 \in \pi_m(S^n)$, $\alpha_2 \in \pi_k(S^m)$, $\alpha_3 \in \pi_l(S^k)$ such that $\alpha_1 \circ \alpha_2 = 0$, $\alpha_2 \circ \alpha_3 = 0$.

Let $f_1\colon S^m \to S^n$, $f_2\colon S^k \to S^m$, $f_3\colon S^l \to S^k$ be representatives of the homotopy classes α_1, α_2, α_3, respectively. Then $f_1 \circ f_2 \simeq 0$, i.e., there exists a contracting homotopy $h_1\colon E^{k+1} \to S^n$. Therefore, we have a map $h_1 \circ Ef_3\colon E^{l+1} \to S^n$. On the other hand, $f_2 \circ f_3 \simeq 0$, i.e., there exists a contracting homotopy $h_2\colon E^{l+1} \to S^m$. Therefore, there is a map $f_1 \circ h_2\colon E^{l+1} \to S^n$. These two maps $h_1 \circ Ef_3, f_1 \circ h_2\colon E^{l+1} \to S^n$ coincide on the boundary of the ball E^{l+1} and determine maps of two $(l+1)$-hemispheres to S^n. Thus they define a map $S^{l+1} \to S^n$, whose homotopy class is denoted by $\langle \alpha_1, \alpha_2, \alpha_3 \rangle$, and which is the value of the sought-for secondary composition product.

Of course, this definition depends on the choice of the representatives f_1, f_2, f_3 and the homotopies h_1, h_2. Hence the secondary composition product is not only a partial operation, but also a multivalued operation.

In 1968 Cohen [**15**] showed that the Whitehead product and the higher composition products of Toda suffice to obtain all elements of the stable homotopy groups of the spheres from the Hopf elements. To be more precise, the Hopf elements are indecomposable elements in the stable homotopy groups of the spheres, and the remaining elements are decomposable, i.e., they can be obtained from the Hopf elements by applying to them the Whitehead and Toda products.

Thus to describe the stable homotopy groups of the spheres, it remains to find the relations between the Whitehead product and the Toda products. Some of these relations for the secondary product

$$\langle\ ,\ ,\ \rangle\colon \sigma_k \times \sigma_m \times \sigma_n \to \sigma_{k+m+n+1}$$

were found by Toda himself. Namely,

1. $\langle \alpha, \beta, \gamma \rangle \circ \delta \subset \langle \alpha, \beta, \gamma \circ \delta \rangle$;

2. $\langle \alpha, \beta, \gamma \circ \delta \rangle \subset \langle \alpha, \beta \circ \gamma, \delta \rangle$;
3. $\langle \alpha \circ \beta, \gamma, \delta \rangle \subset \langle \alpha, \beta \circ \gamma, \delta \rangle$;
4. $\alpha \circ \langle \beta, \gamma, \delta \rangle \subset (-1)^{\alpha} \langle \alpha \circ \beta, \gamma, \delta \rangle$;
5. $0 \in \langle \langle \alpha, \beta, \gamma \rangle, \delta, \epsilon \rangle + (-1)^{\alpha} \langle \alpha \langle \beta, \gamma, \delta \rangle >, \epsilon \rangle + (-1)^{\alpha+\beta} \langle \alpha, \beta, \langle \gamma, \delta, \epsilon \rangle \rangle$;
6. $\langle \alpha, \beta, \gamma \rangle = (-1)^{\alpha\beta+\alpha\gamma+\beta\gamma+1} \langle \gamma, \beta, \alpha \rangle$;
7. $0 \in \langle \alpha, \beta, \gamma \rangle + (-1)^{\alpha\gamma+\beta\gamma} \langle \gamma, \alpha, \beta \rangle + (-1)^{\alpha\gamma+\alpha\beta} \langle \beta, \gamma, \alpha \rangle$.

Relations 1–5 are associativity-type relations for the Whitehead and Toda products. Relations 6–7 are commutativity-type relations for the Toda products.

Actually, it is very difficult to work with the Toda products, which are partial and multivalued, and to find relations between them. A much more convenient language for these purposes is the language of Stasheff's A_∞-structures.

One can try to define an A_∞-algebra structure on the stable homotopy groups of the spheres by repeating Toda's construction. For the operation

$$\pi_1 \colon \pi_m(S^n) \times \pi_k(S^m) \times \pi_l(S^k) \to \pi_{l+1}(S^n),$$

it was done by Baues [**5**] in 1989. Let us recall this construction.

Let $\{S^m, S^n\}$ be the space of base-point-preserving continuous maps from S^m to S^n, and let $\eta \colon \{S^m, S^n\} \to \pi_m(S^n)$ be the projection. Let us fix a map $\xi \colon \pi_m(S^n) \to \{S^m, S^n\}$ of choosing representatives in homotopy classes. Then $\eta \circ \xi = \mathrm{Id}$, and $\xi \circ \eta(f)$ is homotopic to f for all $f \colon S^m \to S^n$. Let us also fix this homotopy $h(f)$.

If $\alpha_1 \in \pi_m(S^n)$ and $\alpha_2 \in \pi_k(S^m)$, then

$$\alpha_1 \circ \alpha_2 = \eta(\xi(\alpha_1) \circ \xi(\alpha_2)).$$

The elements $\xi(\alpha_1 \circ \alpha_2)$ and $\xi(\alpha_1) \circ \xi(\alpha_2)$ are homotopic. The corresponding homotopy is $h(\xi(\alpha_1) \circ \xi(\alpha_2))$. Denote it by $\xi_1(\alpha_1, \alpha_2)$. Now let α_1, α_2 be as above, and let $\alpha_3 \in \pi_l(S^k)$. Consider the homotopies

$$\xi_1(\alpha_1, \alpha_2) \circ \xi(\alpha_3), \quad \xi(\alpha_1) \circ \xi_1(\alpha_2, \alpha_3), \quad \xi_1(\alpha_1 \circ \alpha_2, \alpha_3), \quad \xi_1(\alpha_1, \alpha_2 \circ \alpha_3).$$

They define a loop in the space $\{S^l, S^n\}$, and hence give a map $S^{l+1} \to S^n$. Denote its homotopy class by $\pi_1(\alpha_1, \alpha_2, \alpha_3)$, and the corresponding homotopy between $\xi\pi_1(\alpha_1, \alpha_2, \alpha_3)$ and the initial loop by $\xi_2(\alpha_1, \alpha_2, \alpha_3)$.

Now let $\alpha_1, \alpha_2, \alpha_3$ be as above, and let $\alpha_4 \in \pi_q(S^l)$. Let us construct an element $\pi_2(\alpha_1, \alpha_2, \alpha_3, \alpha_4) \in \pi_{q+2}(S^n)$. To this end, we consider the maps

$$\xi_2(\alpha_1, \alpha_2, \alpha_3) \circ \xi(\alpha_4), \quad \xi_1(\alpha_1, \alpha_2) \circ \xi_1(\alpha_3, \alpha_4), \quad \xi(\alpha_1) \circ \xi_2(\alpha_2, \alpha_3, \alpha_4),$$
$$\xi_2(\alpha_1 \circ \alpha_2, \alpha_3, \alpha_4), \quad \xi_2(\alpha_1, \alpha_2 \circ \alpha_3, \alpha_4), \quad \xi_2(\alpha_1, \alpha_2, \alpha_3 \circ \alpha_4),$$
$$\xi_1(\alpha_1, \pi_1(\alpha_2, \alpha_3, \alpha_4)), \quad \xi_1(\pi_1(\alpha_1, \alpha_2, \alpha_3), \alpha_4).$$

They define a map of a two-dimensional cell whose boundary consists of the maps

$$\xi(\pi_1(\alpha_1, \alpha_2, \alpha_3) \circ \alpha_4), \quad \xi(\alpha_1 \circ \pi_1(\alpha_2, \alpha_3, \alpha_4)),$$
$$\xi\pi_1(\alpha_1 \circ \alpha_2, \alpha_3, \alpha_4), \quad \xi\pi_1(\alpha_1, \alpha_2 \circ \alpha_3, \alpha_4), \quad \xi\pi_1(\alpha_1, \alpha_2, \alpha_3 \circ \alpha_4).$$

Therefore, the sum of these maps is homotopic to zero. But we have fixed all homotopies. Hence we have two maps of the boundary, which together give a map of the two-dimensional sphere to $\{S^q, S^n\}$. It generates a map from S^{l+2} to S^n. Its homotopy class $\pi_2(\alpha_1, \alpha_2, \alpha_3, \alpha_4)$ is the required value of the operation π_2. Unfortunately, the subsequent steps in this construction of the operations $\pi_3, \ldots$ are more and more difficult.

Another problem is the fact that the operations π_1 and π_2 constructed above are not additive, and we cannot directly use the structure of an A_∞-algebra. To make the operations additive, we pass from the homotopy groups to the Adams and Bousfield–Kan spectral sequences.

Recall that the E_∞-term of the Adams spectral sequence for the stable homotopy groups of the spheres is isomorphic to the homology of the cobar construction FK on the Milnor coalgebra, taken with the A_∞-coalgebra structure. In this case, the only primitive elements of the Milnor coalgebra are the elements ξ_1, ξ_1^2, ξ_1^4, and ξ_1^8. They generate the indecomposable elements h_i, $0 \leq i \leq 3$, in the E^∞-term of the Adams and Bousfield–Kan spectral sequences for the homotopy groups of the spheres. Therefore, applying to this case the method of describing A_∞-algebras, we obtain the following theorem.

THEOREM. *Let the E_∞-term of the Adams spectral sequence for the stable homotopy groups of the spheres be regarded as an A_∞-algebra. Then its only indecomposable elements are the elements of the form h_0, h_1, h_2, and h_3.*

This theorem is a reformulation of the result of [**15**] about the description of the stable homotopy groups of the spheres.

Similarly, using the fact that the E_∞-term of the Bousfield–Kan spectral sequence for the unstable homotopy groups of the spheres S^n is isomorphic to the homology of the cobar construction $F(K, S^n)$ and applying to this homology the method of describing A_∞-monads, we obtain the following result.

THEOREM. *Let the term $E_\infty(S^*)$ of the Bousfield–Kan spectral sequence for the unstable homotopy groups of the spheres be regarded as an A_∞-monad. Then its indecomposable elements are the elements of the form $h_i \otimes u^n$, $0 \leq i \leq 3$, such that $2^i \leq n$.*

If the suspension homomorphism, which is defined for spheres, is included in the set of operations under consideration, then only the elements of the form $h_i \otimes u^{2^i}$, $0 \leq i \leq 3$, are indecomposable.

To describe the term $E_\infty(X)$ of the Adams spectral sequence for the stable homotopy groups of a topological space X, it should be endowed with the structure of an A_∞-algebra over the A_∞-monad $E_\infty(S^*)$. In this case the generators are the primitive elements of the homology $H_*(X)$ with respect to the structure of an A_∞-comodule over the Milnor A_∞-coalgebra, i.e., exactly those elements of the homology $H_*(X)$ which survive in the term $E_\infty(X)$. Therefore, in this case the generators are the elements of the homology $H_*(X)$ that belong to the image of the Hurewicz homomorphism $h\colon \sigma_*(X) \to H_*(X)$.

In the case of the Bousfield–Kan spectral sequence for the unstable homotopy groups of a topological space X, the structure of an A_∞-$\vee$-algebra on the term $E_\infty(X)$ should be additionally included into the set of operations under consideration. Then the generators are the primitive elements of the homology $H_*(X)$ with respect to the structure of an A_∞-coalgebra over the Milnor A_∞-coalgebra, i.e., exactly those elements of $H_*(X)$ that survive in the term $E_\infty(X)$. Therefore, in this case the generators are the elements of the homology $H_*(X)$ that belong to the image of the Hurewicz homomorphism $h\colon \pi_*(X) \to H_*(X)$.

Bibliography

1. J. F. Adams, *On the cobar construction*, Proc. Nat. Acad. Sci. U.S.A. **42** (1956), 409–412.
2. ———, *Lectures on generalised cohomology*, Category Theory, Homology Theory and their Applications, III (Battelle Inst. Conf., Seattle, 1968), Lecture Notes in Math., vol. 99, Springer-Verlag, Berlin–New York, 1969, pp. 1–138.
3. ———, *Infinite loop spaces*, Princeton Univ. Press, Princeton, NJ, 1978.
4. H. J. Baues, *The double bar and cobar constructions*, Compositio Math. **43** (1981), 331–341.
5. H. J. Baues and W. Dreckmann, *The cohomology of homotopy categories and the general linear group*, K-Theory **3** (1989), 307–338.
6. N. A. Berikashvili, *Differentials of a spectral sequence*, Akad. Nauk Gruzin. SSR Trudy Tbilis. Mat. Inst. Razmadze **51** (1976), 5–106. (Russian)
7. J. M. Boardman and R. M. Vogt, *Homotopy invariant algebraic structures on topological spaces*, Lecture Notes in Mathematics, vol. 347, Springer-Verlag, Berlin–New York, 1973.
8. A. K. Bousfield, E. B. Curtis, D. M. Kan, D. G. Quillen, D. L. Rector, and J. W. Schlesinger, *The* mod p *lower central series and the Adams spectral sequence*, Topology **5** (1966), 331–342.
9. A. K. Bousfield and D. M. Kan, *The homotopy spectral sequence of a space with coefficients in a ring*, Topology **11** (1972), 79–106.
10. ———, *Homotopy limits, completions and localizations*, Lecture Notes in Math., vol. 304, Springer-Verlag, Berlin–New York, 1972.
11. K. S. Brown, *Cohomology of groups (summary of talks)*, Algebraic K-Theory (Proc. Conf., Evanston, IL, 1976), Lecture Notes in Math., vol 551, Springer-Verlag, Berlin–New York, 1976, pp. 249–259.
12. E. H. Brown, *Twisted tensor products*, Ann. of Math. (2) **69** (1959), 223–246.
13. R. Bruner, J. P. May, J. E. McCluer, and M. Steinberger, *H_∞ ring spectra and their applications*, Lecture Notes in Math., vol. 1176, Springer-Verlag, Berlin–New York, 1986.
14. F. R. Cohen, T. J. Lada, and J. P. May, *The homology of iterated loop spaces*, Lecture Notes in Math., vol. 533, Springer-Verlag, Berlin–New York, 1976.
15. J. Cohen, *The decomposition of stable homotopy*, Ann. of Math. (2) **87** (1968), 305–320.
16. E. B. Curtis, *Some relations between homotopy and homology*, Ann. of Math. (2) **82** (1965), 386–413.
17. E. Dyer and R. Lashof, *Homology of iterated loop spaces*, Amer. J. Math. **84** (1962), 35–88.
18. T. tom Dieck, *Steenrod-Operationen in Kobordismen-Theorien*, Math. Z. **107** (1968), 380–401.
19. H. Freudenthal, *Über die Klassen der Sphärenabbildungen*, Compositio Math. **5** (1937), 299–314.
20. *Homotopy theory of differential forms*, Collection of papers[1], "Mir", Moscow, 1981. (Russian)
21. V. K. Gugenheim, L. A. Lambe, and J. D. Stasheff, *Perturbation theory in differential homological algebra.* II, Illinois J. Math. **35** (1991), 357–373.
22. V. K. Gugenheim and J. P. May, *On the theory and applications of differential torsion products*, Mem. Amer. Math. Soc. **142** (1974).
23. H. Hopf, *Über die Abbildungen von Sphären auf Sphären niedrigerer Dimension*, Fund. Math. **25** (1935), 427–440.
24. T. V. Kadeishvili, *On the theory of homology of fiber spaces*, Uspekhi Mat. Nauk **35** (1980), no. 3, 183–188; English transl., Russian Math. Surveys **35** (1980), no. 3, 231–238.
25. S. O. Kochman, *Homology of the classical groups over the Dyer–Lashof algebra*, Trans. Amer. Math. Soc. **185** (1973), 83–136.

[1]For the contents of this volume, see MR **82k:**55001.

26. P. S. Landweber, *Cobordism operations and Hopf algebras*, Trans. Amer. Math. Soc. **129** (1967), 94–110.
27. S. MacLane, *Homology*, Springer-Verlag, Berlin–New York, 1967.
28. J. P. May, *Simplicial objects in algebraic topology*, Van Nostrand, Princeton, NJ, 1967.
29. ——, *The geometry of iterated loop spaces*, Lecture Notes in Math., vol. 271, Springer-Verlag, Berlin–New York, 1972.
30. ——, *Infinite loop space theory*, Bull. Amer. Math. Soc. **83** (1977), 456–494.
31. J. P. May, F. Quinn, N. Ray, and J. Tornehave, *E_∞ ring spaces and E_∞ ring spectra*, Lecture Notes in Math., vol. 577, Springer-Verlag, Berlin–New York, 1977.
32. W. S. Massey and F. P. Peterson, *The* mod 2 *cohomology structure of certain fibre spaces*, Mem. Amer. Math. Soc. **74** (1967).
33. R. J. Milgram, *Iterated loop spaces*, Ann. of Math. (2) **84** (1966), 386–403.
34. R. E. Mosher and M. C. Tangora, *Cohomology operations and applications in homotopy theory*, Harper & Row, New York–London, 1968.
35. S. P. Novikov, *Rings of operations and spectral sequences of Adams type in extraordinary cohomology theories. U-cobordism and K-theory*, Dokl. Akad. Nauk SSSR **172** (1967), 33–36; English transl., Soviet Math. Dokl. **8** (1967), 27–31.
36. M. M. Postnikov, *Investigations in the homotopy theory of continuous mappings*, Trudy Mat. Inst. Steklov. **46** (1955); English transl., Amer. Math. Soc. Transl. (2) **7** (1957), 1–134.
37. ——, *Lectures on algebraic topology. Homotopy theory of cell complexes*, "Nauka", Moscow, 1985. (Russian)
38. S. B. Priddy, *Dyer-Lashof operations for the classifying spaces of certain matrix groups*, Quart. J. Math. Oxford Ser. (2) **26** (1975), 179–193.
39. D. Quillen, *Homotopical algebra*, Lecture Notes in Math., vol. 43, Springer-Verlag, Berlin–New York, 1967.
40. ——, *Rational homotopy theory*, Ann. of Math. (2) **90** (1969), 205–295.
41. D. C. Ravenel and W. S. Wilson, *The Hopf ring for complex cobordism*, J. Pure Appl. Algebra **9** (1977), 241–280.
42. R. M. Switzer, *Algebraic topology—homotopy and homology*, Springer-Verlag, Berlin–New York, 1975.
43. J. P. Serre, *Lie algebras and Lie groups*, Benjamin, New York–Amsterdam, 1965; *Algèbres de Lie semi-simples complexes*, Benjamin, New York–Amsterdam, 1966.
44. H. E. Spanier, *Algebraic topology*, McGraw Hill, New York–London, 1966.
45. J. D. Stasheff, *Homotopy associativity of H-spaces*, Trans. Amer. Math. Soc. **108** (1963), 275–312.
46. ——, *H-spaces from a homotopy point of view*, Lecture Notes in Math., vol. 161, Springer-Verlag, Berlin–New York, 1970.
47. N. E. Steenrod and D. B. A. Epstein, *Cohomology operations*, Princeton University Press, Princeton, NJ, 1962.
48. R. E. Stong, *Notes on cobordism theory*, Princeton University Press, Princeton, NJ, 1968.
49. H. Toda, *Composition methods in homotopy groups of spheres*, Princeton University Press, Princeton, NJ, 1962.
50. G. W. Whitehead, *The $(n+2)$nd homotopy groups of the n-sphere*, Ann. of Math. (2) **52** (1950), 245-247.
51. ——, *Recent advances in homotopy theory*, CBMS Regional Conf. Ser. Math., vol. 5, Amer. Math. Soc., Providence, RI, 1970.
52. D. B. Fuchs, A. T. Fomenko, and V. L. Gutenmacher, *Homotopic topology*, Izdat. Moskov. Univ., Moscow, 1969; English transl., Akad. Kiadó, Budapest, 1986.
53. A. Ya. Khelemskii, *Banach and polynormed algebras*, "Nauka", Moscow, 1989; English transl., Clarendon Press, Oxford–New York, 1993.
54. P. J. Hilton and S. Wylie, *Homology theory: An introduction to algebraic topology*, Cambridge University Press, Cambridge–New York, 1960.
55. D. Husemoller, *Fibre bundles*, McGraw–Hill, New York–London, 1966.

Index

Selected Titles in This Series

(Continued from the front of this publication)

161 **Michael Blank,** Discreteness and continuity in problems of chaotic dynamics, 1997
160 **V. G. Osmolovskiĭ,** Linear and nonlinear perturbations of the operator div, 1997
159 **S. Ya. Khavinson,** Best approximation by linear superpositions (approximate nomography), 1997
158 **Hideki Omori,** Infinite-dimensional Lie groups, 1997
157 **V. B. Kolmanovskiĭ and L. E. Shaĭkhet,** Control of systems with aftereffect, 1996
156 **V. N. Shevchenko,** Qualitative topics in integer linear programming, 1997
155 **Yu. Safarov and D. Vassiliev,** The asymptotic distribution of eigenvalues of partial differential operators, 1997
154 **V. V. Prasolov and A. B. Sossinsky,** Knots, links, braids and 3-manifolds. An introduction to the new invariants in low-dimensional topology, 1997
153 **S. Kh. Aranson, G. R. Belitsky, and E. V. Zhuzhoma,** Introduction to the qualitative theory of dynamical systems on surfaces, 1996
152 **R. S. Ismagilov,** Representations of infinite-dimensional groups, 1996
151 **S. Yu. Slavyanov,** Asymptotic solutions of the one-dimensional Schrödinger equation, 1996
150 **B. Ya. Levin,** Lectures on entire functions, 1996
149 **Takashi Sakai,** Riemannian geometry, 1996
148 **Vladimir I. Piterbarg,** Asymptotic methods in the theory of Gaussian processes and fields, 1996
147 **S. G. Gindikin and L. R. Volevich,** Mixed problem for partial differential equations with quasihomogeneous principal part, 1996
146 **L. Ya. Adrianova,** Introduction to linear systems of differential equations, 1995
145 **A. N. Andrianov and V. G. Zhuravlev,** Modular forms and Hecke operators, 1995
144 **O. V. Troshkin,** Nontraditional methods in mathematical hydrodynamics, 1995
143 **V. A. Malyshev and R. A. Minlos,** Linear infinite-particle operators, 1995
142 **N. V. Krylov,** Introduction to the theory of diffusion processes, 1995
141 **A. A. Davydov,** Qualitative theory of control systems, 1994
140 **Aizik I. Volpert, Vitaly A. Volpert, and Vladimir A. Volpert,** Traveling wave solutions of parabolic systems, 1994
139 **I. V. Skrypnik,** Methods for analysis of nonlinear elliptic boundary value problems, 1994
138 **Yu. P. Razmyslov,** Identities of algebras and their representations, 1994
137 **F. I. Karpelevich and A. Ya. Kreinin,** Heavy traffic limits for multiphase queues, 1994
136 **Masayoshi Miyanishi,** Algebraic geometry, 1994
135 **Masaru Takeuchi,** Modern spherical functions, 1994
134 **V. V. Prasolov,** Problems and theorems in linear algebra, 1994
133 **P. I. Naumkin and I. A. Shishmarev,** Nonlinear nonlocal equations in the theory of waves, 1994
132 **Hajime Urakawa,** Calculus of variations and harmonic maps, 1993
131 **V. V. Sharko,** Functions on manifolds: Algebraic and topological aspects, 1993
130 **V. V. Vershinin,** Cobordisms and spectral sequences, 1993
129 **Mitsuo Morimoto,** An introduction to Sato's hyperfunctions, 1993
128 **V. P. Orevkov,** Complexity of proofs and their transformations in axiomatic theories, 1993
127 **F. L. Zak,** Tangents and secants of algebraic varieties, 1993

For a complete list of titles in this series, visit the
AMS Bookstore at **www.ams.org/bookstore/**.